ROBOTICS RESEARCH TRENDS

ROBOTICS RESEARCH TRENDS

XING P. GUÔ
EDITOR

Nova Science Publishers, Inc.
New York

Copyright © 2008 by Nova Science Publishers, Inc.

All rights reserved. No part of this book may be reproduced, stored in a retrieval system or transmitted in any form or by any means: electronic, electrostatic, magnetic, tape, mechanical photocopying, recording or otherwise without the written permission of the Publisher.

For permission to use material from this book please contact us:
Telephone 631-231-7269; Fax 631-231-8175
Web Site: http://www.novapublishers.com

NOTICE TO THE READER

The Publisher has taken reasonable care in the preparation of this book, but makes no expressed or implied warranty of any kind and assumes no responsibility for any errors or omissions. No liability is assumed for incidental or consequential damages in connection with or arising out of information contained in this book. The Publisher shall not be liable for any special, consequential, or exemplary damages resulting, in whole or in part, from the readers' use of, or reliance upon, this material. Any parts of this book based on government reports are so indicated and copyright is claimed for those parts to the extent applicable to compilations of such works.

Independent verification should be sought for any data, advice or recommendations contained in this book. In addition, no responsibility is assumed by the publisher for any injury and/or damage to persons or property arising from any methods, products, instructions, ideas or otherwise contained in this publication.

This publication is designed to provide accurate and authoritative information with regard to the subject matter covered herein. It is sold with the clear understanding that the Publisher is not engaged in rendering legal or any other professional services. If legal or any other expert assistance is required, the services of a competent person should be sought. FROM A DECLARATION OF PARTICIPANTS JOINTLY ADOPTED BY A COMMITTEE OF THE AMERICAN BAR ASSOCIATION AND A COMMITTEE OF PUBLISHERS.

LIBRARY OF CONGRESS CATALOGING-IN-PUBLICATION DATA

Robotics research trends / Xing P. Guô, editor.
   p. cm.
 ISBN-13: 978-1-60021-997-9 (hardcover)
 ISBN-10: 1-60021-997-7 (hardcover)
 1. Robotics--Research. I. Guô, Xing P.
TJ211.R5694   2008
629.8'92--dc22
                                                                                                                       2007041223

Published by Nova Science Publishers, Inc. ✢ *New York*

CONTENTS

Preface		vii
Expert Commentary		1
	Robotics Research Trends *Anthony Engwirda*	3
Short Communication		7
	Intelligent Modeling and Adaptive Control of Flexible Robot Manipulators *Chang-Woo Park*	9
Research and Review Studies		31
Chapter 1	Latest Progress of 3-D Reconstruction from Multiple Camera Images *Kenichi Kanatani*	33
Chapter 2	Project Diva: Guidance and Vision Surveillance Techniques for an Autonomous Airship *Alexandra Moutinho, Luiz Mirisola, José Azinheira and Jorge Dias*	77
Chapter 3	An Architecture for Adaptive Swarms *Suranga Hettiarachchi, Paul Maxim and William M. Spears*	121
Chapter 4	Preliminary-Announcement Function of Mobile Robots' Upcoming Operation *Takafumi Matsumaru*	155
Chapter 5	DEVS and Timed Automata for the Design of Control Systems *Norbert Giambiasi and Hernán P. Dacharry*	193
Chapter 6	Noisy Surface Smoothing Using Tsallis Entropy *Hong Zhou, Yonghuai Liu and Xuejun Ren*	223

Chapter 7	An Embedded Real-Time Control Architecture for Unmanned Marine Vehicles *Gabriele Bruzzone and Massimo Caccia*	249
Chapter 8	A Mobile Haptic Interface for Bimanual Manipulations in Extended Remote/Virtual Environments *Angelika Peer, Thomas Schauß, Ulrich Unterhinninghofen and Martin Buss*	267
Chapter 9	Cooperative Multiple Robots Collision Avoidance Problem Based on Bernstein-Bézier Path Tracking *Igor Škrjanc and Gregor Klančar*	289
Chapter 10	ABLE: A Standing Style Transfer System for Disabled Lower Limbs *Yoshikazu Mori*	309
Chapter 11	Stochastic Analysis of a Repairable Standby Robot-Safety System *Shen Cheng and B.S. Dhillon*	333
Index		357

PREFACE

Robotics began as a science fiction creation which has become quite real, first in assembly line operations such as automobile manufacturing, airplane construction etc. They have now reached such areas as the internet, ever-multiplying-medical uses and sophisticated military applications. Control of today's robots is often remote which requires even more advanced computer vision capabilities as well as sensors and interface techniques. Learning has become crucial for modern robotic systems as well. This new book presents the latest research in the field.

In the Short Communication, in order to control an uncertain flexible joint manipulator, an intelligent modeling method and direct model reference adaptive control via Takagi-Sugeno fuzzy modeling based on the Parallel distributed Compensation (PDC) concept are developed for the MIMO plant model. The proposed control scheme is proposed to provide asymptotic tracking of a reference signal for the systems with uncertain parameters. From Lyapunov stability analysis and simulation results, the developed control law and adaptive law guarantee the boundedness of all signals in the closed-loop system. In addition, the plant state tracks the state of the reference model asymptotically with time for any bounded reference input signal.

Chapter 1 summarizes recent progress of the theories and techniques for 3-D reconstruction from multiple images taken by multiple cameras. The authors start with the camera imaging geometry in terms of homogeneous coordinates and the intrinsic and extrinsic parameters. Next, they describe the epipolar geometry for two, three, and four cameras, introducing such concepts as the fundamental matrix, epipolar lines, epipoles, the trifocal tensor, and the quadrifocal tensor. Then, they present the selfcalibration technique using the absolute dual quadric constraint. Finally, they give the definition of the affine camera model and a procedure for 3-D reconstruction based on it. The detailed algorithms are listed in the Appendix.

Unmanned Aerial Vehicles have a wide spectrum of potential civilian applications as observation and data acquisition platforms. Most of the aerial surveillance applications require low altitude, low speed platforms. The vehicle should ideally be able to hover above an area, allow long duration studies, take-off and land vertically without the need for runways. For such a scenario, lighter-than-air (LTA) vehicles are often better suited than airplanes and helicopters. Chapter 2 introduces the Portuguese airship project named DIVA, and the approaches currently under development and that are being implemented in the areas of robotic integration, navigation and guidance, and vision based surveillance. The airship platform is described, along with an overview of the architecture developed, to document the practical experience gained in this field. Looking at the typical autonomous mission objective, the authors present a control approach for airship path-tracking, covering the whole flight envelope from hover to the normal cruise flight. An asymptotically stable backstepping controller is designed from the airship nonlinear dynamics and kinematics. Some practical issues are then considered and the control law is improved to take into account input saturations and wind disturbances, maintaining its asymptotic stability for a bounded wind estimation error. The presented simulation results illustrate the controller performance during a full realistic mission that covers all the usual tasks: vertical take-off and landing, stabilization and route path-tracking. Wind disturbances are also included. In vision systems used in aerial robotics, inertial and earth field magnetic sensors can provide valuable data about the observer ego-motion, as well as an absolute orientation reference. Here, the inertial orientation measurements are used to compensate the rotational degrees of freedom in two different computer vision tasks: first, inertial data is used to project images on a leveled plane, relaxing the demands on interest point matching algorithms when performing image mosaicing; second, in the rotation-compensated, pure translation case, full homographies are reduced to planar homologies, and the heights over the ground plane on two views are calculated more accurately. Visual odometry for the airship 3D trajectory is performed by calculating the focus of expansion during the motion. These results are important in the context of vision based aerial mapping and aerial vision surveillance. A technique is also presented that allows aerial vision tracking of dynamic objects moving in the ground. Experimental results are presented, illustrating the approach and some achievements.

The focus of the research is to design and build rapidly deployable, adaptive, costeffective, and autonomous distributed robot swarms. The objective is to provide a scientific, yet practical, approach to the design and analysis of swarm behaviors. Chapter 3 provides an overview of their work in this area. First, the authors summarize the basis for our robot control algorithms, which they call *artificial physics* or *physicomimetics*. Unlike biomimetic approaches, the authors focus on robotic behaviors that are similar to those shown by solids, liquids, and gases. Solid formations are useful for distributed sensing tasks, while liquids are for obstacle avoidance tasks. Gases are practical for coverage tasks, such as surveillance and sweeping. Physicomimetics is scalable, robust, and fault-tolerant.

Despite the fact that physicomimetics is amenable to theoretical analyses that guide its use, the fact remains that a real-world environment will often have unanticipated qualities that hurt the performance of the robot swarm. Hence, the authors also describe our novel technique for adaptive swarms. Unlike prior off-line approaches that attempt to re-train the behavior of the swarms in a simulation environment, our on-line approach adapts the behavior of the swarm in real time, while the swarm is performing the task.

In order to function properly the robots in the swarm must be able to accurately localize their local neighbors and to share information. Hence they also outline our enabling hardware technology for swarms of robots. Our plug-in hardware module provides the capability to accurately localize neighboring robots, without using global information and/or the use of vision systems. It also couples localization with data exchange, allowing physicomimetics and adaptation to be fully integrated onto physical robots.

In Chapter 4 the authors propose approaches and equipment for preliminarily announcing and indicating to people the speed and direction of movement of mobile robots moving on a twodimensional plane. They introduce the four approaches categorized into (1) announcing the state just after the present and (2) indicating operations from the present to some future time continuously. To realize the approaches, the authors use omni-directional display (PMR-2), flat-panel display (PMR-6), laser pointer (PMR-1), and projector (PMR-5) for the announcement unit of prototype robots. The four robots were exhibited at the 2005 International Robot Exhibition (iREX05). They had visitors answer questionnaires in a 5-stage evaluation. The projection robot PMR-5 received the highest evaluation score among the four. An examination of differences by gender and age suggested that some people prefer friendly expressions, simple method to inform, and a minimum of information to be presented at one time.

The formal verification of temporal properties is a central issue in the design of real-time control systems, in this context, a multi-formalism framework is proposed using the Timed Automata formalism to describe high-level properties and the DEVS formalism to describe the design-level specification of the control system. The framework introduced in Chapter 5 lays on a sound mathematical basis allowing the formal verification of timed properties (described with timed automata) of the design specification (given by a DEVS model). Furthermore, the convenience of the framework is illustrated by presenting a case study of a subsystem of an industrial Production Cell generally used in the literature to compare formal methods.

3D modelling finds a wide range of applications in robotics research from object recognition and robot localization to path planning and obstacle avoidance. However due to the presence of surface scanning noise and range image registration and fusion errors, the finally reconstructed surfaces are often distorted and thus present obstacles to their applications. In Chapter 6, the authors employ the entropy maximization (EntMax) principle in conjunction with the Tsallis entropy from statistical mechanics to optimize the mesh node locations and normals without altering the mesh node connectivity. While the traditional Shannon entropy can only describe extensive systems, the Tsallis entropy can describe a variety of systems: extensive, sub-extensive, and super-extensive. The nodes in the mesh are indeed entangled and interact with each other. The flexibility of the Tsallis entropy in describing different systems is so useful for noisy surface smoothing, since through adjusting the non-extensivity parameter in the Tsallis entropy, it is possible to model various degrees of interaction among neighbouring nodes in the mesh and thus achieve the desired smoothing effect. A comparative study based on real images shows that the proposed algorithm is easy to implement and effectively smoothes the rough surfaces with their geometric details being desirably retained.

Chapter 7 presents an innovative embedded real-time control architecture, which has been designed, implemented and applied to the field of marine robotics. Based on recent enhancements both in computing power of commercially available off-the-shelf (COTS) boards and in performance of free software, the proposed platform is able to manage the different scheduling requirements of the modules constituting advanced intelligent control architectures. The architecture is being satisfactorily used for developing the family of marine robotic vehicles of CNR-ISSIA. In particular, discussion focuses on the application case of the Charlie unmanned surface vehicle (USV) presenting experimental results in typical operating conditions.

In Chapter 8, the concept of a new mobile haptic interface for unconstrained bimanual manipulation is presented. This device, which has been developed at the High-Fidelity Telepresence and Teleaction Research Centre, Munich, Germany, allows locomotion and haptic interaction at the same time. In contrast to most existing haptic interfaces, it is therefore not restricted to desktop applications but also enables bimanual manipulation tasks with high interaction forces in extended remote or virtual environments. The design of this mobile haptic interface is based on a modular system consisting of two components: two haptic interfaces and a mobile platform. While the haptic interfaces only cover parts of the human arm workspace, the mobile platform extends these to arbitrarily large remote environments. A special design and control concept of the haptic interfaces makes it possible to decouple translational from rotational movements. This decoupling helps to significantly simplify the control algorithms which handle the interaction between the single components. The mobile platform which carries the two haptic interfaces must be positioned in such a way that the manipulability of both haptic interfaces is maximized. Different optimization strategies are presented and compared. The motion of the mobile platform must be synchronized with the control of the haptic interfaces in order to hide the platform motion from the operator. Finally, experimental results are presented.

In Chapter 9 a new cooperative collision-avoidance method formultiple nonholonomic robots with constraints and known start and goal velocities based on Bernstein- Bézier curves is presented. In the presented examples the velocities and accelerations of the mobile robots are constrained and the start and the goal velocity are defined for each robot. This means that the proposed method can be used as subroutine in a huge path-planning problem in the real time, in a way to split the whole path in smaller partial paths. The reference path of each robot from the start pose to the goal pose, is obtained by minimizing the penalty function, which takes into account the sum of all the paths subjected to the distances between the robots, which should be bigger than the minimal distance defined as the safety distance, and subjected to the velocities and accelerations which should be lower than the maximal allowed for each robot. When the reference paths are defined the model predictive trajectory tracking is used to define the control. A prediction model derived from linearized tracking-error dynamics is used to predict future system behavior. A control law is derived from a quadratic cost function consisting of the system tracking error and the control effort. The results of the simulation, real experiments and some future work ideas are discussed.

The authors have developed a standing style transfer system "ABLE" for a person with disabled legs. It realizes travel in a standing position even on uneven ground, standing up motion from a chair, and ascending stairs. ABLE consists of three modules: a pair of telescopic crutches, a powered lower extremity orthosis, and a pair of mobile platforms. The authors show the conceptual design of ABLE and the motion of each module. Cooperative operations using three modules are discussed through simulations. The standing up motion from a chair and ascending stairs, however, had a problem with adaptability to the environment and safety because it had executed the movement that relied on telescopic crutches. To solve these problems, they propose a new motion technique and compare it with the previous method. Some experimental results are also shown in Chapter 10.

Chapter 11 presents reliability and availability analyses of a mathematical model representing one robot and (n-1) standby safety units with a perfect switch to replace a failed safety unit. Robot and safety unit failure rates are assumed constant and the failed system repair times are assumed arbitrarily distributed. General expressions for state probabilities, system availabilities, reliability, and mean time to failure were obtained by using Markov and supplementary variable methods. General expressions for the robot - safety system steady state availability were developed for exponential, gamma, Weibull, Rayleigh, and lognormal failed system repair time distributions. Some plots of these expressions for special case models are shown. These plots shows that the robot-safety system availability, state probabilities decrease with time and robot-safety system steady state availability increases with the increasing values of the safety unit repair rate.

EXPERT COMMENTARY

ROBOTICS RESEARCH TRENDS

Anthony Engwirda
Griffith University

Robotics is a young and dynamic field and we have solved challenges over the last two decades that were thought to be intractable. Reactive robotics and subsumption architecture delivered real world and real time machines. The Honda company demonstrated the P2 and P3 biped robots. These robots were able to walk on two legs and even navigate steps. In Japan there are humanoid robots that dance in synchronous harmony. Advances in control algorithms enabled the mars rovers to operate in the obstacle rich real world without instant contact to mission control. Similar advances enabled the robotic space craft, Deep Space 1 to venture into space and take advantage of opportunistic events. Advances in mobile robot cooperation and real time image recognition and response have resulted in ever more popular and successful Robot Soccer competitions. Advances in Modular Robotics are enabling mobile robots with high levels of polymorphism and self repair. Research by NASA JPL into Cooperative Construction Teams are opening new applications and off world capabilities.

The pace of development within mobile robotics will never be fast enough for some. There exists a perception that mobile robotics has not yet arrived until such time as a humanoid robot stands in a kitchen and washes the dishes. Historical precedent demonstrates that complex technologies take significant time to become reliable and find suitable applications. The history of aircraft is a suitable analogy. The first powered flight was in 1904. It was not until 1914 and the advent of the Great War, that aircraft were considered reliable enough to be pressed into military service. The first application was air reconnaissance. However the reports from pilots of enemy troop concentrations were often dismissed as unreliable exaggeration by senior Officers. It was even thought that any person who wanted to be a pilot was of unsound mind. When the aircraft was introduced to the bombing role, the early bombs were hand grenades and had to be thrown from the aircraft while at low altitude. The early dogfights between aircraft were fought with revolvers, grenades and even bricks. It was not until the deflection plate and the interrupter gear that a machine gun could be forward mounted upon an aircraft. Even so, there were still cases as late as 1918 when pilots shot off their own airscrew. The aircraft technology advanced very quickly during the war. There is one account of a German Ace who was forced down behind enemy lines and captured. He subsequently escaped and returned to his squadron only to find that speed and maneuverability of new aircraft was beyond his ability. Applications for the aircraft began to emerge after the war, including barn storming, joy flights, mail delivery and finally passenger service.

Mobile robotics has reached the milestone of real world and time performance with the advent of reactive robotics and subsumption architecture. The entertainment industry has been a driver for the technology with mobile robot toys and pets. Opportunities for industrial applications of mobile robots remain limited due to our failure to address the key issues of automation and self-sustainability identified by David McFarland, a biologist and early pioneer of mobile robotics. Addressing automation and self-sustainability will improve the relationship between quantities of mobile robots and human supervision.

A typical robotics student will ask two sequential questions. How do I build a mobile robot? Now that I have built it, what can I do with it? A mobile robot that bumps into a wall and follows a line is interesting, but how does that translate into a useful application? This approach is fine for teaching the basics of robotics, but we need to go beyond that barrier and look for the applications that will assist the development of the technology. A self-evident technology is one which is part of everyday life, such as the aircraft, automobile, computer, mobile phone, television, etc. Current applications for mobile robotics do not have this profile are therefore not yet self-evident. I would suggest that we need to ask more questions. What are the strengths and weaknesses of the technology? How can we address some of the weaknesses? Can we leverage the strengths to apply the technology to other applications. Increasing the application domain will act as a driver that will yield and increase in funding for mobile robot research. Investigation of the social consequences of mobile robot technology may also steer the research in a particular direction. The history of the Industrial Revolution may provide insights into the effects of a Robotic Revolution. How will the introduction of mobile robots affect the work environment? What employment opportunities will be lost and what new ones created? If we were able to build machines just like people and place them in our homes, how would that affect us? When it comes time to dispose of our obsolete humanoid robot for the new model, would we become less humane and transfer this same philosophy to the elderly? Would the existence of disposable humanoid robots give rise to a resurgence of racism or the reintroduction of human slavery? It is not sufficient for us to solve the technological problem of how to build robots, we must be mindful of the potential consequences. We stand at the birth of a revolutionary new technology, with the opportunity to focus direction of research along a desirable path and history will hold us accountable. The progression of technology means that our success is inevitable. Therefore, let us deliver a beneficial technology.

It can be argued that there exists misdirected research within any field including robotics. We are often tempted to follow a logic where resources are applied to the direction that offers the greatest perceived probability of success. Resources for other research directions tend to be pruned through lack of interest, funding, or publication. The problem with this approach is that the other directions sometimes yield the desired results. One recent example of this phenomena was the medical research into stomache ulcers. It was generally accepted by the medical community that stomache ulcers were caused by stress. One brave medical practitioner had a theory that a bacteria within the stomach was the predominant cause of this condition. Opponents to the theory argued that ulcer causing bacteria could not survive within the hostile environment of the stomach. The researcher examined the stomache lining and found traces of the bacteria within the folds, which offered partial protection. Evidence to support his theory was discounted and it was assumed that the samples were tainted or the results were inconclusive. To support his argument the doctor swallowed a dose of bacteria and then the cure. It is now generally accepted that the bacteria is the leading cause of stomach ulcers and treatment for patients is available. If the doctor had followed the advice from the majority and abandoned his research then this medical breakthrough would not be available today and we would still be treating stomach ulcer patients for stress. We should never give up on a theory that we believe to be correct due to the weight of popular opinion. This is not to suggest that we should open the floodgates for diverse ideas, but that we should enable a degree of alternative thinking. In his IEEE Presidential message, Fukuda suggested that people from diverse backgrounds could generate new ideas to advance robotics.

The future outlook for mobile robotics remains promising. The trend is gradually making the transition away from the limits of the 3ds of robotics, being dirty, dull and dangerous applications. In their place, there is a rise in entertainment robots and the beginnings of new applications in resource gathering and construction. While the current space applications are driven by the hostile nature of the environment, the economical performance of mobile robots will one day surpass an equivalent human workforce. Static robots in factories have long performed applications such as spray painting and welding. The static robots proved to be economical, reliable and faster than their human counterparts. The key issue restraining self-evident mobile robotic technology is economics. We can and will resolve this issue by addressing applications, performance, cost and Self-Reliance. Our goal should be to demonstrate that mobile robots can perform significant applications in a manner which is superior to human workers.

SHORT COMMUNICATION

Intelligent Modeling and Adaptive Control of Flexible Robot Manipulators

Chang-Woo Park*
Korea Electronics Technology Institute, 401-402 B/D 193, Yakdae-Dong,
Wonmi-Gu, Puchon-Si, Kyunggi-Do, 420-734, Korea

Abstract

In this chapter, in order to control an uncertain flexible joint manipulator, an intelligent modeling method and direct model reference adaptive control via Takagi-Sugeno fuzzy modeling based on the Parallel distributed Compensation (PDC) concept are developed for the MIMO plant model. The proposed control scheme is proposed to provide asymptotic tracking of a reference signal for the systems with uncertain parameters. From Lyapunov stability analysis and simulation results, the developed control law and adaptive law guarantee the boundedness of all signals in the closed-loop system. In addition, the plant state tracks the state of the reference model asymptotically with time for any bounded reference input signal.

Keywords: Fuzzy control, flexible joint manipulator, adaptive control, model reference control

1. Introduction

Many of today's robots are driven by actuators with high gear ratios, such as harmonic drivers for high torque and low operation speed. Furthermore, mechanical damage to the robot and the environment can be minimized in an accidental collision involving the arm as the flexible joints and links can absorb a certain amount of impact force due to collision, and this will render the joint compliance sometimes be a desirable feature [1].

To this end, many researches on the control of flexible joint manipulator have been done such as the model based approaches which include feedback linearization scheme [1] and invarient manifold scheme [2-3], robust control [4] and adaptive control [5-6].

---
* E-mail address: drcwpark@keti.re.kr

However, although the joint flexibility has demonstrated some potential merits, the difficulty with modeling and controlling such a flexible mechanical system with high performance made most robot designers prefer to manufacture mechanically rigid arms with stiff joints. Hence, in this chapter, we sill tackle the problem of controlling for flexible joint robots via fuzzy modeling and fuzzy model based controller and propose a complete solution to solving the problem of model uncertainty.

Fuzzy logic controllers are generally considered applicable to plants that are mathematically poorly understood and where the experienced human operators are available. However, the fuzzy control has not been regarded as a rigorous science due to the lack of the guarantee of the global stability and acceptable performance. To overcome this drawback, since Takagi-Sugeno (TS) fuzzy model [7] which can express a highly nonlinear functional relation in spite of a small number of fuzzy implication rules was proposed, there have been significant research on the stability analysis and systematic design of fuzzy controllers [8-10]. In their research, the nonlinear plant is represented by a TS fuzzy model and the control design is carried out based on the fuzzy model via the so-called Parallel Distributed Compensation(PDC) scheme and Linear matrix inequality based optimization.

In order to deal with the uncertainties of nonlinear systems, in the fuzzy control system literature, a considerable amount of adaptive schemes have been suggested [10-15]. An adaptive fuzzy system is a fuzzy logic system equipped with an adaptive law. The major advantage of adaptive fuzzy controller over the conventional fuzzy controller is that the adaptive fuzzy controller is capable of incorporating linguistic fuzzy information from human operators. Most of them were based on the feedback linearization scheme of indirect adaptive approach in which the approximation ability of the fuzzy system was utilized or an online adaptation scheme was usually used to estimate the unknown parameters of the system and an appropriate controller was then designed to control the plant to satisfy a desired performance.

In this chapter, to control the flexible joint manipulator, we present alternative direct adaptive fuzzy controller based on model reference approach, in which the desired process response to a command signal is specified by means of a parametically defined reference model, for MIMO plants with poorly understood dynamics or plants subjected to parameter uncertainties. We utilized TS fuzzy model for uncertain flexible joint manipulator modeling and PDC. The adaptation law for adjusting the parameters in feedback and feedforward gain of PDC controller is designed so that the plant output tracks the reference model output.

2. Takagi-Sugeno Fuzzy Model Based Control

Consider the continuous-time nonlinear system described by the Takagi-Sugeno fuzzy model. The i th rule of continuous-time TS model is of the following form:

$$R^i: \text{ If } x_1(t) \text{ is } M_1^i \cdots \text{ and } x_n(t) \text{ is } M_n^i \\ \text{then } \dot{x}(t) = A_i x(t) + B_i u(t) \tag{2.1}$$

where,
$$x^T(t) = [x_i(t), x_2(t), \cdots, x_n(t)],$$
$$u^T(t) = [u_i(t), u_2(t), \cdots, u_m(t)]$$

Given a pair of input $(x(t), u(t))$, the final output of the fuzzy system is inferred as follows:

$$\dot{x}(t) = \frac{\sum_{i=1}^{l} w_i(t)\{A_i x(t) + B_i u(t)\}}{\sum_{i=1}^{l} w_i(t)} \qquad (2.2)$$

where, $w_i(t) = \prod_{j=1}^{n} M_j^i(x_j(t))$, and $M_j^i(x_j(t))$ is the grade of membership of $x_j(t)$ in M_j^i.

In order to design fuzzy controllers to stabilize fuzzy system (2.2), we utilize the concept of PDC. The PDC controller shares the same fuzzy sets with fuzzy model (2.2) to construct its premise part. That is, the PDC controller is of the following form:

$$R^i : \text{If } x_1(t) \text{ is } M_1^i \text{ and} \cdots \text{and } x_n(t) \text{ is } M_n^i \qquad (2.3)$$
$$\text{then } u(t) = -K_i x(t)$$

Given a state feedback $x(t)$, the final output of the fuzzy PDC controller (2.3) is inferred as follows:

$$u(t) = -\frac{\sum_{i=1}^{l} \omega_i(t) K_i x(t)}{\sum_{i=1}^{l} \omega_i(t)} \qquad (2.4)$$

Where $w_i(t) = \prod_{j=1}^{n} M_j^i(x_j(t))$

By substituting the controller (2.4) into the model (2.2), we can construct the closed-loop fuzzy control system as following:

$$\dot{x}(t) = \frac{\sum_{i=1}^{l}\sum_{j=1}^{l} w_i(t) w_j(t) \{A_i - B_i K_j\} x(t)}{\sum_{i=1}^{l}\sum_{j=1}^{l} w_i(t) w_j(t)} \qquad (2.5)$$

A sufficient condition for ensuring the stability of the closed-loop fuzzy system (2.5) is given in Theorem 1, which was derived in [2].

Theorem 1: The equilibrium of a fuzzy control system (2.5) is asymptotically stable in the large if there exists a common positive definite matrix P such that

$$G_{ij}^T P + P G_{ij} = -Q_{ij} \tag{2.6}$$

for all $i, j = 1, 2, \cdots, l$
where $G_{ij} = A_i - B_i K_j$ and Q_{ij} is apositive definite matrix.

The design problem of model based fuzzy control is to select K_j $(j = 1, 2, \cdots, l)$ which satisfy the stability conditions(2.6). In [9], the common P problem was solved efficiently via convex optimization techniques for LMI's (Linear Matrix Inequality). However, the fuzzy control (2.4) does not guarantee the stability of system in the presence of parameter uncertainty. Moreover, the design of the control parameters is not possible for the systems whose parameters are unknown, In order to overcome these drawbacks, in this research, an adaptive control scheme is developed for the plant models whose parameters are unknown.

3. Adaptive Fuzzy Control Based on Model Reference Approach

In this section, an adaptive fuzzy model reference control scheme for MIMO TS fuzzy system is developed. Consider again the nonlinear plant represented by the TS model (2.1) of (2.2), where state $x \in R^n$ is available for measurement, $A_i \in R^{n \times n}$, $B_i \in R^{n \times q}$ $(i = 1, \cdots, l)$ are unknown constant matrices and (A_i, B_i) are controllable. The control objective is to choose the input vector $u \in R^q$ such that all signals in the closed-loop plant are bounded and the plant state x follows the state $x_m \in R^n$ of a reference model specified by the system

$$\dot{x}_m = \frac{\sum_{i=1}^{l}\sum_{j=1}^{l} w_i(x)\mu_j(x)\{(A_m)_{ij} x_m + (B_m)_{ij} r\}}{\sum_{i=1}^{l}\sum_{j=1}^{l} w_i(x)\mu_j(x)} \tag{3.1}$$

Where $(A_m)_{ij} \in R^{n \times n}$ $(i = 1, \cdots, l)$ satisfy the stability condition of fuzzy system given in Theorem 1, $(B_m)_{ij} \in R^{n \times q}$, and $r \in R^q$ is a bounded reference input vector. The reference model and input r are chosen so that $x_m(t)$ represents a desired trajectory that x has to follow.

3.1. Control Law Design

If the matrices A_i, B_i were known, we could apply the control law

$$u = \frac{\sum_{j=1}^{l} \mu_j(x)\left(-K_j^* x + L_j^* r\right)}{\sum_{j=1}^{l} \mu_j(x)} \tag{3.2}$$

Where $\mu_j(x) = w_j(x)$, and obtain the closed-loop plant

$$\dot{x} = \frac{\sum_{i=1}^{l}\sum_{j=1}^{l} w_i(x)\mu_j(x)\left\{\left(A_i - B_i K_j^*\right)x + B_i L_j^* r\right\}}{\sum_{i=1}^{l}\sum_{j=1}^{l} w_i(x)\mu_j(x)} \tag{3.3}$$

Hence, if $K_j^* \in R^{q \times n}$, and $L_j^* \in R^{q \times q}$ are chosen to satisfy the algebraic equations

$$A_i - B_i K_j^* = (A_m)_{ij}, \qquad B_i L_j^* = (B_m)_{ij} \tag{3.4}$$

Then the transfer matrix of the closed-loop plant is the same as that of the reference model and $x(t) \to x_m(t)$ exponentially fast for any bounded reference input signal $r(t)$. However, the design of the control parameters is not possible for the system whose parameters are unknown. To overcome this drawback, in this research, following controller is developed for the plant models of which parameters are unknown.

Let us assume that K_j^*, L_j^* in (3.4) exist, i.e., that there is sufficient structural flexibility to meet the control objective, and propose the control law

$$u = \frac{\sum_{j=1}^{l} \mu_j(x)\left(-K_j(t)x + L_j(t)r\right)}{\sum_{j=1}^{l} \mu_j(x)} \tag{3.5}$$

Where, $K_j(t)$, $L_j(t)$ are the estimates of K_j^*, L_j^*, respectively, to be generated by an appropriate adaptive law.

3.2. Adaptive Law Design

By adding an subtracting the desired input term, namely,

$$\sum_{j=1}^{l} \mu_j(x) \left\{ -B_i\left(K_j^* x - L_j^* r\right) / \sum_{j=1}^{l} \mu_j(x) \right\}$$

the plant equation and using (3.4), we obtain

$$\dot{x} = \frac{\sum_{i=1}^{l}\sum_{j=1}^{l} w_i(x)\mu_j(x)(A_m)_{ij}}{\sum_{i=1}^{l}\sum_{j=1}^{l} w_i(x)\mu_j(x)} x$$
$$+ \frac{\sum_{i=1}^{l}\sum_{j=1}^{l} w_i(x)\mu_j(x)}{\sum_{i=1}^{l}\sum_{j=1}^{l} w_i(x)\mu_j(x)} r \qquad (3.6)$$
$$+ \frac{\sum_{i=1}^{l}\sum_{j=1}^{l} w_i(x)\mu_j(x) B_i \left(K_j^* x - L_j^* r + u\right)}{\sum_{i=1}^{l}\sum_{j=1}^{l} w_i(x)\mu_j(x)}$$

Furthermore, by adding and subtracting the estimated input term multiplied by $\sum_{i=1}^{l} w_i B_i / \sum_{i=1}^{l} w_i$, that is,

$$\frac{\sum_{i=1}^{l} w_i B_i}{\sum_{i=1}^{l} \omega_i} \left\{ \frac{\sum_{i=1}^{l} \mu_j(x)\left\{\left(K_j(t)x - L_j(t)r\right)\right\}}{\sum_{j=1}^{N} \mu_j(x)} + u \right\}$$

In the reference model (3.1), we obtain

$$\dot{x}_m \frac{\sum_{i=1}^{l}\sum_{j=1}^{l} w_i(x)\mu_j(x)(A_m)_{ij}}{\sum_{i=1}^{l}\sum_{j=1}^{l} w_i(x)\mu_j(x)} x_m$$

$$+ \frac{\sum_{i=1}^{l}\sum_{j=1}^{l} w_i(x)\mu_j(x)(B_m)_{ij}}{\sum_{i=1}^{l}\sum_{j=1}^{l} w_i(x)\mu_j(x)} \qquad (3.7)$$

$$+ \frac{\sum_{i=1}^{l}\sum_{j=1}^{l} w_i(x)\mu_j(x) B_i\left(K_j(t)x - L_j(t)r + u\right)}{\sum_{i=1}^{l}\sum_{j=1}^{l} w_i(x)\mu_j(x)}$$

By using (3.6) and (3.7), we can express the equation of the tracking error defined as $e(t) \overset{\Delta}{=} x(t) - x_m(t)$, I,e.,

$$\dot{e} = \frac{\sum_{i=1}^{l}\sum_{j=1}^{l} w_i(x)\mu_j(x)(A_m)_{ij}}{\sum_{i=1}^{l}\sum_{j=1}^{l} w_i(x)\mu_j(x)} e$$

$$+ \frac{\sum_{i=1}^{l}\sum_{j=1}^{l} w_i(x)\mu_j(x) B_i\left(-\tilde{K}_j x + \tilde{L}_j r\right)}{\sum_{i=1}^{l}\sum_{j=1}^{l} w_i(x)\mu_j(x)} \qquad (3.8)$$

Where $\tilde{K}_j = K_j(t) - K_j^*$ and $\tilde{L}_j = L_j(t) - L_j^*$.

In the dynamic equation (3.8) of tracking error, B_i are unknown, We assume that L_j^* are either positive definite or negative definite and define $\Gamma_j^{-1} = L_j^* \operatorname{sgn}(l_j)$, where $I_j = 1$ if L_j^* is positive definite and $I_j = -1$ if L_j^* is negative definite. Then $B_i = (B_m)_{ij} L_j^{*-1}$ and (3.8) becomes

$$\dot{e} = \frac{\sum_{i=1}^{l}\sum_{j=1}^{l} w_i(x)\mu_j(x)(A_m)_{ij}}{\sum_{i=1}^{l}\sum_{j=1}^{l} w_i(x)\mu_j(x)} e$$
$$+ \frac{\sum_{i=1}^{l}\sum_{j=1}^{l} w_i(x)\mu_j(x)(B_m)_{ij} L_j^{*-1}(-\tilde{K}_j x + \tilde{L}_j r)}{\sum_{i=1}^{l}\sum_{j=1}^{l} w_i(x)\mu_j(x)} \quad (3.9)$$

Now, by using the tracking error dynamics (3.9), we derive that adaptive law for updating the desired control parameters K_j^*, L_j^* so that the closed-loop plant model (3.6) follows the reference model (3.1). We assume that the adaptive law has the general structure

$$\dot{K}_j(t) = F_j(x, x_m, e, r) , \quad \dot{L}_j = G_j(x, x_m, e, r) \quad (3.10)$$

Where F_j and G_j $(i = 1, \cdots, l)$ are functions of known signals that are to be chosen so that the equilibrium

$$K_{je} = K_j^* \quad L_{je} = L_j^*, \quad e_e = 0 \quad (3.11)$$

We propose the following Lyapunov function candidate

$$V(e, \tilde{K}_j, \tilde{L}_j) = e^T p e + \sum_{j=1}^{l} tr(\tilde{K}_j^T \Gamma_j \tilde{K}_j + \tilde{L}_j^T \Gamma_j \tilde{L}_j) \quad (3.12)$$

Where $P = P^T > 0$ is a common positive definite matrix of the Lyapunov equations $(A_m)_{ij}^T P + P(A_m)_{ij} < -Q_{ij}$ for all $Q_{ij} = Q_{ij}^T > 0$ $(i, j = 1, \cdots, l)$ whose existence is guaranteed by the stability assumption for A_m. Then, after some straightforward mathematical manipulations, we obtain the time derivative $\dot{V}$ of V along the trajectory of (3.9), (3.10) as

$$\dot{V} = -e^T \frac{\sum_{i=1}^{l}\sum_{j=1}^{l} w_i(x)\mu_j(x) Q_{ij}}{\sum_{i=1}^{l}\sum_{j=1}^{l} \omega_i(x)\mu_j(x)} e$$

$$+2tr\left\{\frac{\sum_{i=1}^{l}\sum_{j=1}^{l}w_i(x)\mu_j(x)\tilde{K}_j^T\Gamma_j(B_m)_{ij}^T\mathrm{sgn}(l_j)}{\sum_{i=1}^{l}\sum_{j=1}^{l}w_i(x)\mu_j(x)}Pe\ x^T\right.$$

$$+\sum_{j=1}^{l}\tilde{K}_j^T\Gamma_j\tilde{K}_j$$

$$+2tr\left\{\frac{\sum_{i=1}^{l}\sum_{j=1}^{l}w_i(x)\mu_j(x)\tilde{L}_j^T\Gamma_j(B_m)_{ij}^T\mathrm{sgn}(l_j)}{\sum_{i=1}^{l}\sum_{j=1}^{l}w_i(x)\mu_j(x)}Pe\ r^T\right.$$

$$+\sum_{j=1}^{l}\tilde{L}_j^T\Gamma_j L_j\ \}$$

(3.13)

In the last two terms of (3.13), of we let

$$\sum_{j=1}^{l}\tilde{K}_j^T\Gamma_j\tilde{K}_j = \frac{\sum_{i=1}^{l}\sum_{j=1}^{l}w_i(x)\mu_j(x)\tilde{K}_j^T\Gamma_j(B_m)_{ij}^T\mathrm{sgn}(l_j)}{\sum_{i=1}^{l}\sum_{j=1}^{l}w_i(x)\mu_j(x)}Pe\ x^T \quad (3.14a)$$

$$\sum_{j=1}^{l}\tilde{L}_j^T\Gamma_j\tilde{L}_j = -\frac{\sum_{i=1}^{l}\sum_{j=1}^{l}w_i(x)\mu_j(x)\tilde{L}_j^T\Gamma_j(B_m)_{ij}^T\mathrm{sgn}(l_k)}{\sum_{i=1}^{l}\sum_{j=1}^{l}w_i(x)\mu_j(x)}Pe\ r^T \quad (3.14b)$$

We can make V to be negative, i.e.,

$$V = -\ e^T\frac{\sum_{i=1}^{l}\sum_{j=1}^{l}w_i(x)\mu_j(x)Q_{ij}}{\sum_{i=1}^{l}\sum_{j=1}^{l}w_i(x)\mu_j(x)}e \leq 0 \quad (3.15)$$

Hence, the obvious choice for adaptive law to make V negative is

$$\dot{\tilde{K}}_j = \dot{K}_j(t) = \frac{\sum_{i=1}^{l} w_i(x)\mu_j(x)(B_m)_{ij}^T \operatorname{sgn}(l_j)}{\sum_{i=1}^{l}\sum_{j=1}^{l} w_i(x)\mu_j(x)} Pex^T$$

$$= \left\{ \frac{\sum_{i=1}^{l} w_i (B_m)_{ij}^T}{\sum_{i=1}^{l} w_i} \right\} \left\{ \frac{\mu_j}{\sum_{j=1}^{l} \mu_j} \right\} \operatorname{sgn}(l_j) Pex^T$$

(3.16a)

$$\dot{\tilde{L}}_j = \dot{L}_j(t) = \frac{\sum_{i=1}^{l} w_i(x)\mu_j(x)(B_m)_{ij}^T \operatorname{sgn}(l_j)}{\sum_{i=1}^{l}\sum_{j=1}^{l} w_i(x)\mu_j(x)} Per^T$$

$$= \left\{ \frac{\sum_{i=1}^{l} w_i (B_m)_{ij}^T}{\sum_{i=1}^{l} w_i} \right\} \left\{ \frac{\mu_j}{\sum_{j=1}^{l} \mu_j} \right\} \operatorname{sgn}(l_j) Per^T$$

(3.16b)

Theorem 2: Consider the plant model (2.2) and the reference model (3.1) with the control law (3.5) an adaptive law (3.26). Assume that the reference input r and the state x_m of the reference model are uniformly bounded. Then the control law (3.5) and the adaptive law (3.16) and the adaptive law (3.26) guarantee that

(i) $K(t)$, $L(t)$, $e(t)$ are bounded
(ii) $e(t) \to 0$ as $t \to \infty$

Proof

From (3.12) and (3.15), it directly follows that V is a Lyapunov function for the system (3.9), (3.10), which implies that the equilibrium given by (3.11) is uniformly stable, which, in turn, implies that the trajectory $\tilde{K}(t), \tilde{L}(t), e(t)$ is bounded for all $t > 0$. Because $e = x - x_m$ and $x_m \in \pounds_\infty$, we have that $r \in \pounds_\infty$. From (3.5) and $r \in \pounds_\infty$, we also have that $u \in \pounds_\infty$; therefore, all signals in the closed-loop a re bounded.

Now, let us show that $e \in \pounds_2$. From (3.12) and (3.15), we conclude that because V is bounded from below and is nonincreasing with time, it has a limit, i.e.,

$$\lim_{t\to\infty} V(e(t),\widetilde{K}_j(t)\widetilde{L}_j(t)) = V_\infty < \infty \tag{3.17}$$

From (3.15) and (3.17), it follows that

$$\int_0^\infty e^T \left(\frac{\sum_{i,j=1}^l w_i\mu_j Q_{ij}}{\sum_{i,j=1}^l w_i\mu_j} \right) e\, d\tau = -\int_0^\infty (V_0 - V\infty) \tag{3.18}$$

Where

$$V_0 = V(e(0),\widetilde{K}_j(0)\widetilde{L}_j(0))$$

On the other hand, from $0 \le w_i \le 1$,

$$0 \le \mu_j \le 1, \text{ and } \lambda_{\min}(Q_{ij})\|e\|^2 \le e^T Q_{ij} e \le \lambda_{\max}(Q_{ij})\|e\|^2$$

We have

$$\{\lambda_{\min}(Q_{ij})\}_{\min} \|e\|^2 \le e^T \left(\frac{\sum_{i,j=1}^l w_i\mu_j Q_{ij}}{\sum_{i,j=1}^l w_i\mu_j} \right) e$$

$$\le \{\lambda_{\max}(Q_{ij})\}_{\max} \|e\|^2 \tag{3.19}$$

Where

$$\{\lambda_{\min}(Q_{ij})\}_{\min} = \min\{\lambda_{\min}(Q_{11}),\cdots,\lambda_{\min}(Q_{ll})\}$$

$$\{\lambda_{\max}(Q_{ij})\}_{\max} = \max\{\lambda_{\max}(Q_{11}),\cdots,\lambda_{\max}(Q_{ll})\}$$

After inserting (3.19) into (3.18), and straightforward manipulation, we have

$$\frac{(V_0 - V_\infty)}{\{\lambda_{\min}(Q_{ij})\}_{\min}} \le \int_0^\infty \|e\|^2 d\tau \le \frac{(V_0 - V_\infty)}{\{\lambda_{\max}(Q_{ij})\}_{\max}}$$

Which implies that $e \in \pounds_2$. Because $e, \tilde{K}_j, \tilde{L}_j, r \leq \pounds_\infty$, it follows from (3.9) that $\dot{e} \in \pounds_\infty$, which, together with $e \in \pounds_2$, implies that $e(t) \to 0$ as $t \to \infty$.

4. Tracking Control of an Uncertain Flexible Joint Manipulator

In this section, the validity and effectiveness of the proposed controller are examined through the simulation of tracking control for a flexible joint manipulator.

The control objective is to follow a given trajectory $q_d(t)$ and to produce a torque vector u such that the trajectory error approaches 0 as $t \to 0$. In the simulation, we examine the effects of parametric variation on behaviors of the closed-loop systems with the proposed TS model based adaptive control scheme.

4.1. Intelligent Modeling for Flexible Joint Manipulator

In order to apply the suggested AFC, we need a TS fuzzy model representation of the manipulator.

After the T-S fuzzy model was proposed there have been efforts to construct an efficient T-S fuzzy model for a given nonlinear system. If the T-S fuzzy model is not exactly modeling the nonlinear system, the designed controller may not be able to guarantee the control performance and the stability of the closed loop control system.

To develop a systematic procedure, an T-S fuzzy modelling method, exact T-S fuzzy modeling has recently been developed. The basic idea of exact T-S fuzzy modeling for nonlinear systems has first been discussed in [16]. Here, the word "exact" means that the defuzzified output of the constructed T-S fuzzy model is mathematically identical to that of the original nonlinear system.

Consider the single link flexible joint manipulator shown in Fig. 1 whose dynamics can be written as

$$\dot{x}_1 = x_2$$

$$\dot{x}_2 = -\frac{MgL}{I}\sin x_1 - \frac{k}{I}(x_1 - x_3)$$

$$\dot{x}_3 = x_4$$

$$\dot{x}_4 = \frac{K}{J}(x_1 - x_3) + \frac{1}{J}u \quad (4.1)$$

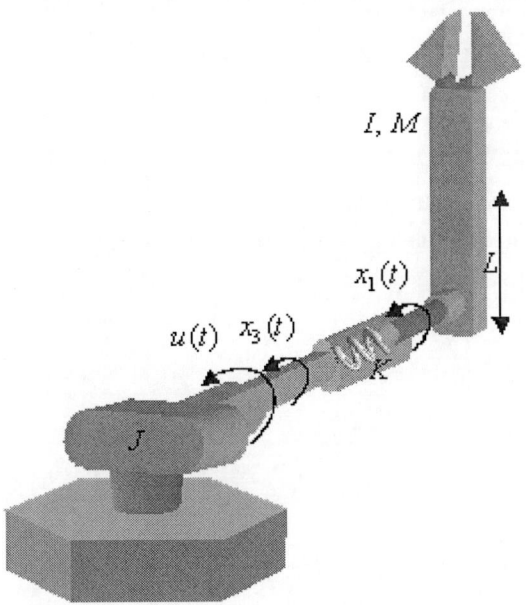

Figure 1. Flexible joint manipulator configuration.

Where, I, J are, respectively, the link and the rotor inertia moments, M is the link mass, k is the joint elastic constant, L is the distance from the axis of the rotation to the link center of mass and g is the gravitational acceleration respectively.

The system (4.1) has a nonlinear term, $\sin(x_1(t))$. If this nonlinear term can be represented as a weighted linear sums of some linear functions, then the TS fuzzy model of (4.1) can be constructed. For this purpose, we first need the following theorem.

Theorem 3 [17]: Consider the following nonlinear term:

$$f_n = x_1 x_2 \cdots x_n \quad , \quad \text{where} \quad x_i \in \left[\Omega_1^i \quad \Omega_2^i \right].$$

It can be exactly represented by a linear weighted sum of the form

$$f_n = (\sum_{i_2, i_3, \cdots i_n = 1}^{2} \mu_{i_2 i_3 \cdots i_n} \bullet g_{i_2 i_3 \cdots i_n}) x_1 \tag{4.2}$$

Where

$g_{i_2 i_3 \cdots i_n} = \prod_{j=2}^{n} \Omega_{i\,ij}^{j,}$ $\mu_{i_2 i_3 \cdots i_n} = \prod_{j=2}^{n} \Gamma_{i_j}^{j}$ in which Γ_{ij}^{j} is positive semi-definite for all $x_j \in \left[\Omega_1 \quad \Omega_2 \right]$, defined as follows.

$$\Gamma_1^j = \frac{-x_j + \Omega_2^j}{\Omega_2^j - \Omega_1^{jj}}, \quad \Gamma_2^j = \frac{x_j - \Omega_2^j}{\Omega_2^j - \Omega_1^j} \tag{4.3}$$

Proof.
Theorem 3 can be proved using inductive reasoning.

If $n=1$, then Theorem 3 is obviously true. When $n=2$, the nonlinear equation is $f_2 = x_1 x_2$, which can be represented as the weighted sum of linear functions of x_1 as follows.

$$f_2 = (\sum_{i_2=1}^{2} \mu_{i_2} q_{i_2}) x_1 = x_2 x_1 \tag{4.4}$$

Where

$$g_1 = \Gamma_1^2, \quad g_2 = \Gamma_2^2, \quad \mu_1 = \frac{-x_2 + \Omega_2^2}{\Omega_2^2 - \Omega_1^2}, \quad \mu_2 = \frac{w_2 + \Omega_2^2}{\Omega_2^2 - \Omega_1^2}$$

Assuming that Theorem 3 holds when $n=k$, then the nonlinear function $f_{k+1} = x_1 x_2 \cdots x_{k+1}$ can be represented by a weighted linear sum of linear functions of x_1 in the following form.

$$\begin{aligned} f_{k+1} &= ((\sum_{i_2,i_3,\cdots,i_k=1}^{2} \mu_{i_2 i_3 \cdots i_k} g_{i_2 i_3 \cdots i_k})(\Gamma_1^{k+1}\Omega_1^{K+1} + \Gamma_2^{K+1}\Omega_2^{K+1}))x_1 \\ &= (\sum_{i_2,i_3,\cdots,i_k=1}^{2} \mu_{i_2 i_3 \cdots i_k 1} g_{i_2 i_3 \cdots i_k 1} + \mu_{i_2 i_3 \cdots i_k 2} g_{i_2 i_3 \cdots i_k 2}) x_1 \\ &= (\sum_{i_2,i_3,\cdots,i_{k+1}=1}^{2} \mu_{i_2 i_3 \cdots i_{k+1}} g_{i_2 i_3 \cdots i_{k+1}}) x_1 \end{aligned} \tag{4.5}$$

Hence, Theorem 3 holds for all n.

Corollary 1 [17]: Assume $x(t) \in [\Omega_1 \quad \Omega_2]$, The nonlinear term

$$f(x(t)) = \sin(x(t)) \tag{4.6}$$

Can be represented by a linear weighted sum of linear functions of the form

$$f(x(t)) = \left(\sum_{i=1}^{2} \mu_i g_i(x(t))\right) x(t) \tag{4.7}$$

Where, $g_1(x(t)) = 1$, $g_2(x)t) = \alpha$ and

$$\mu_1 = \Gamma_1, \quad \mu_2 = \Gamma_2$$

$$\Gamma_1 = \frac{\sin(x(t)) - \alpha x(t)}{(1-\alpha)x(t)}, \quad \Gamma_2 = \frac{x(t) - \sin(x(t))}{(1-\alpha)x(t)},$$

for $x(t) \neq 0$

$$\Gamma_1 = 1, \quad \Gamma_2 = 0 \quad \text{for } x(t) = 0 \text{ and}$$

$$\alpha = \sin^{-1}(\max(\Omega_1, \Omega_2))$$

Proof.
It follows directly from Theorem 3.
Using corollary 1, an exact TS fuzzy model of (4.1) can be represented as follows [17].
Plant rules:

Rule 1: IF $x_1(t)$ is about Ω_1 THEN $\dot{x}(t) = A_1 x(t) + B_1 u(t)$

Rule 2: IF $x_1(t)$ is about Ω_2 THEN $\dot{x}(t) + A_2 x(t) + B_2 u(t)$ (4.8)

Where

$$A_1 = \begin{bmatrix} 0 & 1 & 0 & 0 \\ -\frac{Mgl}{I} & 0 & \frac{k}{I} & 0 \\ 0 & 0 & 0 & 1 \\ \frac{k}{J} & 0 & \frac{-k}{J} & 0 \end{bmatrix}, \quad A_2 = \begin{bmatrix} 0 & 1 & 0 & 0 \\ -\frac{\alpha Mgl}{I} & -\frac{k}{I} & 0 & \frac{k}{I} & 0 \\ 0 & 0 & 0 & 1 \\ \frac{k}{J} & 0 & -\frac{k}{J} & 0 \end{bmatrix}$$

$$B_1 = B_2 = \begin{bmatrix} 0 \\ 0 \\ 0 \\ \frac{k}{J} \end{bmatrix} \quad (4.9)$$

And the membership functions ' *about* Ω_1 ' and ' *about* Ω_2 ' are, respectively,

$$\Gamma_1(x_1) = \frac{\sin(x_1(t)) - \alpha x_1(t)}{(1-\alpha)x_1(t)}, \quad \Gamma_2(x_1) = \frac{x_1(t) - \sin(x_1(t))}{(1-\alpha)x_1(t)}, \quad (4.10)$$

for $x_1(t) \neq 0$

$$\Gamma_1 = 1, \quad \Gamma_2 = 0, \quad \text{for} \quad x_1(t) = 0$$

where $\alpha = \sin^{-1}(\max(\Omega_1, \Omega_2))$ and Γ_i is positive definite for all $x_1(t) \in [\Omega_1 \quad \Omega_2]$. In the simulation, $[\Omega_1 \quad \Omega_2]$ was chosen as $[-2.85 \quad 2.85]$.

Although the exact fuzzy model of the flexible joint manipulator does not have any modeling uncertainties since the deffuzzified output of the TS fuzzy model is exactly same to that of original nonlinear flexible joint manipulator (4.1), the exact modeling scheme may have some demerit. If the nonlinearities in the system model have very complicated form of the number of them is very large, the methodology presented in Theorem 3 can not be applied easily.

An alternatice TS fuzzy modeling technique, the linearization method is often utilized to construct a T-S fuzzy model for a nonlinear system. The linearization based TS fuzzy modeling technique is the most popular as it is simple and the consequent rule base becomes intuitive although the modeling error inevitably exists.

By applying the Lyapunov linearization method [17] at operating points $x_1 = -\pi, 0, \pi$, we obtain the TS fuzzy model for the robot manipulator as followings.

Rule 1 : IF x_1 is about $-\pi$ THEN $\dot{x} = A_1 x + B_1 u$
Rule 2 : IF x_1 is about 0 THEN $\dot{x} = A_2 x + B_2 u$
Rule 3 : IF $x_1 \in$ about π THEN $\dot{x} = A_3 x + B_3 u$ (4.11)

$$A_1 = A_3 = \begin{bmatrix} 0 & 1 & 0 & 0 \\ \frac{Mgl}{I} - \frac{k}{I} & 0 & \frac{k}{I} & 0 \\ 0 & 0 & 0 & 1 \\ \frac{k}{J} & 0 & -\frac{k}{J} & 0 \end{bmatrix}, \quad A_2 = \begin{bmatrix} 0 & 1 & 0 & 0 \\ \frac{Mgl}{I} - \frac{k}{I} & 0 & \frac{k}{I} & 0 \\ 0 & 0 & 0 & 1 \\ \frac{k}{J} & 0 & -\frac{k}{j} & 0 \end{bmatrix} \quad \text{and}$$

$$B_1 = B_2 = B_3 = \begin{bmatrix} 0 \\ 0 \\ 0 \\ \frac{1}{J} \end{bmatrix}$$

The whole state space formed by state vector of the original nonlinear equations is partitioned into three different fuzzy subspaces whose center is located at the center of corresponding membership functions shown is Fig. 2

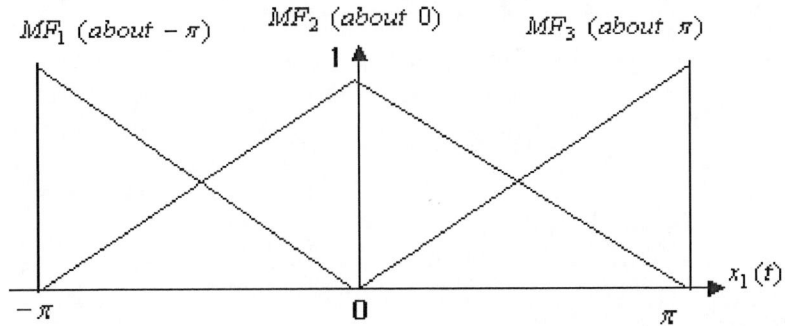

Figure 2. Membership functions.

4.2. Control Results

In order to apply the proposed adaptive fuzzy control scheme, the reference model for the plant state x to follow should be specified. In this simulation, the closed-loop eigenvalues for each subsystem are chosen to be the same, which in turn make the reference model for each fuzz subspace to be the same and linear one as following:

$$\dot{x}_m = \begin{bmatrix} 0 & 1 & 0 & 0 \\ 0 & 0 & 1 & 0 \\ 0 & 0 & 0 & 1 \\ -4 & -10 & -10 & -5 \end{bmatrix} x_m + \begin{bmatrix} 0 \\ 0 \\ 0 \\ 1 \end{bmatrix} r \qquad (4.12)$$

The PDC controller shares the same fuzzy sets with fuzzy model to construct its premise part. That is, the PDC controller is of the following form:

$$R^i : If \ x_1 is MF_i,$$
$$then \quad u(t) = -K_i [x_1 x_2 x_3 x_4]^T + L_i r(t) \qquad (4.13)$$

The feedback control gains K_i and L_i of each fuzzy state feedback controller is updated by adaptive law so that the closed-loop plant follows the reference model (4.12).

Now by using (3.16), we derive the adaptive law for updating the elements of K_j and L_j so that the closed-loop plant follows the reference model.

$$\dot{K}_j(t) = \left\{ \frac{\mu_j}{\sum_{j=1}^{2} \mu_j} \right\} sgn(l_j) B_m^T Pex^T$$

$$\dot{L}_j(t) = \left\{ \frac{\mu_j}{\sum_{j=1}^{2} \mu_j} \right\} sgn(l_j) B_m^T Per^T \quad (4.14)$$

Where $B_m^T = \begin{bmatrix} 0 & 0 & 0 & 1 \end{bmatrix}$.

The parameters of nominal plant model used in this simulation are as follows.

$$M = 0.2687 Kg, \quad I = 0.03 Kg.m^2, \quad L = 1m, \quad k = 31 N/m,$$

$$J = 0.004 Kg.m^2, and\, g = 9.8 m/s^2 \quad (4.15)$$

To test the adaptation abilities of the proposed scheme, the mass of link is varied with time as $m = 0.2687 + 0.15 \sin 3\pi t$ and the initial value for state x_1 is assumed as $x_1 = \frac{\pi}{6}$.

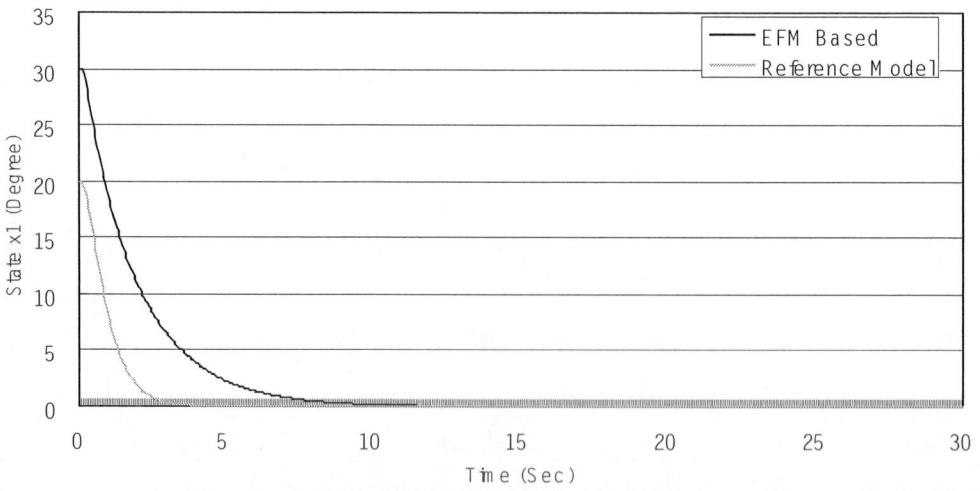

Figure 3. Regulation based on EFM.

The designed adaptive fuzzy controller was applied to the original nonlinear model of flexible joint manipulator (4.1) in the simulation. Figure 3,4 and 5 show the simulation results of regulation of joint angle with exact fuzzy model (EFM) and linearization based fuzzy model (LFM). From these figures, It is shown that the regulation problem can be solved under parametric uncertainties. Figure 6, 7 and 8 show the tracking control results with both EFM and LFM. In both cases, the tracking can be accomplished successfully. The response characteristics of EFM based control such as response time is better than that of LFM based control. This is due to the fuzzy modeling ability of LFM. If more fuzzy rules, that is, linearization at more operating points can be possible, the difference between both the models can be reduced.

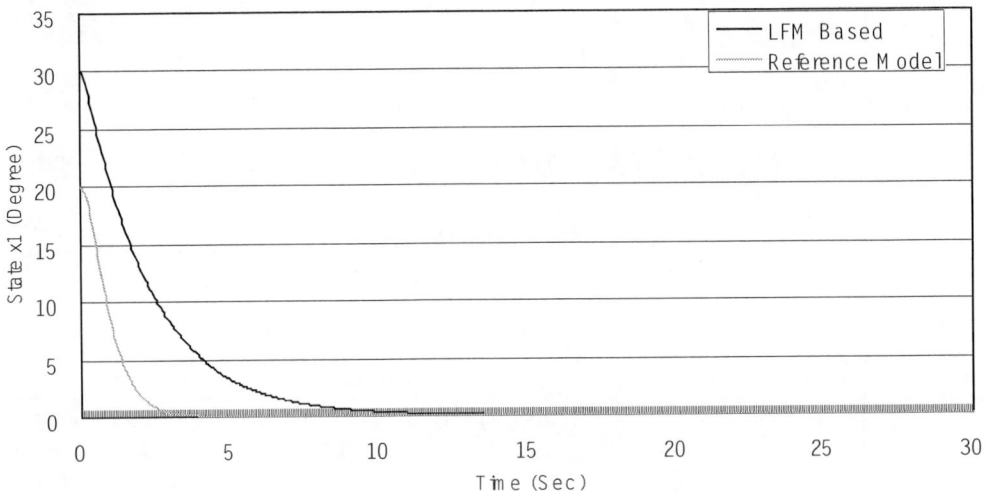

Figure 4. Regulation based on LFM.

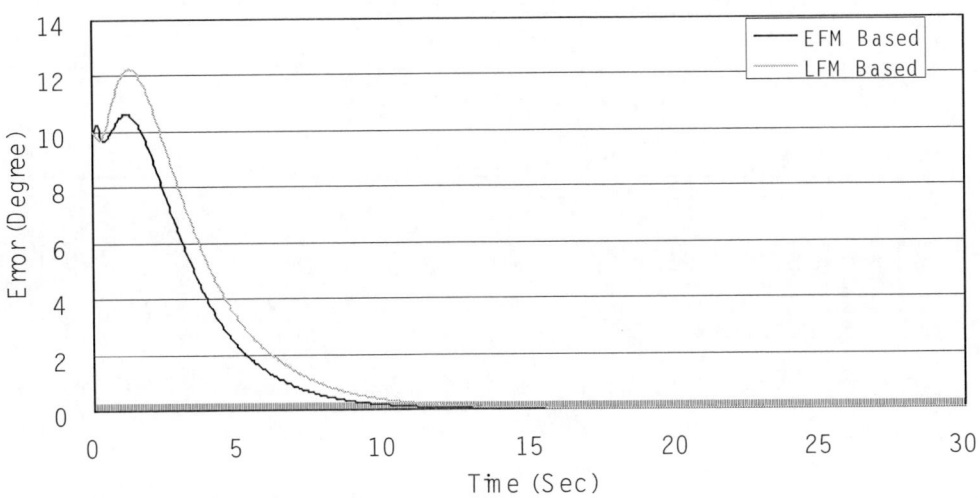

Figure 5. Regulation error.

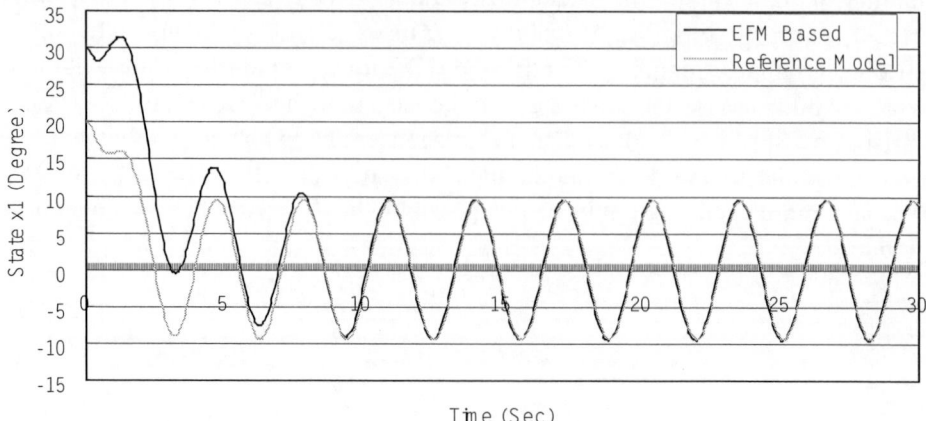

Figure 6. Tracking based on EFM.

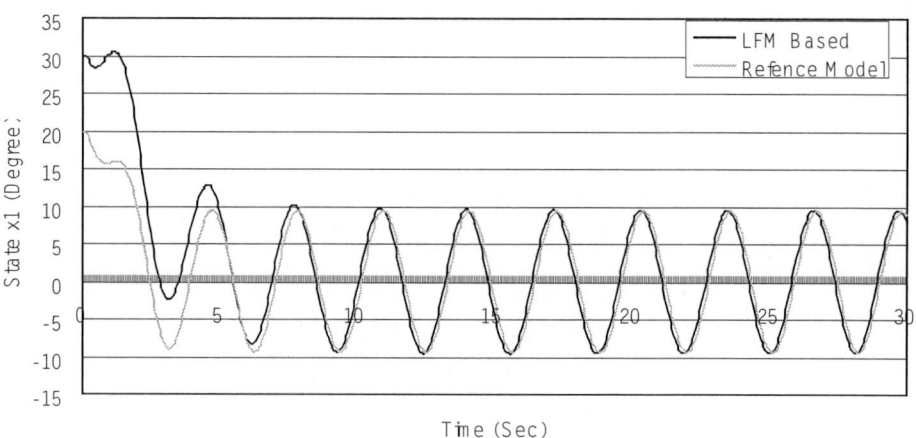

Figure 7. Tracking based on LFM.

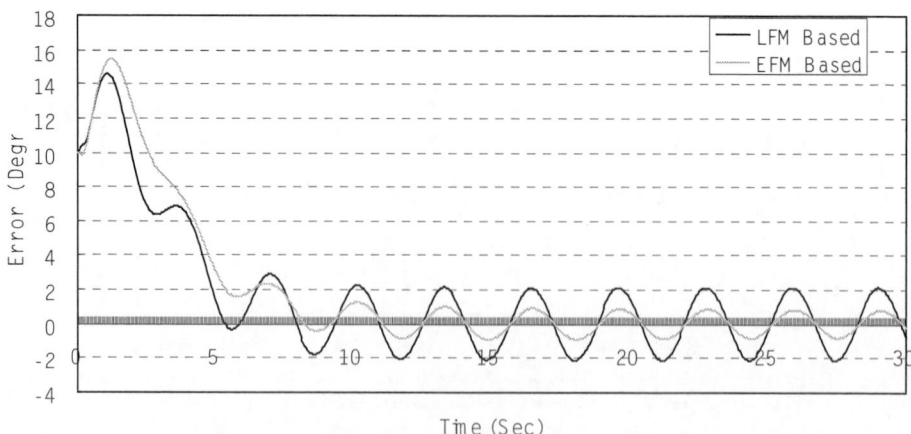

Figure 8. Tracking error.

5. Conclusion

In this chapter, we have developed an alternative TS fuzzy model based adaptive control scheme via model reference approach for flexible joint manipulators with parameter uncertainty in their model. We have used exact fuzzy modeling method and linearization based modeling method to represent the flexible joint manipulator. The adaptation law adjusts the controller parameters on-line so that the plant output tracks the reference model output, The developed adaptive law guarantees the boundedness of all signals in the closed-loop system and ensures that the plant state tracks the state of the reference model asymptotically with time for any bounded reference input signal. The proposed adaptive fuzzy control scheme was applied to tracking control of a single link flexible joint manipulator to verify the validity and effectiveness of the control scheme. From the simulation results, we conclude that the suggested scheme can effectively achieve the trajectory tracking in spite of parameter perturbation.

References

[1] Khorasani, K.: 'Nonlinear feedback control of flexible joint manipulator: a single link case study,' *IEEE trans. Automatic Control*, 1990, 35, (10), pp.1145-1149
[2] Khorasani, K. and Spong, M. W.: 'Invariant manifolds and their application to robot manipulators with slexible joints,' *IEEE Int. Conf. of Robotics and Automation*, 1985, St. Louis, p.110-116
[3] Khorasani, K. and Kokotovic, P. V.: 'Feedback linearization of a flexible manipulator near its rigid body manifold.' *Systems and Control Letters*, 1985, 6, pp.187-192
[4] Spong, M. W.: 'Modeling and control of elastic joint manipulators,' *ASME Journal of Dynamic Systems, Measurement and Control*, 1987, 109, pp.310-319
[5] Lozano, R. and Brogliato, B.: 'Adptive control of robot manipulators with flexible joints,' *IEEE Trans. Automatic Control*, 1992, AC-37, pp.174-181
[6] Chen, K. P. and Fu, L. C.: 'Nonlinear adaptive motion control for a manipulator with flexible joints,' *Proc. 1989 IEEE Int. Conf. on Robotics and Automation*, 1989, Phoenix, AZ, pp.1201-1207
[7] Takagi, T. and Sugeno, M.: 'Fuzzy identification of systems and its applications to modeling and control,' *IEEE Trans. Syst., Man, Cybern.*, 1985, 15, (1), pp.116-132
[8] Tanaka, K. and Sugeno, M.: 'Stability analysis and design of fuzzy control systems,' *Fuzzy Sets and Syst.*, 1992, 45, (2), pp.135-156
[9] Wang, H. O., Tanaka, K. and Griffin, M. F.: 'An approach to fuzzy control of nonlinear systems: stability and design issues,' *IEEE Tans. Fuzzy Syst.*, 1996, 4, (1), pp.14-23
[10] Chen, B. S., Lee, C. H. and Cang, Y. C.: 'H tracking design of uncertain nonlinear SISO systems: adaptive fuzzy approach,' *IEEE Tans, Fuzzy Syst.*, 1996, 4, (1), pp.32-42
[11] Spooner, J. T. and Passino, K. M.: 'Stable adaptive control using fuzzy systems and neural networks,' *IEEE Tans. Fuzzy Syst.*, 1996, 4, (3), pp.339-359
[12] Wang, L. X.: 'Stable adaptive fuzzy controllers with application to inverted pendulum tracking,' *IEEE Tans. Fuzzy Syst.*, 1996, 26, (5), pp.667-691
[13] Tsay, D. L., Chung, H. Y. and Lee, C. J.: 'The adaptive control of nonlinear systems using the Sugeon-type of fuzzy logic,' *IEEE Tans. Fuzzy Syst.*, 1999, 7, (2), pp.225-229

[14] Fischle, K. and Schroder, D.: 'An improved stable adaptive fuzzy control method,' *IEEE Tans. Fuzzy Syst.*, 1999, 7, (1), pp.27-40

[15] Leu, Y. G., Wang, W. Y. and Lee, T. T.: 'Robust adaptive fuzzy-neural controllers for uncertain nonlinear systems,' *IEEE Tans. Robot. And Automat.*, 1999, 15, (5), pp.805-817

[16] Kawamoto, S.,: 'Nonlinear control and rigorous stability analysis based on fuzzy system for inverted pendulum,' *Porc, IEEE Int. Conf. Fuzzy Systems*, Sep. 1996, New Orleans, pp.1427-1432

[17] Lee, H. J., Park, J. B. and Chen G.: 'Robust fuzzy control of nonlinear systems with parametric uncertainties,' *IEEE Tans. Fuzzy* Syst., 2001, 9, (2), PP.369-379

RESEARCH AND REVIEW STUDIES

Chapter 1

LATEST PROGRESS OF 3-D RECONSTRUCTION FROM MULTIPLE CAMERA IMAGES

Kenichi Kanatani
Dept. of Information Technology,
Okayama Univ., Okayama

Abstract

This chapter summarizes recent progress of the theories and techniques for 3-D reconstruction from multiple images taken by multiple cameras. We start with the camera imaging geometry in terms of homogeneous coordinates and the intrinsic and extrinsic parameters. Next, we describe the epipolar geometry for two, three, and four cameras, introducing such concepts as the fundamental matrix, epipolar lines, epipoles, the trifocal tensor, and the quadrifocal tensor. Then, we present the self-calibration technique using the absolute dual quadric constraint. Finally, we give the definition of the affine camera model and a procedure for 3-D reconstruction based on it. The detailed algorithms are listed in the Appendix.

1. Introduction

Analyzing camera or video images for understanding the 3-D meaning of the captured scene is generally known as *computer vision* (also *machine vision*, *robot vision*, or *image understanding*, depending on the emphasis of the researchers), which is one of the most crucial elements of autonomous robotic operations. In general terms, the procedure consists of the following three stages:

- Image processing for detecting, extracting, and matching *features*, which can be points, lines, regions, or anything that is characteristic to that scene.

- Acquiring *metric* information such as locations, orientations, distances, sizes, and motions of the objects in the scene.

- Obtaining *semantic* information such as classification, recognition, labeling, indexing, and retrieval of specific objects in the scene.

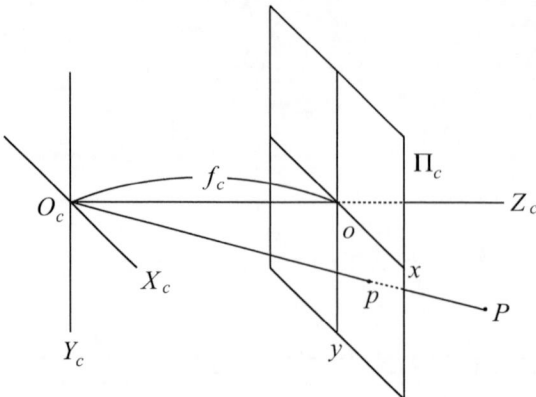

Figure 1. Perspective projection.

These three stages roughly correspond to what has historically been known as *early* (or *low-level*) *vision*, *intermediate-level vision*, and *high-level vision*, respectively [22]. However, these are not necessarily treated separately. In fact, these stages are closely and interactively interwoven in most real computer vision systems.

One of the essential techniques for the second stage is to compute the 3-D shape of the scene or objects from multiple images, know as *3-D reconstruction* or *structure from motion* (*SFM*). This computation critically depends on the *camera imaging geometry*, i.e., the geometric relationship between a 3-D scene and its projection onto a 2-D image. In contrast, analysis for the third stage crucially relies on the *domain knowledge* specific to individual applications such as faces, gestures, gaits, traffic, aerial photographs, and medical images.

Although the third stage is the ultimate goal of computer vision, it is still a very challenging task, and no universally satisfactory technologies have yet been established. However, the 3-D reconstruction technique for the second stage has been extensively studied in the last few decades to arrive at almost definitive conclusions. The aim of this chapter is to present thus established latest technologies of 3-D reconstruction from multiple images. Standard textbooks on this subject are, for example, [4, 5, 6, 8, 13, 14, 15, 23, 44].

2. Camera Imaging Geometry

2.1. Perspective Projection

We identify an image, or a photograph, with a mapping from a 3-D scene onto a 2-D plane and call this mapping the *camera model*. The standard model is *perspective projection* (Fig. 1): we imagine in the scene a point O_c, called the *viewpoint*, and a plane Π_c, called the *image plane* or *retina*, and assume that a point P in the scene is mapped to the intersection p of the image plane Π_c with the line O_cP, called the *line of sight*. This models an ideal *pin-hole camera* and is known to describe real cameras with sufficient accuracy.

The line starting from the viewpoint O_c and perpendicularly passing through the image plane Π_c is called the *optical axis*. We define an $X_cY_cZ_c$ coordinate system with the origin at the viewpoint O_c and the Z_c-axis along the optical axis. The intersection o of the optical

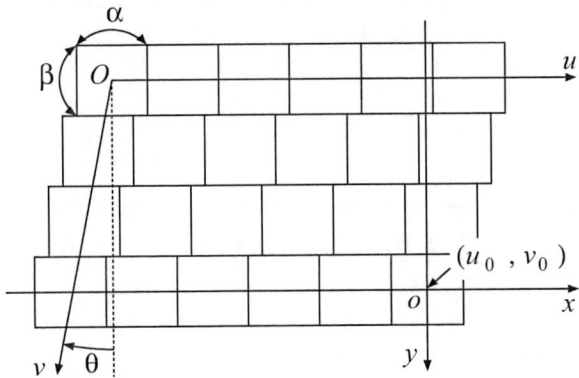

Figure 2. Pixels and the image coordinate system.

axis with the image plane Π_c is called the *principal point*. We define an xy coordinate system with the origin at the principal point o and the x- and the y-axes parallel to the X_c- and the Y_c-axes, respectively (Fig. 1). Then, a point (X_c, Y_c, Z_c) in the scene is projected onto a point (x, y) in the image plane given by

$$x = f_c \frac{X_c}{Z_c}, \qquad y = f_c \frac{Y_c}{Z_c}, \qquad (1)$$

where f_c, called the *focal length*, is the distance from the viewpoint O_c to the image plane Π_c.

2.2. Pixel Coordinates

In real cameras, the image plane corresponds to the array of photo-cells, or *pixels*. The physical photo-cell configuration, in particular the configuration of the R-G-B (red, green, and blue) photocells, may differ depending on the type of the camera. Conceptually, however, we can think of pixels capable of perceiving R, G, and B placed in parallel rows at equal intervals in horizontal and vertical directions, but the vertical columns of pixels are not necessarily orthogonal to the horizontal rows. Also, the inter-pixel distance may not be the same in horizontal and vertical directions. Labeling the upper-left pixel $(u, v) = (0, 0)$, we count the pixels $u = 1, 2, \ldots$ rightward and $v = 1, 2, \ldots$ downward. Thus, the integer pair (u, v) is identified with the position at the center of that pixel. Inter-pixel, or *subpixel*, positions are specified with real number pairs (u, v) by linear interpolation. This defines a continuous *pixel coordinate system* of the image plane (Fig. 2).

If the xy coordinate system is oriented so that the x-axis is directed rightward in parallel to the horizontal pixel rows and the y-axis downward, the pixel coordinates (u, v) and the image coordinates (x, y) are related by

$$u = \frac{x}{\alpha} + \frac{y}{\alpha} \tan \theta + u_0, \qquad v = \frac{y}{\beta} + v_0, \qquad (2)$$

where (u_0, v_0) are the pixel coordinates of the principal point o, and α and β are, respectively, the distances between consecutive pixels in the horizontal and vertical directions. We define the angle between the horizontal and vertical pixel directions to be $\pi/2 + \theta$ and call θ the *skew angle*.

Remark 1 The xy coordinate system as defined above is "reversed" as compared with the usual sense. This convention originates from the human intuition that a hypothetical z-axis extends "away" from the viewer toward the scene, making the x-, y- and z-axes a right-handed system.

Remark 2 In most textbooks, the angle between the horizontal and vertical pixel directions is defined to be θ. Then, the first of Eqs. (2) becomes $u = x/\alpha + (y/\beta)\cot\theta + u_0$. We prefer our convention, because the skewless camera corresponds to $\theta = 0$ rather than $\theta = \pi/2$.

2.3. Intrinsic Parameters

Combining Eqs. (1) and (2), we have

$$\begin{pmatrix} u \\ v \\ 1 \end{pmatrix} \simeq K \begin{pmatrix} X_c \\ Y_c \\ Z_c \end{pmatrix}, \qquad (3)$$

where and throughout this chapter the symbol $\simeq$ means that one side is a multiple of the other by a nonzero constant. The matrix K is defined by

$$K = \begin{pmatrix} f\gamma & f\gamma\tan\theta & u_0 \\ 0 & f & v_0 \\ 0 & 0 & 1 \end{pmatrix}, \qquad (4)$$

where we put $f = f_c/\beta$, the normalized focal length so that the vertical distance between pixel rows is 1. Customarily, it is simply called the "focal length". We also define $\gamma = \beta/\alpha$, called the *aspect ratio*. The constants f, γ, θ, u_0, and v_0 are called the *intrinsic parameters* of the camera, and the matrix K the *intrinsic parameter matrix*.

Remark 3 For digital cameras today, we can expect $\gamma \approx 1$ and $\theta \approx 0$ with high precision and the principal point (u_0, v_0) is nearly at the center of the photo-cell array.

Remark 4 In some textbooks, the vertical interval β is defined not as the distance between consecutive "rows" but as the distance between consecutive "pixels" in the vertical direction. In that case, the second of Eqs. (2) becomes $v = y/\beta\cos\theta + v_0$, so the (22) element of the matrix K in Eq. (4) is $f/\cos\theta$. If we use the skew angle convention mentioned in Remark 2, $\cos\theta$ is replaced by $\sin\theta$. However, precise interpretation of the matrix K is not essential. Many recent textbooks simply write

$$K = \begin{pmatrix} f_1 & s & u_0 \\ 0 & f_2 & v_0 \\ 0 & 0 & 1 \end{pmatrix}, \qquad (5)$$

emphasizing the fact that it is *an upper triangular matrix with 1 in the (33) element*.

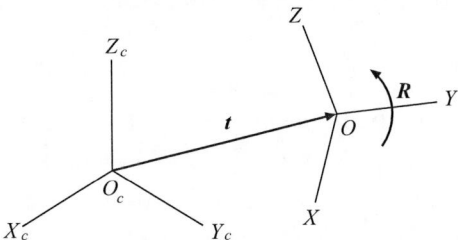

Figure 3. The camera coordinate system and the world coordinate system.

2.4. Motion Parameters

Since the $X_c Y_c Z_c$ coordinate system is defined with respect to the camera (i.e., the viewpoint O_c and the optical axis), it is called the *camera coordinate system*. We also define an XYZ coordinate system fixed to the scene and call it the *world coordinate system*. Let t be its origin described with respect to the camera coordinate system. If the world coordinate system is rotated by R relative to the camera coordinate system, a point in the scene with world coordinates (X, Y, Z) has the following camera coordinates (X_c, Y_c, Z_c) (Fig. 3):

$$\begin{pmatrix} X_c \\ Y_c \\ Z_c \end{pmatrix} = R \begin{pmatrix} X \\ Y \\ Z \end{pmatrix} + t. \tag{6}$$

We call $\{R, t\}$ the *motion parameters* or the *extrinsic parameters* of the camera.

Remark 5 The above motion parameters $\{R, t\}$ are a description with respect to the *camera coordinate system*. Alternatively, they can be described with respect to the world coordinate system. Let t_c be the origin of the camera coordinate system described with respect to the world coordinate system. If the camera coordinate system is rotated by R_c relative to the world coordinate system, we obtain instead of Eq. (6)

$$\begin{pmatrix} X \\ Y \\ Z \end{pmatrix} = R_c \begin{pmatrix} X_c \\ Y_c \\ Z_c \end{pmatrix} + t_c, \tag{7}$$

and the two descriptions $\{R, t\}$ and $\{R_c, t_c\}$ are related by

$$R = R_c^\top, \qquad t = -R_c^\top t_c. \tag{8}$$

2.5. Projection Matrix

From Eqs. (3) and (6), we can see that the pixel coordinates (u, v) are related to the world coordinates (X, Y, Z) in the form

$$u \simeq PX, \tag{9}$$

where we put

$$u = \begin{pmatrix} u \\ v \\ 1 \end{pmatrix}, \qquad X = \begin{pmatrix} X \\ Y \\ Z \\ 1 \end{pmatrix}, \tag{10}$$

and
$$P = K \begin{pmatrix} R & t \end{pmatrix}. \qquad (11)$$

This 3×4 matrix P is called the *projection matrix* or the *camera matrix*. The vectors in Eqs. (10) represent the *homogeneous coordinates* of the point (u, v) in the image and the point (X, Y, Z) in the scene. Hereafter, we refer to points represented by vectors u and X simply as "point u" and "point X", respectively.

Remark 6 Homogeneous coordinates are used not only for points in 2-D and 3-D but also for lines in 2-D and planes in 3-D, as we will see later. They are the description of points, lines, and planes with a set of real numbers, not all zero, defined up to a nonzero multiplier. For example, triples x^1, x^2, x^3 and cx^1, cx^2, cx^3 for an arbitrary $c \neq 0$ describe the same point in 2-D (the superscripts are indices, not powers). If $x^3 \neq 0$, the usual coordinates, or the *inhomogeneous coordinates*, are

$$x = \frac{x^1}{x^3}, \qquad y = \frac{x^2}{x^3}. \qquad (12)$$

If $x^3 = 0$, the point is interpreted to be at infinity; such a point is called an *ideal point*. Similarly, quadruples X^1, X^2, X^3, X^4 and cX^1, cX^2, cX^3, cX^4 for an arbitrary $c \neq 0$ describe the same point in 3-D. If $X^4 \neq 0$, its inhomogeneous coordinates are

$$X = \frac{X^1}{X^4}, \qquad Y = \frac{X^2}{X^4}, \qquad Z = \frac{X^3}{X^4}. \qquad (13)$$

If $X^4 = 0$, the point is an *ideal point* at infinity. The symbol $\simeq$ in Eqs. (3) and (9) reflects the indeterminacy of the absolute scale of homogeneous coordinates.

Remark 7 If we use the motion parameters $\{R_c, t_c\}$ described with respect to the world coordinate system, Eq. (11) becomes

$$P = K \begin{pmatrix} R_c^\top & -R_c^\top t \end{pmatrix} = K R_c^\top \begin{pmatrix} I & -t \end{pmatrix}. \qquad (14)$$

(I denotes the unit matrix.) In this chapter, we adopt the description with respect to the camera coordinate system. Generally, the expressions become simpler if described with respect to the camera coordinate system, because the camera imaging geometry is usually defined with respect to the camera.

2.6. Absolute Conic

Since Eq. (9) is a relationship between homogeneous coordinates, it also holds for ideal points. In other words, Eq. (9) defines a mapping from the 3-D *projective space* $\mathcal{P}^3$ obtained by adding all ideal points in 3-D to $\mathcal{R}^3$ onto the 2-D *projective space* $\mathcal{P}^2$ obtained by adding all ideal points in 2-D to $\mathcal{R}^2$.

The set Π_∞ of points X^1, X^2, X^3, X^4 in $\mathcal{P}^3$ with $X^4 = 0$ is called the *ideal plane*. The set Ω_∞ of (imaginary) points in Π_∞ that satisfy

$$(X^1)^2 + (X^2)^2 + (X^3)^2 = 0 \qquad (15)$$

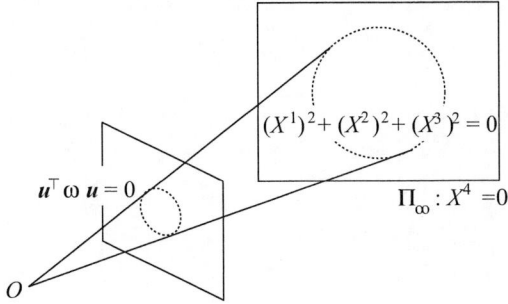

Figure 4. The absolute conic and its projection.

is called the *absolute conic*. It can be shown that any projection u of a point $X \in \Omega_\infty$ in the form of Eq. (9) satisfies, *irrespective* of the motion parameters $\{R, t\}$,

$$u^\top \omega u = 0, \qquad \omega \equiv (K^{-1})^\top K^{-1}. \qquad (16)$$

The set of (imaginary) points u that satisfy this equation is interpreted to be the camera projection of the absolute conic Ω_∞ (Fig. 4).

Remark 8 If we are given camera images of objects in the scene with known 3-D information, we can determine the intrinsic parameters and the motion parameters of the camera in many different ways, depending on the type of the available 3-D information about the scene. Such a procedure is called *camera calibration*, and most known calibration procedures can be given projective geometric interpretations in terms of the absolute conic [45].

3. Epipolar Geometry

3.1. Multilinear Constraints

When geometric primitives such as points, lines, and planes in the scene are viewed by multiple cameras located in different positions, description of the relationships among their projection images is called *epipolar geometry* (typically for two cameras) or *multilinear geometry* (typically for more than two cameras).

Suppose we observe a point X in the scene by M cameras. Let u_κ be its projection onto the κth image, $\kappa = 1, ..., M$, and P_κ the projection matrix of the κth camera. For each camera, the relationship of Eq. (9) holds. If we introduce an indeterminate nonzero constant λ_κ instead of the relation $\simeq$, we have

$$\lambda_\kappa u_\kappa = P_\kappa X. \qquad (17)$$

The constant λ_κ is called the *projective depth*. Rearranging all the equations of this form for $\kappa = 1, ..., M$ in a matrix form, we obtain

$$\begin{pmatrix} P_1 & u_1 & 0 & \cdots & 0 \\ P_2 & 0 & u_2 & \cdots & 0 \\ \vdots & 0 & 0 & \ddots & \vdots \\ P_M & 0 & 0 & \cdots & u_M \end{pmatrix} \begin{pmatrix} X \\ -\lambda_1 \\ \vdots \\ -\lambda_M \end{pmatrix} = \begin{pmatrix} 0 \\ 0 \\ \vdots \\ 0 \end{pmatrix}. \qquad (18)$$

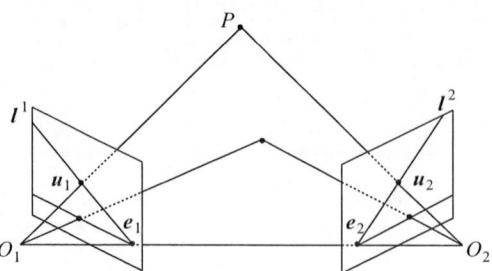

Figure 5. Epipolar lines and epipoles.

Since some X ($\neq 0$) and λ_κ, $\kappa = 1, ..., M$, that satisfy this equation should exist, the $3M \times (M+4)$ matrix on the left-hand side has at most rank $M+3$. Hence, all $(M+4) \times (M+4)$ minors should vanish. This leads to constraints on projection images in M (= 2, 3, 4) images [12].

Remark 9 It is easy to see that unless the chosen $(M+4) \times (M+4)$ minor contains two or more columns of P_κ, we cannot obtain a meaningful constraint on the projection in the κth image. In fact, if only one column of P_κ is included, the resulting minor is linear in its elements, so its vanishing does not give any information about P_κ. Hence, if M projection matrices are to be constrained by the vanishing of a $(M+4) \times (M+4)$ minor, we need $2M \leq M+4$, or $M \leq 4$. Thus, we can obtain constraints on only two, three, and four images.

3.2. Fundamental Matrix

For $M = 2$ (two images), the matrix on the left-hand side of Eq. (18) is 6×6, so we obtain only one constraint: the matrix has determinant 0. This is rewritten as

$$u_1^\top F u_2 = 0, \tag{19}$$

where F is a 3×3 matrix called the *fundamental matrix*. Its (ij) element is

$$F_{ij} = \sum_{k,l,m,n=1}^{3} \epsilon_{ikl}\epsilon_{jmn} \det P^{klmn}_{1122}, \tag{20}$$

where P^{klmn}_{1122} is the 4×4 matrix consisting of the kth row of P_1, the lth row of P_1, the mth row of P_2, and the nth row of P_2. From Eq. (20), it can be shown that the fundamental matrix F has rank 2.

Remark 10 The symbol ϵ_{ijk} denotes the signature of the permutation (ijk). Namely, it takes on 1 if (ijk) is an even permutation of (123), -1 if it is an odd permutation, and 0 otherwise. This symbol is called the *Levi-Civita* (or *Eddington*) *epsilon*.

3.3. Epipolar Constraint

The line starting from the viewpoint O_1 of the first camera and passing through the point u_1 in the image plane of the first camera is called the *line of sight* of u_1. The line of sight of u_2 is similarly defined. Geometrically, Eq. (19) describes the requirement that the line of sight of u_1 and the line of sight of u_2 should intersect at a point (it may be at infinity) in the scene. (Fig. 5). The set of points u that satisfy $l^\top u = 0$ for some l defines a line in the image. The vector l labels this line up to a nonzero multiplier (i.e., l and cl defines the same line for $c \neq 0$). The three components of l define the homogeneous coordinates of this line. Henceforth, we abbreviate the line represented by vector l simply as "line l".

Eq. (19) implies that the point u_1 is on the line $l^1 = Fu_2$, which is called the *epipolar line* of point u_2. Eq. (19) also implies that the point u_2 is on the line $l^2 = F^\top u_1$, called the *epipolar line* of point u_1. Thus, Eq. (19) states that *a point in one image should be on the epipolar line of the corresponding point in the other image*. This requirement is called the *epipolar constraint*. If follows that if the fundamental matrix F is known, one can find point correspondence by searching the other image along the epipolar line of u (Fig. 5).

3.4. Epipoles

Since the fundamental matrix F has rank 2, it has a null vector. So does $F^\top$, too. In other words, there exist vectors e_1 and e_1 such that $F^\top e_1 = 0$ and $Fe_2 = 0$. Identifying e_1 and e_2 with homogeneous coordinates of points in the image, we call them the *epipoles*. Geometrically, the epipole e_1 is *the projection of the viewpoint O_2 of the second camera onto the first image*, and the epipole e_2 is *the projection of the viewpoint O_1 of the first camera onto the second image* (Fig. 5). From Eq. (19), we can see that in the first image the epipolar line $l^1 = Fu_2$ of any point u_2 passes through the epipole e_1, i.e., $l^{1\top} e_1 = 0$. Similarly, in the second image, the epipolar line $l^2 = F^\top u_1$ of any point u_1 passes through the epipole e_2, i.e, $l^{2\top} e_2 = 0$.

It follows that epipolar lines of all points in the other image pass through the epipole, defining a *pencil of lines* (Fig. 5). This is easily understood if we note that the epipolar line of a point u_2 of the second image is nothing but the intersection of the first image plane with the plane defined by u_2 and the viewpoints O_1 and O_2 of the two cameras. This plane is called the *epipolar plane* of u_2 (and hence of the corresponding point u_1). The line connecting the two viewpoints O_1 and O_2 is called the *baseline*. All epipolar planes contain the baseline, defining a *pencil of planes* (Fig. 5).

3.5. Three-View Geometry

For $M = 3$ (three images), we obtain from Eq. (18) the following *trilinear constraint*:

$$\sum_{i,j,k,l,m=1}^{3} \epsilon_{jlp}\epsilon_{kmq} T_i^{jk} u_1^i u_2^l u_3^m = 0. \tag{21}$$

Here, u_κ^i denotes the ith component of u_κ, and

$$T_i^{jk} = \sum_{l,m=1}^{3} \epsilon_{ilm} \det P_{1123}^{lmjk} \tag{22}$$

is called the *trifocal tensor*.

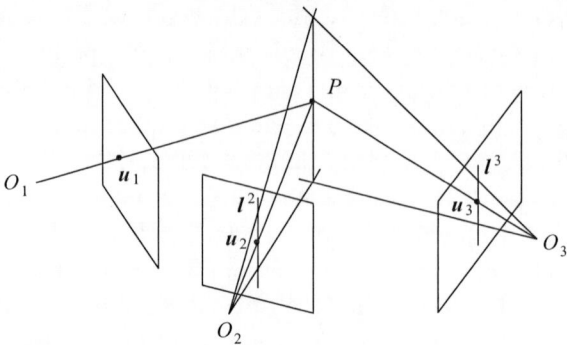

Figure 6. Trifocal constraint.

Given a line l in the image plane, the plane Π_l defined by the viewpoint O_c and the line l is called the *back projection* of the line l. Let Π_{l^2} be the back projection of an arbitrary line l^2 passing through u_2 in the second image, and Π_{l^3} the back projection of an arbitrary line l^3 passing through u_3 in the third image. Geometrically, Eq. (21) describes the requirement that *the line of sight of u_1 in the first image should meet the intersection of the two planes Π_{l^2} and Π_{l^3} at a single point* (it may be at infinity) (Fig. 6).

Remark 11 Take an arbitrary point v_2 ($\neq u_2$) in the second image and an arbitrary point v_3 ($\neq u_3$) in the third image. Multiplying Eq. (21) by $v_2^p v_3^q$ and summing it over p and q, we obtain

$$\sum_{i,j,k=1}^{3} T_i^{jk} u_1^i \Big(\sum_{l,p=1}^{3} \epsilon_{jlp} u_2^l v_2^p\Big) \Big(\sum_{m,q=1}^{3} \epsilon_{kmq} u_3^m v_3^q\Big) = 0. \qquad (23)$$

If we define lines

$$l^2 = u_2 \times v_2, \qquad\qquad l^3 = u_3 \times v_3, \qquad (24)$$

Eq. (23) is rewritten as

$$\sum_{i,j,k=1}^{3} T_i^{jk} u_1^i l_j^2 l_k^3 = 0, \qquad (25)$$

which describe the geometric relationship mentioned earlier.

3.6. Four-View Geometry

For $M = 4$ (four images), we obtain from Eq. (18) the *quadrilinear constraint*

$$\sum_{i,j,k,l,m,n,p,q=1}^{3} \epsilon_{ima}\epsilon_{jnb}\epsilon_{kpc}\epsilon_{lqd} Q^{ijkl} u_1^m u_2^n u_3^p u_4^q = 0, \qquad (26)$$

where

$$Q^{ijkl} = \det \boldsymbol{P}_{1234}^{ijkl}, \qquad (27)$$

is called the *quadrifocal tensor*. Geometrically, Eq. (26) describes the requirement that *the back projections Π_{l^1}, ..., Π_{l^4} of arbitrary lines l^1, ..., l^4 in each image passing through points u_1, ..., u_4, respectively, should meet at a single point* (Fig. 7).

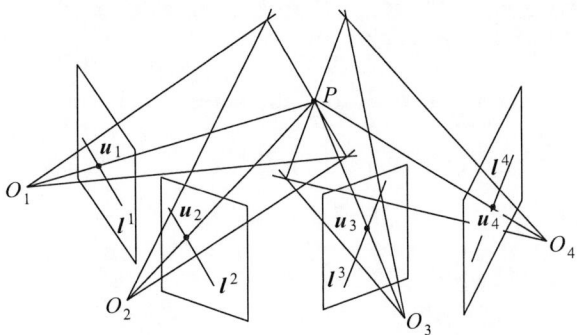

Figure 7. Quadrifocal constraint.

Remark 12 Take an arbitrary point v_κ ($\neq u_\kappa$) in the κth image, $\kappa = 1, 2, 3, 4$. Multiplying Eq. (26) with $v_1^a v_2^b v_3^c v_4^d$ and summing it over a, b, c, and d, we obtain

$$\sum_{i,j,k,l=1}^{3} Q^{ijkl} \Big(\sum_{m,a=1}^{3} \epsilon_{ima} u_1^m v_1^a \Big) \Big(\sum_{n,b=1}^{3} \epsilon_{jnb} u_2^n v_2^b \Big) \Big(\sum_{p,c=1}^{3} \epsilon_{kpc} u_3^p v_3^c \Big) \Big(\sum_{q,d=1}^{3} \epsilon_{lqd} u_4^q v_4^d \Big) = 0. \tag{28}$$

If we define lines

$$\begin{aligned} l^1 &= u_1 \times v_1, & l^2 &= u_2 \times v_2, \\ l^3 &= u_3 \times v_3, & l^4 &= u_4 \times v_4, \end{aligned} \tag{29}$$

Eq. (28) is rewritten as

$$\sum_{i,j,k,l=1}^{3} Q^{ijkl} l_i^1 l_j^2 l_k^3 l_m^4 = 0, \tag{30}$$

which describe the geometric relationship mentioned earlier.

4. 3-D Reconstruction from Images

4.1. Classification of the Problem

Suppose we observe N points X_α, $\alpha = 1, ..., N$, in the scene by M cameras having projection matrices P_κ, $\kappa = 1, ..., M$. Equivalently, we may move one camera, changing its parameters and taking pictures at M different instances, which is also equivalent to fix the camera position and move the scene relative to it. In whichever interpretation, let $u_{\kappa\alpha}$ the projection of point X_α onto the κth image. For each point and each image, we have the relationship described in the form of Eq. (9):

$$u_{\kappa\alpha} \simeq P_\kappa X_\alpha. \tag{31}$$

Given projection images $u_{\kappa\alpha}$, $\kappa = 1, ..., M$, $\alpha = 1, ..., N$, the task of computing X_α, $\alpha = 1, ..., N$, is called *3-D reconstruction* or *structure from motion* (*SFM*). The problem is classified into the following three cases (we adopt the multiple camera interpretation for simplicity):

(i) The projection matrix P of each camera is known.

(ii) The intrinsic parameter matrix K of each camera is known (but the motion parameters $\{R, t\}$ are not).

(iii) The projection matrix P of each camera is unknown.

In Case (i), Eq. (31) determines the 3-D coordinates $(X_\alpha, Y_\alpha, Z_\alpha)$ of point $\boldsymbol{X}_\alpha$ up to one degree of freedom, which corresponds to the *depth* of the point $\boldsymbol{X}_\alpha$ along the line of sight. In order to determine it uniquely, we need to observer two or more images. Computing the depths of points in the scene in this way is called (*multi-camera*) *stereo vision*.

In Case (ii), the cameras are said to be *calibrated*. In this case, we first compute the fundamental matrix $\boldsymbol{F}$ from point correspondences between two images. Then, the motion parameters $\{R, t\}$ are determined by solving Eq. (20), and the problem reduces to stereo vision of Case (ii).

In Case (iii), the cameras are said to be *uncalibrated*. 3-D reconstruction in this case is called *self-calibration* or *autocalibration*.

Remark 13 In Cases (ii) and (iii), the positions of the points in the scene and the camera motion parameters are determined only up to an unknown scale factor. This is because small camera motions relative to a small object located nearby cannot be distinguished from large camera motions relative to a large object located far away, as long as projection images are the only available information.

Remark 14 For calibrated cameras (Case (ii)), the motion parameters computed from the fundamental matrix $\boldsymbol{F}$ has ambiguity of "mirror image". This is because we only require the 3-D positions of observed points to be on the lines of sight that they defines. As a result, the reconstructed shape can be a mirror image "behind" the camera. Mirror image solutions can be removed by imposing the constraint that observed points be in front of the cameras, which Hartley [7] called *cheirality* (or *chirality*) (see [14, 15] for the actual procedure).

4.2. Self-calibration

In Case (iii) (self-calibration), the projection matrices $\boldsymbol{P}_\kappa$ and the 3-D points $\boldsymbol{X}_\alpha$ in Eq. (31) are both unknown. It is immediately seen from Eq. (31) that the solution is indeterminate if there is no constraint on the cameras or the 3-D points. In fact, if $\boldsymbol{X}_\alpha$ and $\boldsymbol{P}_\kappa$ are a solution, we have another solution

$$\tilde{\boldsymbol{X}}_\alpha \simeq \boldsymbol{H}\boldsymbol{X}_\alpha, \qquad \tilde{\boldsymbol{P}}_\kappa \simeq \boldsymbol{P}_\kappa \boldsymbol{H}^{-1} \qquad (32)$$

for an arbitrary nonsingular 4×4 matrix $\boldsymbol{H}$.

The first of Eqs. (32) can be regarded as applying a *projective transformation* (or a *homography*) $\boldsymbol{H}$ to the 3-D projective space $\mathcal{P}^3$ (Fig. 8). Accordingly, the points $\boldsymbol{X}_\alpha$ and $\tilde{\boldsymbol{X}}_\alpha$ have the same *projective structure*. For example, collinear points are mapped to collinear points, coplanar points are mapped to coplanar points, and their *incidence relationships*, such as "on ...", "passing through ..." and "meeting at ...", are preserved. However, *metric*

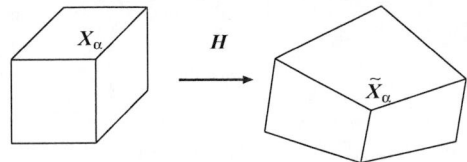

Figure 8. Projective transformation.

properties such as lengths and angles are not preserved. 3-D reconstruction determined up to an arbitrary projective transformation is called *projective reconstruction*.

In order to select a correct solution, one needs some constraint on either the cameras or the points. Selecting a unique solution by imposing such constraint is termed *upgrading* of projective reconstruction into *Euclidean* (or *metric*) *reconstruction*.

Note that Eqs. (32) are rewritten as

$$\boldsymbol{X}_\alpha \simeq \boldsymbol{H}^{-1}\tilde{\boldsymbol{X}}_\alpha, \qquad \boldsymbol{P}_\kappa \simeq \tilde{\boldsymbol{P}}_\kappa \boldsymbol{H}. \qquad (33)$$

If, for example, we know the true 3-D positions $\boldsymbol{X}_\alpha$ of five (or more) points in general position, we can uniquely determine the projective transformation $\boldsymbol{H}$ that maps, or "rectifies", the five points $\tilde{\boldsymbol{X}}_\alpha$ to their true positions $\boldsymbol{X}_\alpha$. Applying the computed $\boldsymbol{H}$ to the remaining points, we obtain the Euclidean reconstruction $\boldsymbol{X}_\alpha$ of all points. If no such five points are known, we need to assume some constraints on cameras and find an appropriate projective transformation $\boldsymbol{H}$ such that the projection matrices $\boldsymbol{P}_\kappa$ rectified by the second of Eqs. (33) satisfy the assumed constraints. This approach is called the *stratified reconstruction*.

Remark 15 Points in 3-D are said to be *in general position* if no three of them are coplanar. If we are given five (or more) points in general position for which we only know their relative configuration up to a scale factor, we can reconstruct the 3-D shape up to position, orientation, and scale by arbitrarily normalizing the position, the orientation, and the scale.

Remark 16 If no 3-D information is given about the scene, the absolute scale cannot be determined from images alone, as pointed out in Remark 13. Hence, all that can be obtained is, strictly speaking, "similarity" reconstruction rather than "Euclidean" or "metric". However, the terms "Euclidean" and "metric" are commonly used to mean "up to similarity".

4.3. Stratified Reconstruction

Eliminating the rotation $\boldsymbol{R}$ from Eq. (11) by using the identity $\boldsymbol{R}\boldsymbol{R}^\top = \boldsymbol{I}$, we obtain for each image

$$\boldsymbol{P}_\kappa \mathrm{diag}(1,1,1,0)\boldsymbol{P}_\kappa^\top = \boldsymbol{\omega}_\kappa^*, \qquad (34)$$

where $\mathrm{diag}(a, b, c, ...)$ denotes the diagonal matrix with diagonal elements $a, b, c, ...$ in that order. The 3×3 matrix $\boldsymbol{\omega}_\kappa^*$ is defined by

$$\boldsymbol{\omega}_\kappa^* \equiv \boldsymbol{K}_\kappa \boldsymbol{K}_\kappa^\top. \qquad (35)$$

Substituting $\boldsymbol{P}_\kappa$ in the second of Eqs. (33) into Eq. (34), we obtain

$$\tilde{\boldsymbol{P}}_\kappa \boldsymbol{\Omega}_\infty^* \tilde{\boldsymbol{P}}_\kappa^\top \simeq \boldsymbol{\omega}_\kappa^*, \qquad (36)$$

where we define the 4×4 matrix Ω_∞^* by

$$\Omega_\infty^* \equiv H \operatorname{diag}(1,1,1,0) H^\top. \tag{37}$$

If the intrinsic parameter matrix K_κ is known (i.e., the camera is calibrated), we can determine ω_κ^* from Eq. (35). Even if ω_κ^* is not completely known, we can obtain constraints on the elements of Ω_∞^* from Eq. (36) if we have some knowledge about ω_κ^*, such as a particular element being 0 or two particular elements being equal (we are assuming that $\tilde{P}_\kappa$ are given). If the number M of images is sufficiently large to give a sufficient number of such constraints on Ω_∞^*, we can determine Ω_∞^*. Frequently used assumptions about the cameras are:

- All cameras have the same intrinsic parameters.
- The location of the principal point is known for all cameras.
- The skew angle θ is 0 for all cameras.
- The aspect ratio γ is 1 for all cameras.

For example, if all cameras have the same intrinsic parameters (i.e., one camera is moved to take multiple pictures without changing its parameters), the unknown is one intrinsic parameter matrix K, so $\omega_1^* = ... = \omega_M^* = \omega^*\ (\equiv KK^\top)$. Hence, Eq. (36) gives $5(M-1)$ equations of Ω_∞^*. If the principal point is known, we can translate the coordinate system so that $u_0 = v_0 = 0$. Then, the (13) and (23) elements of K in Eq. (4) are 0, and hence the (13) and (23) elements of $\omega_\kappa^* = K_\kappa K_\kappa^\top$ are also 0. In this case, Eq. (36) gives $2M$ equations of Ω_∞^*. If the skew angle is zero in addition, the (12) element of ω_κ^* is also zero, so we obtain $3M$ equations of Ω_∞^*. If furthermore the aspect ratio γ is 1, the (11) element and the (22) element are equal, giving M additional equations. If we obtain nine or more such equations, we can solve them for Ω_∞^* up to a scale factor. If Ω_∞^* is determined, ω_κ^* is determined from Eq. (36). Then, the projective transformation H is determined from Eq. (37). The intrinsic parameter matrix K_κ is obtained by solving Eq. (35).

Remark 17 From Eq. (4), the matrix ω_κ^* in Eq. (35) has the form

$$\omega_\kappa^* = \begin{pmatrix} f_\kappa^2 \gamma_\kappa^2 + s_\kappa^2 + u_{0\kappa}^2 & f_\kappa s_\kappa^2 + u_{0\kappa} v_{0\kappa} & u_{0\kappa} \\ f_\kappa s_\kappa^2 + u_{0\kappa} v_{0\kappa} & f_\kappa^2 + v_{0\kappa}^2 & v_{0\kappa} \\ u_{0\kappa} & v_{0\kappa} & 1 \end{pmatrix}, \tag{38}$$

where we put $s_\kappa = f_\kappa \gamma_\kappa \tan \theta$. This is a 3×3 symmetric matrix with six different elements. Hence, if all the intrinsic parameters are known, Eq. (36) gives five constraints for each κ (one degree of freedom is lost for the indeterminate scale factor). The unknown is the 4×4 symmetric matrix Ω_∞^* with ten independent elements, but it has scale indeterminacy. Hence, two views are sufficient.

If the intrinsic parameters are all unknown but are the same for all cameras (or one camera is moved), we need to observe M views such that $5(M-1) \geq 9$, or $M \geq 3$. If the principal point $(u_{0\kappa}, u_{0\kappa})$ is known but other parameters can vary from frame to frame, the number M of necessary views is such that $2M \geq 9$, or $M \geq 5$. If the skew s_κ is 0 in addition, this is relaxed to $3M \geq 9$, or $M \geq 3$ views. If furthermore the aspect ratio γ_κ is 1, this becomes $4M \geq 9$, so we still need to observe $M \geq 3$ views.

Remark 18 If we have more equations than the number of unknowns, inconsistencies arise among these equations in the presence of noise in the data. Theoretically, we can determine the unknowns in a statistically optimal ways [15], but this is too complicated to carry out. So, a simple least-squares minimization is conducted in practice. This, however, causes another problem: Ω_∞^* should have rank 3 from the definition of Eq. (37), but it has generally rank 4 if computed by least squares. Ad-hoc treatments, such as computing the singular value decomposition (SVD) of the obtained Ω_∞^* and replacing the smallest singular value by 0, are widely employed.

Remark 19 If Ω_∞^* is obtained, Eq. (37) does not completely determine the projective transformation H: it has rotational ambiguity, and its fourth column is arbitrary. This corresponds to the fact that the orientation and the location of the world coordinate system can be arbitrarily defined. The details of the computation is given in Appendix A.

Remark 20 From the computed Ω_∞^*, Eq. (36) determines ω_κ^* up to a scale factor. Then, Eq. (35) must be solved for K_κ, which should be an upper triangular matrix. A standard procedure, called the *Cholesky factorization*, is well known for decomposing a given positive semi-definite symmetric matrix into the product of an upper triangular matrix and its transpose. The indeterminate scale of K_κ is fixed so that its (33) element becomes 1.

Remark 21 The stratified reconstruction approach was proposed by Faugeras [4] and others. First, the constant camera constraint was used by many researchers. Later, Heyden and Åström [9, 10] showed that Euclidean reconstruction is possible using as few constraints as zero skew alone if a sufficient number of images and point correspondences are available. The constraint in the form of Eq. (36) was first formulated by Triggs [42]. Pollefeys et al. [28] demonstrated that accurate reconstruction is indeed possible by this approach. Since then, various modifications and simplifications have been devised for imposing the constraint. Many researchers used nonlinear optimization in one form or another, but later simple formulations using linear computations have been found in many forms; see [30, 31, 32]. The actual procedure of one such approach is given in Appendix A.

4.4. Dual Absolute Quadric Constraint

Comparing the second of Eqs. (16) and Eq. (35), we can see that the matrix ω_κ^* is the inverse of ω_κ, which represents the projection, onto the κth image, of the absolute conic Ω_∞. This means that the set of lines l that satisfy $l^\top \omega_\kappa^* l = 0$ is the *envelope* of, or the set of tangent lines to, the (imaginary) conic defined by the first of Eqs. (16). In projective geometry, this is called the *line pencil of second class* dual to the conic $u^\top \omega_\kappa u = 0$.

Eq. (36) states that the line pencil of second class represented by ω_κ^* is the projection, onto the κth image, of the *plane pencil of second class* represented by Ω_∞^*, i.e., the set of planes with homogeneous coordinates π that satisfy $\pi^\top \Omega_\infty^* \pi = 0$. This is the envelope of, or the set of tangent planes to, the absolute conic Ω_∞ regarded as a degenerate (imaginary) quadric surface (a 2-D "disk") (Fig. 9). This envelope is called the *dual absolute quadric*. From this projective geometric interpretation, Eq. (36) is called the *dual absolute quadric constraint*.

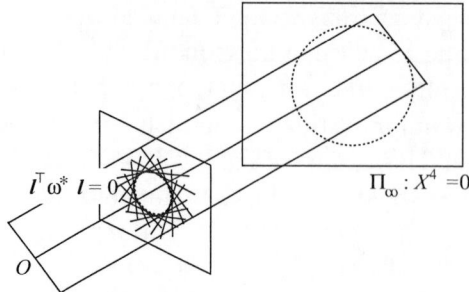

Figure 9. Dual absolute quadric constraint.

Remark 22 The fact that the constraint for Euclidean reconstruction can be given a projective geometric interpretation in terms of the dual absolute quadric is one of the greatest theoretical advances of 3-D reconstruction from images. For this reason, almost all papers, articles and books on 3-D reconstruction now start with theorems of projective geometry involving the absolute conic. At the cost of this elegance, however, this projective geometric interpretation makes the reconstruction procedure incomprehensible to average computer vision researchers, who tend to shy away from such mathematical sophistication involving imaginary quantities. In reality, the actual reconstruction procedure can be described without any reference to projective geometry, as we showed in Section 4.3. It is still being debated among researchers whether the projective geometric interpretation helps or prevents people's understanding of this method.

4.5. Projective Reconstruction

In order to start stratified reconstruction, we need an initial projective reconstruction. The most frequently used method for it is called *factorization*. If the projective depth $\lambda_{\kappa\alpha}$ is introduced as in Eq. (17), Eq. (31) is rewritten as the following equality:

$$\lambda_{\kappa\alpha} \boldsymbol{u}_{\kappa\alpha} = \boldsymbol{P}_\kappa \boldsymbol{X}_\alpha. \tag{39}$$

Let $\tilde{\boldsymbol{u}}_\alpha$ be the $3M$-D vector obtained by vertically stacking $\lambda_{1\alpha}\boldsymbol{u}_{1\alpha}, \lambda_{2\alpha}\boldsymbol{u}_{2\alpha}, ..., \lambda_{M\alpha}\boldsymbol{u}_{M\alpha}$, and $\tilde{\boldsymbol{p}}_i$ the $3M$-D vector obtained by vertically stacking the ith columns of $\boldsymbol{P}_1, \boldsymbol{P}_2, ..., \boldsymbol{P}_M$. Then, Eq. (39) is expressed in the form

$$\tilde{\boldsymbol{u}}_\alpha = X_\alpha^1 \tilde{\boldsymbol{p}}_1 + X_\alpha^2 \tilde{\boldsymbol{p}}_2 + X_\alpha^3 \tilde{\boldsymbol{p}}_3 + X_\alpha^4 \tilde{\boldsymbol{p}}_4, \tag{40}$$

where X_α^i is the ith component of the vector $\boldsymbol{X}_\alpha$. Eq. (40) states that the N vectors $\tilde{\boldsymbol{u}}_\alpha$ are all constrained to be in the *4-D subspace* $\mathcal{L}$ of $\mathcal{R}^{3M}$ spanned by $\{\tilde{\boldsymbol{p}}_1, \tilde{\boldsymbol{p}}_2, \tilde{\boldsymbol{p}}_3, \tilde{\boldsymbol{p}}_4\}$. This fact is called the *subspace constraint*.

We can see that Eq. (39) holds if we multiply both the projective depth $\lambda_{\kappa\alpha}$ and the homogeneous coordinates $\boldsymbol{X}_\alpha$ by a common nonzero constant c_α. As a result, the vector $\tilde{\boldsymbol{u}}_\alpha$ is multiplied by c_α. In order to remove this indeterminacy, we normalize $\tilde{\boldsymbol{u}}_\alpha$ to be a unit vector: $\|\tilde{\boldsymbol{u}}_\alpha\| = 1$. Then, we obtain the following iterative procedure for computing $\boldsymbol{X}_\alpha$:

1. Give initial values for the projective depths $\lambda_{\kappa\alpha}$.

Latest Progress of 3-D Reconstruction from Multiple Camera Images

2. Compute the $3M$-D vectors $\tilde{\boldsymbol{u}}_\alpha$ and fit a 4-D subspace $\mathcal{L}$ to the resulting $\tilde{\boldsymbol{u}}_\alpha$ by least squares.

3. Adjust the projective depths $\lambda_{\kappa\alpha}$ so that the square distance from each $\tilde{\boldsymbol{u}}_\alpha$ to the fitted subspace $\mathcal{L}$ is minimized.

4. Go back to Step 2, and repeat this until the computation converges.

5. Letting an arbitrary orthonormal basis of the converged subspace $\mathcal{L}$ be $\tilde{\boldsymbol{p}}_i$, determine $\boldsymbol{X}_\alpha$ by expanding $\tilde{\boldsymbol{u}}_\alpha$ in the form of Eq. (40) by least squares.

Remark 23 In Step 1, the initial values of the projective depths $\lambda_{\kappa\alpha}$ can be set to 1. If all the cameras are "affine cameras" (to be defined in the next section), it can be shown that a solution such that $\lambda_{\kappa\alpha} = 1$ exists.

Remark 24 The least-squares solution in Step 2 can be immediately obtained by solving an eigenvalue problem. In fact, if we let

$$C = \sum_{\alpha=1}^{N} \tilde{\boldsymbol{u}}_\alpha \tilde{\boldsymbol{u}}_\alpha^\top, \tag{41}$$

the subspace $\mathcal{L}$ is spanned by the eigenvectors of C for the largest four (positive) eigenvalues; the rest of the eigenvalues should vanish if the solution is exact. Alternatively, we may compute the singular value decomposition (SVD) in the form

$$\begin{pmatrix} \tilde{\boldsymbol{u}}_1 & \cdots & \tilde{\boldsymbol{u}}_N \end{pmatrix} = \boldsymbol{U}\boldsymbol{\Lambda}\boldsymbol{V}^\top, \tag{42}$$

where $\boldsymbol{U}$ is a $3M \times 3M$ orthogonal matrix, $\boldsymbol{V}$ is a $N \times N$ orthogonal matrix, and $\boldsymbol{\Lambda}$ is a diagonal matrix. The diagonal elements of $\boldsymbol{\Lambda}$ consist of singular values in descending order; only four are nonzero if the solution is exact. The basis of the $\mathcal{L}$ is given by the first four columns of $\boldsymbol{U}$. Usually, the use of SVD is computationally more efficient than the eigenvalue computation of Eq. (41).

Remark 25 The factorization approach to projective reconstruction was first introduced by Sturm and Triggs [34] and Triggs [41] with the observation that Eq. (39) for all κ and α can be rearranged in the form

$$\begin{pmatrix} \lambda_{11}\boldsymbol{u}_{11} & \cdots & \lambda_{1N}\boldsymbol{u}_{1N} \\ \vdots & \ddots & \vdots \\ \lambda_{M1}\boldsymbol{u}_{M1} & \cdots & \lambda_{MN}\boldsymbol{u}_{MN} \end{pmatrix} = \begin{pmatrix} \boldsymbol{P}_1 \\ \vdots \\ \boldsymbol{P}_M \end{pmatrix} \begin{pmatrix} \boldsymbol{X}_1 & \cdots & \boldsymbol{X}_N \end{pmatrix}. \tag{43}$$

In our notation, the vector $\tilde{\boldsymbol{u}}_\alpha$ is the αth column of the matrix on the left-hand side, and $\tilde{\boldsymbol{p}}_i$ is the ith column of the first matrix on the right-hand side. Sturm and Triggs [34] and Triggs [41] determined the projective depths $\lambda_{\kappa\alpha}$ so that the matrix on the left-hand side of Eq. (43) can be factorized into two matrices, hence the name "factorization". To do this, they determined the projective depths $\lambda_{\kappa\alpha}$ by using the epipolar constraints (Section 3.3) on pairwise images, computing the fundamental matrices of image pairs in advance. See

Deguchi [2] for more details. Ueshiba and Tomita [43] did direct numerical search for $\lambda_{\kappa\alpha}$ based on the perturbation theorem of SVD. It was Heyden et al. [11] who explicitly stated the subspace constraint and reduced the problem to eigenvalue problem solving. However, they considered the space of the vectors constructed from "all projected points in each image", rather than the vectors constructed from "each projected point in all images", as in the above formulation. In this sense, their method is "dual" to the above treatment, which is based on Mahumud and Herbert [25]. Mahumud et al. [26] also presented an alternative update strategy.

Remark 26 In Step 3, it is easy to see that the square distance is a quadratic form in $\lambda_{\kappa\alpha}$ [25]. So, the solution that minimizes this subject to the normalization $\|\tilde{u}_\alpha\|^2 = \sum_{\kappa=1}^{M} \|u_{\kappa\alpha}\|^2 \lambda_{\kappa\alpha}^2 = 1$ is directly obtained by solving a generalized eigenvalue problem [15]. In Appendix B, the detailed procedure of Steps 1 – 5 ("primal method") is described together with its dual form ("dual method").

Remark 27 Iterations of Steps 2 – 4 are guaranteed to converge, because the sum of square distances of $\tilde{u}_\alpha$ to the fitted subspace $\mathcal{L}$ monotonically decreases due to the minimization in Step 3. This type of iteration is a special variant of the *EM algorithm* [3]. However, the convergence is, though guaranteed, very slow in general.

5. 3-D Reconstruction from Affine Cameras

5.1. Affine Cameras

In terms of homogeneous coordinates, perspective projection can be written as a linear equation in the form of Eq. (9), but this is in appearance only; the relationship is essentially nonlinear, as can be seen from Eq. (3), which makes the subsequent analysis very difficult. The analysis is made much easier if Eq. (3) is approximated by a linear relationship in the form

$$\begin{pmatrix} u \\ v \end{pmatrix} = \Pi \begin{pmatrix} X_c \\ Y_c \\ Z_c \end{pmatrix} + \pi, \qquad (44)$$

where Π is a 2×3 matrix, π is a 2-D vector, and (X_c, Y_c, Z_c) is a point in the scene described with respect to the camera coordinate system. This approximation holds up to reasonable accuracy if

1. the object of our interest is localized around the world coordinate origin t, and

2. the size of the object is small as compared with $\|t\|$.

The approximate imaging geometry in the form of Eq. (44) is called an *affine camera*. Unlike the perspective camera model, the elements of the matrix Π and the vector π in Eq. (44) are now some functions of the motion parameters $\{R, t\}$. In order that Eq. (44) well mimic the perspective projection of Eq. (1), we require the following:

(i) The camera imaging is symmetric around the Z-axis.

(ii) The camera imaging does not depend on R.

(iii) The frontal parallel plane passing through the world coordinate origin is projected as if by perspective projection.

Requirement (i) states that if the scene is rotated around the optical axis by an angle θ, the resulting image should also rotate around the image origin by the same angle θ, a very natural requirement. Requirement (ii) is also natural, since the orientation of the world coordinate system can be defined arbitrarily, and such indeterminate parameterization should not affect the actual observation. Requirement (iii) corresponds to the assumption that the object of our interest is small and localized around the world coordinate origin t. It can be shown that in order that Requirements (i) – (iii) be satisfied, Eq. (44) must have the following form [20]:

$$\begin{pmatrix} u \\ v \end{pmatrix} = \frac{1}{\zeta}\left(\begin{pmatrix} X_c \\ Y_c \end{pmatrix} + \beta(t_z - Z_c)\begin{pmatrix} t_x \\ t_y \end{pmatrix}\right). \tag{45}$$

Here, t_x, t_y, and t_z are the three components of t, and $\{\zeta, \beta\}$ are arbitrary functions of $\sqrt{t_x^2 + t_y^2}$ and t_z; function ζ determines the size of the projected image, while function β describes the deformation of the projection image as the point moves away from the plane $Z_c = t_z$. Typical examples are the following three (Fig. 10):

Orthographic projection

$$\zeta = 1, \qquad \beta = 0. \tag{46}$$

Weak perspective (or *scaled orthographic*) *projection* [27, 40]

$$\zeta = \frac{t_z}{f_c}, \qquad \beta = 0. \tag{47}$$

Paraperspective projection [27]

$$\zeta = \frac{t_z}{f_c}, \qquad \beta = \frac{1}{t_z}. \tag{48}$$

Remark 28 The concept of affine camera and its epipolar geometry were presented by Shapiro et al. [33]. It was also shown that any affine camera can be interpreted to be paraperspective projection by appropriately adjusting the scale, the position, and the orientation of the world coordinate system [1]. This fact was exploited for object recognition from a single image [39]. The weak perspective and paraperspective models were introduced by Tomasi and Kanade [40] and Poelman and Kanade [27]. The generic form of Eq. (45) was derived by Kanatani et al. [20].

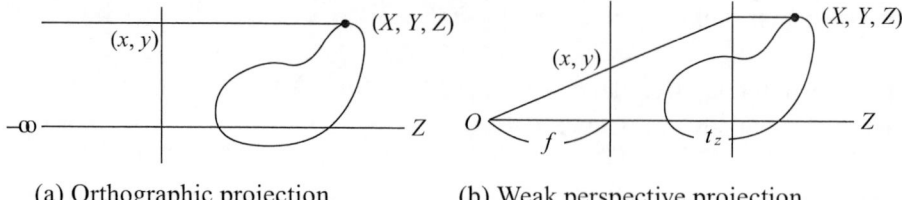

(a) Orthographic projection. (b) Weak perspective projection.

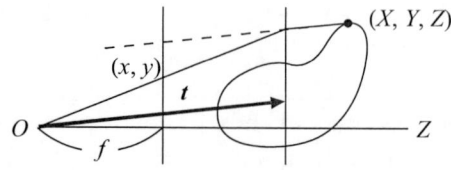

(c) Paraperspective projection.

Figure 10. Affine camera models.

5.2. Affine Space Constraint

If a point in the scene is represented by a vector X_α of homogeneous coordinates with the fourth component 1, Eqs. (6) and (44) imply that its projection onto the κth image is represented by a vector $u_{\kappa\alpha}$ with the third component 1 in the form

$$u_{\kappa\alpha} = \begin{pmatrix} \Pi_\kappa R_\kappa & \Pi_\kappa t_\kappa + \pi_\kappa \\ 0\ 0\ 0 & 1 \end{pmatrix} X_\alpha, \qquad (49)$$

where Π_κ and π_κ are, respectively, the values of the matrix Π and the vector π in Eq. (44) for the κth image, and $\{R_\kappa, t_\kappa\}$ are the motion parameters of the κth camera. Eq. (49) shows that an affine camera is a special case of the general projection in the form of Eq. (39) with the conditions that

- the third row of the projection matrix P_κ is (0 0 0 1), and
- the projective depths $\lambda_{\kappa\alpha}$ are all 1 (Remark 23).

It follows that, corresponding to Eq. (40), the following relationship holds:

$$\tilde{u}_\alpha = X_\alpha \tilde{p}_1 + Y_\alpha \tilde{p}_2 + Z_\alpha \tilde{p}_3 + \tilde{p}_4. \qquad (50)$$

As in Section 4.5, $\tilde{u}_\alpha$ is a vector, which we call the *trajectory* of the αth point, obtained by vertically stacking $u_{1\alpha}, u_{2\alpha}, ..., u_{M\alpha}$, while $\tilde{p}_i$ is a vector obtained by vertically stacking the ith columns of the matrix on the right-hand side of Eq. (49) for $\kappa = 1, ..., M$. We call $\tilde{p}_1, ..., \tilde{p}_4$ the *motion vectors*. We can see that every third component of the vector equation in Eq. (50) gives the identity $1 = 1$, so they can be removed. As a result, all the trajectories $\tilde{u}_\alpha$ and the motion vectors $\tilde{p}_i$ become $2M$-D vectors. Eq. (50) states that all the trajectories $\tilde{u}_\alpha$ are constrained to be in the *3-D affine space* $\mathcal{A}$ of $\mathcal{R}^{2M}$ passing through $\tilde{p}_4$ and spanned by the motion vectors $\{\tilde{p}_1, \tilde{p}_2, \tilde{p}_3\}$. This fact is called the *affine space constraint*.

Remark 29 The affine space constraint is not only a basis for 3-D reconstruction from affine camera images but also the core principle of *multibody motion segmentation* from images.

This is because if we observe multiple objects that are moving independently in the scene, the affine space constraint should hold for each rigid motion. Hence, if we track feature points that belong to multiple objects, classifying them into different motions is equivalent to classifying their trajectories, regarded as $2M$-D vectors, into different affine spaces in $\mathcal{R}^{2M}$. See [16, 17, 18, 35, 36, 37, 38] for actual applications.

5.3. Affine Reconstruction and Metric Constraints

The standard procedure for 3-D reconstruction based on the affine space constraint is called *factorization* for the reason explained shortly.

First, we fit a 3-D affine space $\mathcal{A}$ to the trajectories $\tilde{u}_\alpha$ by least squares. It is specified by a particular point $\tilde{p}_C \in \mathcal{A}$ and orthonormal vectors $\{\tilde{q}_1, \tilde{q}_2, \tilde{q}_3\}$ that span $\mathcal{A}$ at $\tilde{p}_C$. If we identify $\{\tilde{q}_1, \tilde{q}_2, \tilde{q}_3\}$ with $\{\tilde{p}_1, \tilde{p}_2, \tilde{p}_3\}$ in Eq. (50), we can determine $(X_\alpha, Y_\alpha, Z_\alpha)$ by expanding each $\tilde{u}_\alpha$ over them in the same way we did in Section 4.5. However, $\tilde{p}_C$ can be anywhere in $\mathcal{A}$, and $\{\tilde{q}_1, \tilde{q}_2, \tilde{q}_3\}$ can be any three linearly independent vectors, not necessarily orthonormal. Hence, the 3-D shape reconstructed from $\tilde{p}_C$ and $\{\tilde{q}_1, \tilde{q}_2, \tilde{q}_3\}$ has ambiguity up to an affine transformation. Such a reconstruction is called *affine reconstruction*. In order to upgrade the solution to Euclidean, we need to rectify the basis correctly by an affine transformation in the form

$$\tilde{p}_i = \sum_{j=1}^{3} A_{ji} \tilde{q}_j. \tag{51}$$

The translational ambiguity due to the arbitrariness of $\tilde{p}_C$ has no effect on the reconstructed 3-D shape. The rectifying transformation matrix $\boldsymbol{A} = (A_{ij})$ is determined by the condition that each $\tilde{p}_i$ consists of coordinates of points in the scene viewed by an affine camera that has the form of Eq. (49). This condition, known as the *metric constraint*, is obtained, as in the case of the dual absolute quadric constraint, by eliminating $\boldsymbol{R}_\kappa$ from the projection relation of Eq. (49) by using the identity $\boldsymbol{R}_\kappa \boldsymbol{R}_\kappa^\top = \boldsymbol{I}$.

Let $\boldsymbol{Q}$ be the $2M \times 3$ matrix with columns $\tilde{q}_1$, $\tilde{q}_2$, and $\tilde{q}_3$ in that order. Let $\boldsymbol{q}_{\kappa(1)}^\dagger$ and $\boldsymbol{q}_{\kappa(2)}^\dagger$ be the $(2\kappa - 1)$th and the 2κth columns of $\boldsymbol{Q}^\top$, respectively. We define the 3×2 matrix $\boldsymbol{Q}_\kappa^\dagger$ by

$$\boldsymbol{Q}_\kappa^\dagger = \left(\boldsymbol{q}_{\kappa(1)}^\dagger \quad \boldsymbol{q}_{\kappa(2)}^\dagger \right). \tag{52}$$

It can be shown (see Appendix C for the derivation) that if we let

$$\boldsymbol{T} = \boldsymbol{A} \boldsymbol{A}^\top, \tag{53}$$

the metric constraint is written in the following form [20]:

$$\boldsymbol{Q}_\kappa^{\dagger\top} \boldsymbol{T} \boldsymbol{Q}_\kappa^\dagger = \boldsymbol{\Pi}_\kappa \boldsymbol{\Pi}_\kappa^\top. \tag{54}$$

As in the stratified reconstruction, we can obtain from Eq. (54) a set of equations for $\boldsymbol{T}$ from the knowledge about the camera model, i.e., the relationships among the elements of the matrix $\boldsymbol{\Pi}_\kappa \boldsymbol{\Pi}_\kappa^\top$ on the right-hand side of Eq. (54). After that, we can obtain the rectifying matrix $\boldsymbol{A}$ by decomposing the computed $\boldsymbol{T}$ in the form of Eq. (53). The computational details for the typical models of Eqs. (46) – (48) and the general affine camera model of Eq. (45) are described in Appendix C.

Remark 30 As mentioned in Section 5.1, the affine camera model is a good approximation when the object of our interest is localized around the world coordinate origin. In such a situation, the world coordinate origin (which can be defined arbitrarily) can be located at the centroid of points $(X_\alpha, Y_\alpha, Y_\alpha)$, which means

$$\sum_{\alpha=1}^{N} X_\alpha = \sum_{\alpha=1}^{N} Y_\alpha = \sum_{\alpha=1}^{N} Z_\alpha = 0. \tag{55}$$

Let $\tilde{\bm{u}}_C$ be the centroid of the trajectories $\tilde{\bm{u}}_\alpha$:

$$\tilde{\bm{u}}_C = \frac{1}{N} \sum_{\alpha=1}^{N} \tilde{\bm{u}}_\alpha. \tag{56}$$

From Eqs. (50) and (55), we can see that the centroid $\tilde{\bm{u}}_C$ coincide with $\tilde{\bm{p}}_4$: $\tilde{\bm{u}}_C = \tilde{\bm{p}}_4$. As in the case of stratified reconstruction, the basis of the affine space $\mathcal{A}$ that optimally fits the trajectories $\tilde{\bm{u}}_\alpha$ and passes through their centroid $\tilde{\bm{u}}_C$ is given by the eigenvectors of the matrix

$$\bm{C} = \sum_{\alpha=1}^{N} (\tilde{\bm{u}}_\alpha - \tilde{\bm{u}}_C)(\tilde{\bm{u}}_\alpha - \tilde{\bm{u}}_C)^\top, \tag{57}$$

for the largest three eigenvalues. Alternatively, we may compute the singular value decomposition (SVD) in the form

$$\begin{pmatrix} \tilde{\bm{u}}_1 - \tilde{\bm{u}}_C & \cdots & \tilde{\bm{u}}_N - \tilde{\bm{u}}_C \end{pmatrix} = \bm{U}\bm{\Lambda}\bm{V}^\top, \tag{58}$$

where $\bm{U}$ is a $2M \times 2M$ orthogonal matrix, $\bm{V}$ is a $N \times N$ orthogonal matrix, and $\bm{\Lambda}$ is a diagonal matrix. The basis of the $\mathcal{A}$ is given by the first three columns $\bm{U}$.

Remark 31 If we let $\tilde{\bm{u}}'_\alpha = \tilde{\bm{u}}'_\alpha - \tilde{\bm{u}}_C$, Eq. (50) for $\alpha = 1, \ldots, N$ can be rearranged in the following form:

$$\begin{pmatrix} \tilde{\bm{u}}'_1 & \cdots & \tilde{\bm{u}}'_N \end{pmatrix} = \begin{pmatrix} \tilde{\bm{p}}_1 & \tilde{\bm{p}}_2 & \tilde{\bm{p}}_3 \end{pmatrix} \begin{pmatrix} X_1 & Y_1 & Z_1 \\ \vdots & \vdots & \vdots \\ X_N & Y_N & Z_N \end{pmatrix}. \tag{59}$$

Hence, computing the solution $\{X_\alpha, Y_\alpha, Z_\alpha\}$ can be given the interpretation that we are *factorizing* the *measurement* (or *observation*) matrix $\bm{W} = \begin{pmatrix} \tilde{\bm{u}}'_1 & \cdots & \tilde{\bm{u}}'_N \end{pmatrix}$ into the product of two matrices: the first describes the motion; the second the shape. This is the origin of the term *factorization*, named by Tomasi and Kanade [40], and the subsequent papers [24, 27] adopt this interpretation. Sturm and Triggs [34] and Triggs [41] presented a projective reconstruction procedure in a similar formalism, and this lead to the term "factorization" also for the approach described in Section 4.5 (Remark 25).

Remark 32 Since the factorization gives the solution by linear computation alone without any iterative search, it is widely used for many applications, such as object recognition and classification, which do not require so very high accuracy of the 3-D shape. Also, this method can be used to obtain a good initial guess of projective reconstruction for the stratified reconstruction.

Remark 33 When we say that we obtain "affine reconstruction" if the metric constraint is not imposed, we must keep in mind that an affine camera is a hypothetical concept; it only approximates existing cameras, which are modeled as perspective projection. Hence, if we use perspectively projected images as input, the resulting shape is not exactly affine reconstruction and is not exactly Euclidean even if the metric constraint is imposed.

Remark 34 The 3-D shape reconstructed by factorization is not unique, having the following ambiguity:

(i) The absolute scale is indeterminate.

(ii) The orientation of the world coordinate system is indeterminate.

(iii) Mirror image ambiguity exists.

The absolute scale indeterminacy is unavoidable as long as images are only available information (Remark 13). In fact, we can see from Eq. (50) that multiplying $\{\tilde{p}_1, \tilde{p}_2, \tilde{p}_3\}$ by a nonzero constant c gives rise to the same effect as dividing $\{X_\alpha, Y_\alpha, Z_\alpha\}$ by c. The orientation of the world coordinate system is indeterminate, because it can be arbitrarily defined in the scene. The mirror image ambiguity arises from the fact that the rectifying matrix A is determined by Eq. (53), which can be rewritten as $T = (\pm AR)(\pm AR)^\top$ for an arbitrary rotation matrix R. The indeterminacy of the rotation R corresponds to the orientation ambiguity; the indeterminacy of the sign corresponds to the mirror image ambiguity.

6. Concluding Remarks

This chapter has summarized recent advancements of the theories and techniques for 3-D reconstruction from multiple images. We started with the description of the camera imaging geometry as perspective projection in terms of homogeneous coordinates. We defined the intrinsic and extrinsic (motion) parameters of the camera by introducing the camera coordinate system and the world coordinate system.

It was shown that the camera imaging is regarded as a mapping from the 3-D projective space $\mathcal{P}^3$ onto the 2-D projective space $\mathcal{P}^2$ and that the absolute conic is invariant to camera motions, providing projective geometric interpretations to camera calibration procedures. Next, we described the epipolar geometry for two, three, and four cameras, introducing such concepts as the fundamental matrix, epipolar lines, epipoles, the trifocal tensor, and the quadrifocal tensor.

We then described the self-calibration technique based on the stratified reconstruction approach, using the absolute dual quadric constraint. We showed that an elegant projective geometric interpretation can be given but that it is not essential or even necessary for actually doing 3-D reconstruction computations. We also described the procedure for computing a projective reconstruction by the factorization based on the subspace constraint.

Finally, we gave the definition of the affine camera model and a procedure for 3-D reconstruction based on it. We discussed possible forms of the affine camera, described the affine space constraint, and introduced the metric constraint that is necessary for Euclidean reconstruction. The detailed procedures for 3-D reconstruction are given in the Appendix.

A. Euclidean Upgrading from Projective Reconstruction

Here, we describe the computational procedure for computing Ω_∞^* in Eq. (35) and H in Eq. (37), given the projection matrices $\tilde{P}_\kappa$ of projective reconstruction (the projective reconstruction procedure is given in Appendix B). We then give the procedure for computing the 3-D shape X_α and the motion parameters $\{R_\kappa, t_\kappa\}$ using the computed projective transformation H.

The following is a modification of the scheme proposed by the method of Seo and Heyden [31], to which several techniques are introduced for increasing robustness. The basic assumption here is that the skew angles θ_κ are all 0 and the aspect ratios γ_κ are all 1 (Sections 2.2 and 2.3). Hence, unknown camera parameters are the focal lengths f_κ and the principal points $(u_{\kappa 0}, v_{\kappa 0})$ for M frames.

A.1. Computation of Ω

Substituting Eq. (35) into Eq. (36), we have

$$\tilde{P}_\kappa \Omega_\infty^* \tilde{P}_\kappa^\top \simeq K_\kappa K_\kappa^\top. \tag{60}$$

Suppose we have an estimate f_κ of the focal length and an estimate $(u_{\kappa 0}, v_{\kappa 0})$ of the principal point for each frame. We tentatively let

$$K_\kappa = \begin{pmatrix} f_\kappa & 0 & u_{\kappa 0} \\ 0 & f_\kappa & v_{\kappa 0} \\ 0 & 0 & 1 \end{pmatrix}. \tag{61}$$

(See Eq. (4). Keep in mind that we are assuming that the skew angle is 0 and the aspect ratio is 1). Multiplying Eq. (60) by K_κ^{-1} from left and $K_\kappa^{-1\top}$ from right, we have

$$Q_\kappa \Omega_\infty^* Q_\kappa^\top \approx \text{scalar} \times I, \tag{62}$$

where we define

$$Q_\kappa \equiv K_\kappa^{-1} \tilde{P}_\kappa. \tag{63}$$

Eq. (62) implies that the (11) and (22) elements of $Q_\kappa \Omega_\infty^* Q_\kappa^\top$ are approximately equal, and its (12), (23), and (31) elements are approximately 0. Namely,

$$\sum_{i,j=1}^{4} Q_{\kappa(1i)} Q_{\kappa(1j)} \Omega_{\infty(ij)}^* - \sum_{i,j=1}^{4} Q_{\kappa(2i)} Q_{\kappa(2j)} \Omega_{\infty(ij)}^* \approx 0, \tag{64}$$

$$\sum_{i,j=1}^{4} Q_{\kappa(1i)} Q_{\kappa(2j)} \Omega_{\infty(ij)}^* \approx 0, \tag{65}$$

$$\sum_{i,j=1}^{4} Q_{\kappa(2i)} Q_{\kappa(3j)} \Omega_{\infty(ij)}^* \approx 0, \tag{66}$$

$$\sum_{i,j=1}^{4} Q_{\kappa(3i)} Q_{\kappa(1j)} \Omega^*_{\infty(ij)} \approx 0, \tag{67}$$

where $Q_{\kappa(ij)}$ and $\Omega^*_{\infty(ij)}$ are the (ij) elements of the matrices $\boldsymbol{Q}_\kappa$ and $\boldsymbol{\Omega}^*_\infty$, respectively. We determine $\boldsymbol{\Omega}^*_\infty$ by minimizing

$$K = \sum_{\kappa=1}^{M} W_\kappa \Big((\text{Eq. (64)})^2 + (\text{Eq. (65)})^2 + (\text{Eq. (66)})^2 + (\text{Eq. (64)})^2 \Big)$$

$$= \sum_{i,j,k,l=1}^{4} A_{ijkl} \Omega^*_{\infty(ij)} \Omega^*_{\infty(kl)}, \tag{68}$$

where W_κ is an appropriate weight (initially we set $W_\kappa = 1$). The $3 \times 3 \times 3 \times 3$ tensor $\mathcal{A} = (A_{ijkl})$ has the form

$$A_{ijkl} = \sum_{\kappa=1}^{M} W_\kappa \Big(Q_{\kappa(1i)} Q_{\kappa(1j)} Q_{\kappa(1k)} Q_{\kappa(1l)} - Q_{\kappa(1i)} Q_{\kappa(1j)} Q_{\kappa(2k)} Q_{\kappa(2l)}$$

$$- Q_{\kappa(2i)} Q_{\kappa(2j)} Q_{\kappa(1k)} Q_{\kappa(1l)} + Q_{\kappa(2i)} Q_{\kappa(2j)} Q_{\kappa(2k)} Q_{\kappa(2l)} + \frac{1}{4}(Q_{\kappa(1i)} Q_{\kappa(2j)} Q_{\kappa(1k)} Q_{\kappa(2l)}$$

$$+ Q_{\kappa(2i)} Q_{\kappa(1j)} Q_{\kappa(1k)} Q_{\kappa(2l)} + Q_{\kappa(1i)} Q_{\kappa(2j)} Q_{\kappa(2k)} Q_{\kappa(1l)} + Q_{\kappa(2i)} Q_{\kappa(1j)} Q_{\kappa(2k)} Q_{\kappa(1l)})$$

$$+ \frac{1}{4}(Q_{\kappa(2i)} Q_{\kappa(3j)} Q_{\kappa(2k)} Q_{\kappa(3l)} + Q_{\kappa(3i)} Q_{\kappa(2j)} Q_{\kappa(2k)} Q_{\kappa(3l)} + Q_{\kappa(2i)} Q_{\kappa(3j)} Q_{\kappa(3k)} Q_{\kappa(2l)}$$

$$+ Q_{\kappa(3i)} Q_{\kappa(2j)} Q_{\kappa(3k)} Q_{\kappa(2l)}) + \frac{1}{4}(Q_{\kappa(3i)} Q_{\kappa(1j)} Q_{\kappa(3k)} Q_{\kappa(1l)} + Q_{\kappa(1i)} Q_{\kappa(3j)} Q_{\kappa(3k)} Q_{\kappa(1l)}$$

$$+ Q_{\kappa(3i)} Q_{\kappa(1j)} Q_{\kappa(1k)} Q_{\kappa(3l)} + Q_{\kappa(1i)} Q_{\kappa(3j)} Q_{\kappa(1k)} Q_{\kappa(3l)}) \Big). \tag{69}$$

The absolute scale of $\boldsymbol{\Omega}^*_\infty$ cannot be determined from Eq. (62), so we tentatively adopt normalization $\sum_{i,j=1}^{4} \Omega^{*2}_{\infty(ij)} = 1$. Since $\boldsymbol{\Omega}^*_\infty$ is a symmetric matrix, we can write

$$\boldsymbol{\Omega}^*_\infty = \begin{pmatrix} w_1 & w_5/\sqrt{2} & w_6/\sqrt{2} & w_7/\sqrt{2} \\ w_5/\sqrt{2} & w_2 & w_8/\sqrt{2} & w_9/\sqrt{2} \\ w_6/\sqrt{2} & w_8/\sqrt{2} & w_3 & w_{10}/\sqrt{2} \\ w_7/\sqrt{2} & w_9/\sqrt{2} & w_{10}/\sqrt{2} & w_4 \end{pmatrix}. \tag{70}$$

Then, the normalization $\sum_{i,j=1}^{4} \Omega^{*2}_{\infty(ij)} = 1$ is equivalent to $\sum_{i=1}^{10} w_i^2 = 1$. If we define the 10×10 matrix

$$\boldsymbol{A}^\dagger = \begin{pmatrix} A_{1111} & A_{1122} & A_{1133} & A_{1144} & \sqrt{2}A_{1112} \\ A_{2211} & A_{2222} & A_{2233} & A_{2244} & \sqrt{2}A_{2212} \\ A_{3311} & A_{3322} & A_{3333} & A_{3344} & \sqrt{2}A_{3312} \\ A_{4411} & A_{4422} & A_{4433} & A_{4444} & \sqrt{2}A_{4412} \\ \sqrt{2}A_{1211} & \sqrt{2}A_{1222} & \sqrt{2}A_{1233} & \sqrt{2}A_{1244} & 2A_{1212} \\ \sqrt{2}A_{1311} & \sqrt{2}A_{1322} & \sqrt{2}A_{1333} & \sqrt{2}A_{1344} & 2A_{1312} \\ \sqrt{2}A_{1411} & \sqrt{2}A_{1422} & \sqrt{2}A_{1433} & \sqrt{2}A_{1444} & 2A_{1412} \\ \sqrt{2}A_{2311} & \sqrt{2}A_{2322} & \sqrt{2}A_{2333} & \sqrt{2}A_{2344} & 2A_{2312} \\ \sqrt{2}A_{2411} & \sqrt{2}A_{2422} & \sqrt{2}A_{2433} & \sqrt{2}A_{2444} & 2A_{2412} \\ \sqrt{2}A_{3411} & \sqrt{2}A_{3422} & \sqrt{2}A_{3433} & \sqrt{2}A_{3444} & 2A_{3412} \end{pmatrix}$$

$$\begin{pmatrix}
\sqrt{2}A_{1113} & \sqrt{2}A_{1114} & \sqrt{2}A_{1123} & \sqrt{2}A_{1124} & \sqrt{2}A_{1134} \\
\sqrt{2}A_{2213} & \sqrt{2}A_{2214} & \sqrt{2}A_{2223} & \sqrt{2}A_{2224} & \sqrt{2}A_{2234} \\
\sqrt{2}A_{3313} & \sqrt{2}A_{3314} & \sqrt{2}A_{3323} & \sqrt{2}A_{3324} & \sqrt{2}A_{3334} \\
\sqrt{2}A_{4413} & \sqrt{2}A_{4414} & \sqrt{2}A_{4423} & \sqrt{2}A_{4424} & \sqrt{2}A_{4434} \\
2A_{1213} & 2A_{1214} & 2A_{1223} & 2A_{1224} & 2A_{1234} \\
2A_{1313} & 2A_{1314} & 2A_{1323} & 2A_{1324} & 2A_{1334} \\
2A_{1413} & 2A_{1414} & 2A_{1423} & 2A_{1424} & 2A_{1434} \\
2A_{2313} & 2A_{2314} & 2A_{2323} & 2A_{2324} & 2A_{2334} \\
2A_{2413} & 2A_{2414} & 2A_{2423} & 2A_{2424} & 2A_{2434} \\
2A_{3413} & 2A_{3414} & 2A_{3423} & 2A_{3424} & 2A_{3434}
\end{pmatrix}, \tag{71}$$

Eq. (68) is written as

$$K = \sum_{i,j=1}^{10} A^\dagger_{ij} w_i w_j. \tag{72}$$

Hence, minimization of Eq. (68) subject to $\sum_{i,j=1}^{4} \Omega^{*2}_{\infty(ij)} = 1$ reduces to minimization of Eq. (72) subject to $\sum_{i=1}^{10} w_i^2 = 1$. The solution is given by the unit eigenvector $\boldsymbol{w} = (w_i)$ of the matrix $\boldsymbol{A}^\dagger = (A^\dagger_{ij})$ (alternatively, we can use SVD, but explicit expressions are cumbersome to write down). The computed $\boldsymbol{w} = (w_i)$ is then converted to a 4×4 matrix in the form of Eq. (70). However, the sign of the eigenvector $\boldsymbol{w}$, hence of $\boldsymbol{\Omega}^*_\infty$, is indeterminate. Also, $\boldsymbol{\Omega}^*_\infty$ must be positive-semi definite with rank 3. So, we redefine $\boldsymbol{\Omega}^*_\infty$ as follows. Let $\sigma_1 \geq \cdots \geq \sigma_4$ be the eigenvalues of $\boldsymbol{\Omega}^*_\infty$, and $\boldsymbol{u}_1, ..., \boldsymbol{u}_4$ the corresponding unit eigenvectors. We let

$$\boldsymbol{\Omega}^*_\infty = \begin{cases} \sigma_1 \boldsymbol{u}_1 \boldsymbol{u}_1^\top + \sigma_2 \boldsymbol{u}_2 \boldsymbol{u}_2^\top + \sigma_3 \boldsymbol{u}_3 \boldsymbol{u}_3^\top & \sigma_3 > 0 \\ -\sigma_4 \boldsymbol{u}_4 \boldsymbol{u}_4^\top - \sigma_3 \boldsymbol{u}_3 \boldsymbol{u}_3^\top - \sigma_2 \boldsymbol{u}_2 \boldsymbol{u}_2^\top & \sigma_2 < 0 \end{cases}. \tag{73}$$

A.2. Update of K_κ

Suppose the left-hand side of Eq. (62) for the computed $\boldsymbol{\Omega}^*_\infty$ has the form

$$\boldsymbol{Q}_\kappa \boldsymbol{\Omega}^*_\infty \boldsymbol{Q}_\kappa^\top = \begin{pmatrix} c_\kappa(11) & c_\kappa(12) & c_\kappa(13) \\ c_\kappa(21) & c_\kappa(22) & c_\kappa(23) \\ c_\kappa(31) & c_\kappa(32) & c_\kappa(33) \end{pmatrix}. \tag{74}$$

If this is not a scalar multiple of the unit matrix $\boldsymbol{I}$, we update $\boldsymbol{K}_\kappa$ in the form of $\boldsymbol{K}_\kappa \leftarrow \delta \boldsymbol{K}_\kappa \boldsymbol{K}_\kappa$, where we let

$$\delta \boldsymbol{K}_\kappa = \begin{pmatrix} \delta f_\kappa & 0 & \delta u_{\kappa 0} \\ 0 & \delta f_\kappa & \delta v_{\kappa 0} \\ 0 & 0 & 1 \end{pmatrix}. \tag{75}$$

The increment $\delta \boldsymbol{K}_\kappa$ is determined in such a way that Eq. (74) is approximated by $\delta \boldsymbol{K}_\kappa \delta \boldsymbol{K}_\kappa^\top$. From Eqs. (74) and (75), we find that

$$\delta u_{\kappa 0} = \frac{c_\kappa(13)}{c_\kappa(33)}, \qquad \delta v_{\kappa 0} = \frac{c_\kappa(23)}{c_\kappa(33)},$$

$$\delta f_\kappa = \sqrt{\frac{1}{2}\left(\frac{c_{\kappa(11)} + c_{\kappa(22)}}{c_{\kappa(33)}} - \delta u_{\kappa 0}^2 - \delta v_{\kappa 0}^2\right)}. \tag{76}$$

Since the projection matrix $\tilde{P}_\kappa$ can be defined only up to scalar multiplication (see Eq. (31)), the matrix Q_κ in Eq. (63) also has scale indeterminacy. So, we normalize Q_κ by dividing it by $\sqrt{c_{\kappa(33)}}$ so that Eq. (74) has approximately the same scale as I for all κ. However, $c_{\kappa(33)}$ can be negative in the presence of extremely large noise, and the inside of the square root in Eqs. (76) may also become negative. In such a case, we skip that frame in the computation. To do this systematically, we make the weight W_κ reflect the closeness of Eq. (74) to a scalar multiple of I. We also measure the goodness of estimation not by totaling the goodness measures of individual frames but by their "median" so that exceptional frames are not counted (see Section A.4).

A.3. Computation of H

Since Ω_∞^* has the form of Eq. (73), a 4×4 matrix H that satisfies Eq. (37) is given up to a rotation by $(\sqrt{\sigma_1}u_1 \; \sqrt{\sigma_2}u_2 \; \sqrt{\sigma_3}u_3 \; v)$ for $\sigma_3 > 0$ and $(\sqrt{-\sigma_4}u_4 \; \sqrt{-\sigma_3}u_3 \; \sqrt{-\sigma_2}u_2 \; v)$ for $\sigma_2 < 0$, where v is an arbitrary vector. The indeterminate freedom of rotation and the arbitrariness of the vector v correspond to the fact that the orientation and the location of the world coordinate system are arbitrary. However, the matrix H must be nonsingular, which means that v must be linearly independent of the first, the second, and the third columns of H. So, we choose as v a unit vector orthogonal to them. This means that we can take as v the remaining unit eigenvector of Ω_∞^*.

A.4. Computational Procedure

The above computation is summarized as follows:

Input:

- Approximate principal points $(u_{\kappa 0}, v_{\kappa 0})$ and the focal lengths f_κ, $\kappa = 1, ..., M$.
- Projection matrices $\tilde{P}_\kappa$, $\kappa = 1, ..., M$.

Output:

- Rectifying projective transformation H.
- Intrinsic parameter matrices K_κ, $\kappa = 1, ..., M$.

Computation:

1. Let
$$\hat{H} = I_{4 \times 4}, \qquad \hat{K} = I_{3 \times 3}, \qquad \hat{J}_{\text{med}} = \infty, \tag{77}$$
where the subscript of I denotes its size (omitted if understood), and ∞ means a sufficiently large number.

2. Initialize K_κ in the form of Eq. (61), and let $W_\kappa = 1$ and $\gamma_\kappa = 1$.

3. Let
$$Q_\kappa = \gamma_\kappa K_\kappa^{-1} \tilde{P}_\kappa. \tag{78}$$

4. Compute the tensor $\mathcal{A} = (A_{ijkl})$ in Eq. (69).

5. Compute the 10-D unit eigenvector w of the 10×10 matrix A in Eq. (71) for the smallest eigenvalue.

6. Compute the tentative matrix Ω_∞^* in Eq. (70).

7. Compute the eigenvalues $\sigma_1 \geq \cdots \geq \sigma_4$ of Ω_∞^* and the corresponding unit eigenvectors $u_1, ..., u_4$.

8. Compute
$$H = \begin{cases} (\sqrt{\sigma_1} u_1 \quad \sqrt{\sigma_2} u_2 \quad \sqrt{\sigma_3} u_3 \quad u_4) & \sigma_3 > 0 \\ (\sqrt{-\sigma_4} u_4 \quad \sqrt{-\sigma_3} u_3 \quad \sqrt{-\sigma_2} u_2 \quad u_1) & \sigma_2 < 0 \end{cases}. \tag{79}$$

9. Do the following computation for $\kappa = 1, ..., M$:

    (a) Compute $c_{\kappa(ij)}$ by Eq. (74), and let
    $$F_\kappa = \frac{c_{\kappa(11)} + c_{\kappa(22)}}{c_{\kappa(33)}} - \left(\frac{c_{\kappa(13)}}{c_{\kappa(33)}}\right)^2 - \left(\frac{c_{\kappa(23)}}{c_{\kappa(33)}}\right)^2. \tag{80}$$

    (b) If $c_{\kappa(33)} > 0$ and $F_\kappa > 0$, compute $\delta u_{\kappa 0}$, $\delta v_{\kappa 0}$, and δf_κ in Eqs. (76) and let
    $$J_\kappa = \left(\frac{c_{\kappa(11)}}{c_{\kappa(33)}} - 1\right)^2 + \left(\frac{c_{\kappa(22)}}{c_{\kappa(33)}} - 1\right)^2 + 2\frac{c_{\kappa(12)}^2 + c_{\kappa(23)}^2 + c_{\kappa(31)}^2}{c_{\kappa(33)}^2}. \tag{81}$$

    Then, update K_κ and γ_κ as follows:
    $$K_\kappa \leftarrow K_\kappa \delta K_\kappa, \qquad \gamma_\kappa \leftarrow \frac{\gamma_\kappa}{\sqrt{c_{\kappa(33)}}}. \tag{82}$$

    (c) Else, let $J_\kappa = \infty$.

10. Compute the following median:
$$J_{\text{med}} = \text{med}_{\kappa=1}^M J_\kappa. \tag{83}$$

11. If $J_{\text{med}} \approx 0$, return H and K_κ and stop.

12. If $J_{\text{med}} \geq \hat{J}_{\text{med}}$, return $\hat{H}$ and $\hat{K}_\kappa$ as H and K_κ and stop.

13. Go back to Step 3 after letting
$$\hat{J}_{\text{med}} \leftarrow J_{\text{med}}, \qquad \hat{H} \leftarrow H, \qquad \hat{K}_\kappa \leftarrow K_\kappa, \qquad W_\kappa \leftarrow e^{-J_\kappa/J_{\text{med}}}. \tag{84}$$

Note that this algorithm does not compute the matrix Ω_∞^* in Eq. (73); it directly outputs the rectifying projective transformation H and the intrinsic parameter matrix K_κ.

A.5. 3-D Positions and Motion Parameters

Using the computed H, we can rectify the projection matrices $\tilde{P}_\kappa$ and the 3-D positions X_α as follows:
$$\bar{P}_\kappa = P_\kappa H, \qquad \bar{X}_\alpha = H^{-1} X_\alpha. \tag{85}$$

The 3-D coordinates $(X_\alpha, Y_\alpha, Z_\alpha)$ are given by Eqs. (13). From the computed K_κ, the motion parameters $\{R_\kappa, t_\kappa\}$ are to be determined such that

$$K_\kappa^{-1} \bar{P}_\kappa \simeq \begin{pmatrix} R_\kappa & t_\kappa \end{pmatrix}. \tag{86}$$

So, we adjust the scale of $K_\kappa^{-1} \bar{P}_\kappa$ so that its first three columns are all unit vectors (in practice, their average norm is made 1). We choose the sign of $K_\kappa^{-1} \bar{P}_\kappa$ so that its first three columns define a rotation matrix R_κ of determinant 1. Then, the fourth column gives the translation t_κ. The resulting R_κ may not be strictly orthonormal in the presence of noise, so we enforce the orthonormality by computing the singular value decomposition

$$R_\kappa = U \mathrm{diag}(\lambda_1, \lambda_2, \lambda_3) V^\top, \tag{87}$$

and letting $R_\kappa = UV^\top$ [15].

A.6. Mirror Image Solution Removal

Now, we remove the mirror image solution (Remark 14). If a point is at $(X_\alpha, Y_\alpha, Z_\alpha)$, its coordinates $(X^c_{\kappa\alpha}, Y^c_{\kappa\alpha}, Z^c_{\kappa\alpha})$ with respect to the κth camera coordinate system are given by

$$\begin{pmatrix} X^c_{\kappa\alpha} \\ Y^c_{\kappa\alpha} \\ Z^c_{\kappa\alpha} \end{pmatrix} = t_\kappa + R_\kappa \begin{pmatrix} X_\alpha \\ Y_\alpha \\ Z_\alpha \end{pmatrix}. \tag{88}$$

We can judge that it is in front of the camera if

$$\sum_{\alpha=1}^{N} \mathrm{sgn}(Z^c_{1\alpha}) > 0, \tag{89}$$

where $\mathrm{sgn}(x)$ returns 1, 0, and -1 for $x > 0$, $x = 0$, and $x < 0$, respectively. If Eq. (89) is not satisfied, we reverse the signs of X_α, Y_α, Z_α, and t_κ. We introduce $\mathrm{sgn}(x)$ because if we require $\sum_{\alpha=1}^{N} Z^c_{1\alpha} > 0$, the judgment may be reversed when a very large depth $Z^c_{1\alpha} \approx \infty$ may be computed to be $Z^c_{1\alpha} \approx -\infty$ in the presence of noise. Theoretically, we should require $\sum_{\kappa=1}^{M} \sum_{\alpha=1}^{N} \mathrm{sgn}(Z^c_{\kappa\alpha}) > 0$, but considering the first camera alone is sufficient in practice.

B. Procedure for Projective Reconstruction

Here, we give two algorithms for projective reconstruction. One is the method of Mahamud and Hebert [25], which we call the *primal method*. The other, which we call the *dual method*, is based on Heyden et al. [11]. We modify these, using corresponding symbols and notations so that their mutual relationships become clear.

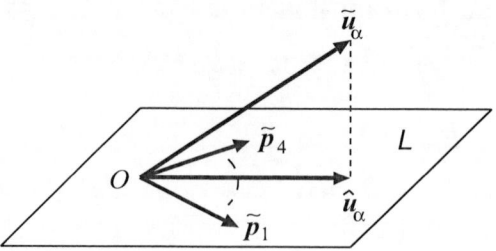

Figure 11. Orthogonal projection of $\tilde{u}_\alpha$ onto $\mathcal{L}$.

B.1. Primal Method

Eq. (40) indicates that vectors $\tilde{u}_\alpha$ (= the columns of the matrix on the left-hand side of Eq. (43)) are constrained to be in the 4-D subspace $\mathcal{L}$ spanned by $\{\tilde{p}_1, \tilde{p}_2, \tilde{p}_3, \tilde{p}_4\}$ if the projective depths $z_{\kappa\alpha}$ are all correct. This does not hold if $z_{\kappa\alpha}$ are not correct, so we update $z_{\kappa\alpha}$ so that each $\tilde{u}_\alpha$ is as close to $\mathcal{L}$ as possible, identifying $\{\tilde{p}_1, \tilde{p}_2, \tilde{p}_3, \tilde{p}_4\}$ with the unit eigenvectors of C in Eq. (41) for the largest four eigenvalues (or the first four columns of U in Eq. (42)). The orthogonal projection of $\tilde{u}_\alpha$ onto $\mathcal{L}$ is (Fig. 11)

$$\hat{u}_\alpha = \sum_{i=1}^{4} (\tilde{u}_\alpha, \tilde{p}_i) \tilde{p}_i, \tag{90}$$

where and hereafter we denote the inner product of vectors a and b by (a, b). Since $\tilde{u}_\alpha$ is normalized to unit norm (Section 4.5), the distance of $\tilde{u}_\alpha$ from the subspace $\mathcal{L}$ is

$$\sqrt{\|\tilde{u}_\alpha\|^2 - \|\hat{u}_\alpha\|^2} = \sqrt{1 - \sum_{i=1}^{4} (\tilde{u}_\alpha, \tilde{p}_i)^2}. \tag{91}$$

Minimizing this is equivalent to maximizing

$$J_\alpha = \sum_{i=1}^{4} (\tilde{u}_\alpha, \tilde{p}_i)^2 = \sum_{i=1}^{4} \Big(\sum_{\kappa=1}^{M} (z_{\kappa\alpha} x_{\kappa\alpha}, p_{i\kappa}) \Big)^2$$

$$= \sum_{\kappa,\lambda=1}^{M} \Big(\sum_{i=1}^{4} (x_{\kappa\alpha}, p_{i\kappa})(x_{\lambda\alpha}, p_{i\lambda}) \Big) z_{\kappa\alpha} z_{\lambda\alpha}, \tag{92}$$

where $p_{i\kappa}$ is the 3-D vector consisting of the $3(\kappa-1)+1$th, $3(\kappa-1)+2$th, and $3(\kappa-1)+3$th components of $\tilde{p}_i$. Thus, Eq. (92) is to be maximized subject to

$$\|\tilde{u}_\alpha\|^2 = \sum_{\kappa=1}^{M} z_{\kappa\alpha}^2 \|x_{\kappa\alpha}\|^2 = 1. \tag{93}$$

Define new variables $\xi_{\kappa\alpha}$ by

$$\xi_{\kappa\alpha} = \|x_{\kappa\alpha}\| z_{\kappa\alpha}, \tag{94}$$

and consider the M-D vector $\boldsymbol{\xi}_\alpha$ with components $\xi_{1\alpha}, ..., \xi_{M\alpha}$. Then, Eq. (93) means $\|\boldsymbol{\xi}_\alpha\| = 1$, and Eq. (92) is rewritten as

$$J_\alpha = \sum_{\kappa,\lambda=1}^{M} A_{\kappa\lambda}^\alpha \xi_{\kappa\alpha} \xi_{\lambda\alpha} = (\boldsymbol{\xi}_\alpha, \boldsymbol{A}^\alpha \boldsymbol{\xi}_\alpha), \qquad (95)$$

where we define the $M \times M$ matrix $\boldsymbol{A}^\alpha = (A_{\kappa\lambda}^\alpha)$ by

$$A_{\kappa\lambda}^\alpha = \frac{\sum_{i=1}^{4} (\boldsymbol{x}_{\kappa\alpha}, \boldsymbol{p}_{i\kappa})(\boldsymbol{x}_{\lambda\alpha}, \boldsymbol{p}_{i\lambda})}{\|\boldsymbol{x}_{\kappa\alpha}\| \cdot \|\boldsymbol{x}_{\lambda\alpha}\|}. \qquad (96)$$

Eq. (95) is maximized by the unit eigenvector $\boldsymbol{\xi}_\alpha$ of the matrix $\boldsymbol{A}^\alpha$ for the largest eigenvalue. The sign is chosen so that

$$\sum_{\kappa=1}^{M} \xi_{\kappa\alpha} \geq 0. \qquad (97)$$

The corresponding projective depths $z_{\kappa\alpha}$ are determined from Eq. (94). The procedure is summarized as follows:

Input: $x_{\kappa\alpha}$, $\kappa = 1, ..., M$, $\alpha = 1, ..., N$.

Output: $\boldsymbol{P}_\kappa$, $\kappa = 1, ..., M$, $\boldsymbol{X}_\alpha$, $\alpha = 1, ..., N$.

Computation:

1. Initialize the projective depths to $z_{\kappa\alpha} = 1$ (Remark 23).

2. Compute $\tilde{\boldsymbol{u}}_\alpha$ and normalize them into unit norm.

3. Fit a 4-D subspace $\mathcal{L}$ to $\tilde{\boldsymbol{u}}_\alpha$ by least squares (Remark 24).

4. Do the following computations for $\alpha = 1, ..., N$.

    (a) Compute the unit eigenvector $\boldsymbol{\xi}_\alpha$ of the matrix $\boldsymbol{A}^\alpha$ defined by Eq. (96) for the largest eigenvalue, and choose the sign as in Eq. (97).

    (b) Determine the projective depths $z_{\kappa\alpha}$ according to Eq. (94).

    (c) Recompute the vector $\tilde{\boldsymbol{u}}_\alpha$.

5. Go back to Step 3, and repeat this until the iterations converge.

6. Compute $\boldsymbol{X}_\alpha = (X_\alpha^i)$ by
$$X_\alpha^i = (\tilde{\boldsymbol{u}}_\alpha, \tilde{\boldsymbol{p}}_i). \qquad (98)$$

7. Determine the projection matrix $\boldsymbol{P}_\kappa$ by
$$\boldsymbol{P}_\kappa = \begin{pmatrix} \tilde{\boldsymbol{p}}_{1\kappa} & \tilde{\boldsymbol{p}}_{2\kappa} & \tilde{\boldsymbol{p}}_{3\kappa} & \tilde{\boldsymbol{p}}_{4\kappa} \end{pmatrix}, \qquad (99)$$

where $\tilde{\boldsymbol{p}}_{i\kappa}$ is a 3-D vector whose first, second, and third components are, respectively, the $(3(\kappa-1)+1)$st, $(3(\kappa-1)+2)$nd, and $(3(\kappa-1)+3)$rd components of $\tilde{\boldsymbol{p}}_i$.

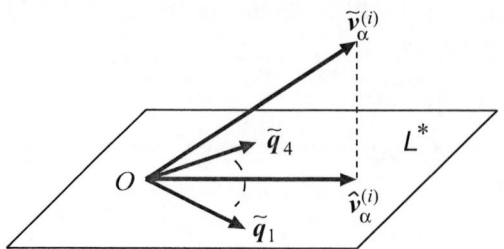

Figure 12. Orthogonal projection of $\tilde{v}_\kappa^{(i)}$ onto $\mathcal{L}^*$.

B.2. Dual Method

Consider the following N-D vectors:

$$\tilde{v}_\kappa^{(1)} = \begin{pmatrix} z_{\kappa 1} x_{\kappa 1} \\ z_{\kappa 2} x_{\kappa 2} \\ \vdots \\ z_{\kappa 1} x_{\kappa 1} \end{pmatrix}, \quad \tilde{v}_\kappa^{(2)} = \begin{pmatrix} z_{\kappa 1} y_{\kappa 1} \\ z_{\kappa 2} y_{\kappa 2} \\ \vdots \\ z_{\kappa 1} y_{\kappa 1} \end{pmatrix}, \quad \tilde{v}_\kappa^{(3)} = \begin{pmatrix} z_{\kappa 1} \\ z_{\kappa 2} \\ \vdots \\ z_{\kappa 1} \end{pmatrix}. \quad (100)$$

Note that the transpose $\tilde{v}_\kappa^{(i)\top}$ is the $(3(\kappa-1)+i)$th row of the matrix on the left-hand side of Eq. (43)), which is written as $\left(\tilde{v}_1^{(1)} \ \tilde{v}_1^{(2)} \ \tilde{v}_1^{(3)} \ \tilde{v}_2^{(1)} \ \cdots \ \tilde{v}_M^{(3)} \right)^\top$. For the scale normalization, we impose

$$\sum_{i=1}^{3} \|\tilde{v}_\kappa^{(i)}\|^2 = \sum_{\alpha=1}^{N} z_{\kappa\alpha}^2 \|x_\kappa\|^2 = 1. \quad (101)$$

If we take out the ith component of Eq. (39) and vertically align it for $\alpha = 1, ..., N$, we obtain

$$\tilde{v}_\kappa^{(i)} = P_{\kappa(i1)} X^1 + P_{\kappa(i2)} X^2 + P_{\kappa(i3)} X^3 + P_{\kappa(i4)} X^4, \quad (102)$$

where $P_{\kappa(ij)}$ is the (ij) element of P_κ, and X^k is the N-D vector consisting of X_α^k (= the kth component of X_α), $\alpha = 1, ..., N$. Eq. (102) implies that the $3M$ vectors $\tilde{v}_\kappa^{(i)}$ belong to the 4-D subspace $\mathcal{L}^*$ spanned by X^1, X^2, X^3, and X^4. The orthonormal basis $\{\tilde{q}_1, ..., \tilde{q}_4\}$ of the subspace $\mathcal{L}^*$ is given by the first four columns of the matrix V in Eq. (42). The orthogonal projection of $\tilde{v}_\kappa^{(i)}$ onto $\mathcal{L}^*$ is (Fig. 12)

$$\hat{v}_\kappa^{(i)} = \sum_{k=1}^{4} (\tilde{v}_\kappa^{(i)}, \tilde{q}_k) \tilde{q}_k. \quad (103)$$

We update $z_{\kappa\alpha}$ so that the sum of squares of the distances of $\tilde{v}_\kappa^{(1)}$, $\tilde{v}_\kappa^{(2)}$, and $\tilde{v}_\kappa^{(3)}$ from the subspace $\mathcal{L}^*$

$$\sum_{i=1}^{3} \left(\|\tilde{v}_\kappa^{(i)}\|^2 - \|\hat{v}_\kappa^{(i)}\|^2 \right) = \sum_{i=1}^{3} \|\tilde{v}_\kappa^{(i)}\|^2 - \sum_{i=1}^{3} \sum_{k=1}^{4} (\tilde{v}_\kappa^{(i)}, \tilde{q}_k)^2 \quad (104)$$

is minimized for each α. Consider the N-D vector $\boldsymbol{\xi}_\kappa$ with components $\xi_{\kappa 1}$, ..., $\xi_{\kappa N}$ defined by Eq. (94). Then, minimizing Eq. (104) is equivalent to maximizing

$$J_\kappa^* = \sum_{i=1}^{3}\sum_{k=1}^{4}(\tilde{\boldsymbol{v}}_\kappa^{(i)}, \tilde{\boldsymbol{q}}_k)^2 = (\boldsymbol{\xi}_\kappa, \boldsymbol{B}^\kappa \boldsymbol{\xi}_\kappa), \qquad (105)$$

where we define the $N \times N$ matrix $\boldsymbol{B}^\kappa = (B_{\alpha\beta}^\kappa)$ by

$$B_{\alpha\beta}^\kappa = \frac{(\boldsymbol{q}_\alpha, \boldsymbol{q}_\beta)(\boldsymbol{x}_{\kappa\alpha}, \boldsymbol{x}_{\kappa\beta})}{\|\boldsymbol{x}_{\kappa\alpha}\| \cdot \|\boldsymbol{x}_{\kappa\beta}\|}. \qquad (106)$$

Here, $\boldsymbol{q}_\alpha$ is the 4-D vector consisting of the αth components of the basis vectors $\tilde{\boldsymbol{q}}_1$, ..., $\tilde{\boldsymbol{q}}_4$. Eq. (105) is maximized by the unit eigenvector $\boldsymbol{\xi}_\kappa$ of the matrix $\boldsymbol{B}^\kappa$ for the largest eigenvalue. The sign is chosen so that

$$\sum_{\alpha=1}^{N} \xi_{\kappa\alpha} \geq 0, \qquad (107)$$

and the corresponding projective depths $z_{\kappa\alpha}$ are determined from Eq. (94). The procedure is summarized as follows:

Input: $x_{\kappa\alpha}$, $\kappa = 1, ..., M$, $\alpha = 1, ..., N$.

Output: $\boldsymbol{P}_\kappa$, $\kappa = 1, ..., M$, $\boldsymbol{X}_\alpha$, $\alpha = 1, ..., N$.

Computation:

1. Initialize the projective depths to $z_{\kappa\alpha} = 1$.

2. Compute the vectors $\tilde{\boldsymbol{v}}_\kappa^{(i)}$ in Eqs. (100), and normalize them as in Eqs. (101).

3. Fit a 4-D subspace $\mathcal{L}^*$ to $\tilde{\boldsymbol{v}}_\kappa^{(i)}$ by least squares.

4. Do the following computations for $\kappa = 1, ..., M$.

    (a) Compute the unit eigenvector $\boldsymbol{\xi}_\kappa$ of the matrix $\boldsymbol{B}^\alpha$ defined by Eq. (106) for the largest eigenvalue, and choose the sign as in Eq. (107).

    (b) Determine the projective depths $z_{\kappa\alpha}$ according to Eq. (107).

    (c) Recompute the vectors $\tilde{\boldsymbol{v}}_\kappa^{(i)}$.

5. Go back to Step 3, and repeat this until the iterations converge.

6. Compute $\boldsymbol{X}_\alpha = (X_\alpha^i)$ by

$$X_\alpha^i = (\text{the } \alpha\text{th component of } \tilde{\boldsymbol{q}}_i). \qquad (108)$$

7. Determine the projection matrix $\boldsymbol{P}_\kappa = (P_{\kappa(ij)})$ by

$$P_{\kappa(ij)} = (\tilde{\boldsymbol{v}}_\kappa^{(i)}, \tilde{\boldsymbol{q}}_j). \qquad (109)$$

C. Affine Camera Factorization

Here, we give the details of the 3-D reconstruction procedure described in Section 5.3. The actual computation depends on what affine camera model we use, so we first describe the general framework that does not depend on specific camera models and then add details that depend on individual models, for which we consider (i) orthographic projection of Eqs. (46) (Fig. 10(a)), (ii) weak perspective projection of Eqs. (47) (Fig. 10(b)), (iii) paraperspective projection of Eqs. (48) (Fig. 10(c)), and (iv) the generic model of Eq. (45). Whichever model we use, we obtain "two" solutions that are mirror images of each other, which cannot be distinguished as long as we use affine camera modeling.

C.1. General Framework

Suppose we track N points over M frames. Let $(x_{\kappa\alpha}, y_{\kappa\alpha})$ be the image coordinates of the αth point in the κth image. The algorithm for affine camera 3-D reconstruction has the following structure [19, 20]. Items with $*$ depend on the camera model we use. The detailed procedure for them is given later.

Input:

- $2M$-D trajectory vectors

$$\tilde{\boldsymbol{u}}_\alpha = \begin{pmatrix} x_{1\alpha} & y_{1\alpha} & x_{2\alpha} & y_{2\alpha} & \cdots & x_{M\alpha} & y_{M\alpha} \end{pmatrix}^\top, \quad \alpha = 1, ..., N. \tag{110}$$

- Focal lengths f_κ, $\kappa = 1, ..., M$ (arbitrary if unknown).

Output:

- Translations $\boldsymbol{t}_\kappa$ (= the world coordinate origin for the κth view).

- Shape vectors, i.e., 3-D positions $\boldsymbol{s}_\alpha$ and $\boldsymbol{s}'_\alpha$ (mirror images of each other) of the points relative to the world coordinate system centered on their centroid.

- Corresponding rotations $\boldsymbol{R}_\kappa$ and $\boldsymbol{R}'_\kappa$ that specify the world coordinate axis orientations.

Computation:

1. Compute the centroid $\tilde{\boldsymbol{u}}_C$ of the trajectory vectors $\tilde{\boldsymbol{u}}_\alpha$ by Eq. (56).

2. Let $\tilde{t}_{x\kappa}$ and $\tilde{t}_{y\kappa}$ be the $(2(\kappa - 1) + 1)$th and $(2(\kappa - 1) + 2)$th components of $\tilde{\boldsymbol{u}}_C$, respectively.

3. Fit a 3-D affine space to the trajectory vectors $\tilde{\boldsymbol{u}}_\alpha$, and let $\{\tilde{\boldsymbol{q}}_1, \tilde{\boldsymbol{q}}_2, \tilde{\boldsymbol{q}}_3\}$ be its basis.

4. Let $\boldsymbol{Q}$ be the $2M \times 3$ matrix having $\tilde{\boldsymbol{q}}_1$, $\tilde{\boldsymbol{q}}_2$, and $\tilde{\boldsymbol{q}}_3$ as its columns, and let $\boldsymbol{q}^\dagger_{\kappa(a)}$ be the $(2(\kappa - 1) + a)$th column of $\boldsymbol{Q}^\top$, $\kappa = 1, ..., M$, $a = 1, 2$.

5. *Compute the 3×3 metric matrix $\boldsymbol{T}$.

6. Compute the eigenvalues $\{\lambda_1, \lambda_2, \lambda_3\}$ of T and the corresponding orthonormal system $\{v_1, v_2, v_3\}$ of unit eigenvectors.

7. *Compute the translation vectors $t_\kappa = (t_{x\kappa}, t_{y\kappa}, t_{z\kappa})^\top$.

8. Compute the following $2M$-D vectors:

$$m_i = \sqrt{\lambda_i} \begin{pmatrix} (q^\dagger_{1(1)}, v_i) \\ (q^\dagger_{1(2)}, v_i) \\ (q^\dagger_{2(1)}, v_i) \\ \vdots \\ (q^\dagger_{M(2)}, v_i) \end{pmatrix}, \qquad i = 1, 2, 3. \tag{111}$$

9. Let M be the $2M \times 3$ motion matrix having m_1, m_2, and m_3 as its columns, and let $m^\dagger_{\kappa(a)}$ be the the $(2(\kappa - 1) + a)$th column of $M^\top$, $\kappa = 1, ..., M$, $a = 1, 2$.

10. *Compute the rotations R_κ.

11. *Recompute the motion matrix M by

$$M = \sum_{\kappa=1}^{M} \Pi_\kappa^\top R_\kappa, \tag{112}$$

where $\Pi_\kappa = (\Pi_{\kappa(ij)})$ is a $3 \times 2M$ matrix that depends on the assumed camera model.

12. Compute the 3-D shape vectors s_α by

$$s_\alpha = (M^\top M)^{-1} M^\top (\tilde{u}_\alpha - \tilde{u}_C). \tag{113}$$

13. *Compute s'_α and R'_κ by

$$s'_\alpha = -s_\alpha, \qquad R'_\kappa = \Omega_\kappa R_\kappa, \tag{114}$$

where Ω_κ is a rotation matrix that depends on the assumed camera model.

C.2. Metric Constraint

The metric constraint of Eq. (54) is derived as follows. By definition, the three columns i_κ, j_κ, and k_κ of the rotation R_κ are the world coordinate axis directions for the κth view. Their homogeneous coordinate representations are $(1\ 0\ 0\ 0)^\top$, $(0\ 1\ 0\ 0)^\top$, and $(0\ 0\ 1\ 0)^\top$, respectively (they define "orientations" in the projective space $\mathcal{P}^3$). Hence, according to Eq. (49), their image projections are represented by the first, the second, and the third columns of $\Pi_\kappa R_\kappa$, respectively, if the third components are removed, i.e., if expressed in inhomogeneous (or usual) coordinates. From Eq. (50), on the other hand, these vectors are, respectively,

$$\begin{pmatrix} \tilde{p}_{1(3(\kappa-1)+1)} \\ \tilde{p}_{1(3(\kappa-1)+2)} \end{pmatrix}, \quad \begin{pmatrix} \tilde{p}_{2(3(\kappa-1)+1)} \\ \tilde{p}_{2(3(\kappa-1)+2)} \end{pmatrix}, \quad \begin{pmatrix} \tilde{p}_{3(3(\kappa-1)+1)} \\ \tilde{p}_{3(3(\kappa-1)+2)} \end{pmatrix}, \tag{115}$$

where $\tilde{p}_{i(j)}$ is the jth component of $\tilde{\boldsymbol{p}}_i$. Thus, we have

$$\boldsymbol{\Pi}_\kappa \boldsymbol{R}_\kappa = \begin{pmatrix} \tilde{p}_{1(3(\kappa-1)+1)} & \tilde{p}_{2(3(\kappa-1)+1)} & \tilde{p}_{3(3(\kappa-1)+1)} \\ \tilde{p}_{1(3(\kappa-1)+2)} & \tilde{p}_{2(3(\kappa-1)+2)} & \tilde{p}_{3(3(\kappa-1)+2)} \end{pmatrix}. \tag{116}$$

The right-hand side equals $\boldsymbol{Q}_\kappa^{\dagger\top} \boldsymbol{A}$ from the definition of $\boldsymbol{A}$ in Eq. (51) and $\boldsymbol{Q}_\kappa^\dagger$ in Eq. (52). Hence, we have

$$\boldsymbol{\Pi}_\kappa \boldsymbol{R}_\kappa = \boldsymbol{Q}_\kappa^{\dagger\top} \boldsymbol{A}. \tag{117}$$

It follows that

$$\boldsymbol{Q}_\kappa^{\dagger\top} \boldsymbol{A} \boldsymbol{A}^\top \boldsymbol{Q}_\kappa^\dagger = \boldsymbol{\Pi}_\kappa \boldsymbol{R}_\kappa \boldsymbol{R}_\kappa^\top \boldsymbol{\Pi}_\kappa^\top = \boldsymbol{\Pi}_\kappa \boldsymbol{\Pi}_\kappa^\top, \tag{118}$$

or Eq. (54) if the metric matrix $\boldsymbol{T}$ is defined by Eq. (53).

C.3. Orthographic Projection

If the orthographic projection model of Eqs. (46) is assumed, (Fig. 10(a)), the metric constraint of Eq. (54) takes the following form [19]:

$$(\boldsymbol{q}_{\kappa(1)}^\dagger, \boldsymbol{T}\boldsymbol{q}_{\kappa(1)}^\dagger) = (\boldsymbol{q}_{\kappa(2)}^\dagger, \boldsymbol{T}\boldsymbol{q}_{\kappa(2)}^\dagger) = 1, \qquad (\boldsymbol{q}_{\kappa(1)}^\dagger, \boldsymbol{T}\boldsymbol{q}_{\kappa(2)}^\dagger) = 0. \tag{119}$$

From these, we determine the metric matrix $\boldsymbol{T}$ by least squares. The computation of Step 5 goes as follows [19]. First, we define the $3 \times 3 \times 3 \times 3$ tensor $\mathcal{B} = (B_{ijkl})$ by

$$B_{ijkl} = \sum_{\kappa=1}^M \Big[(\boldsymbol{q}_{\kappa(1)}^\dagger)_i (\boldsymbol{q}_{\kappa(1)}^\dagger)_j (\boldsymbol{q}_{\kappa(1)}^\dagger)_k (\boldsymbol{q}_{\kappa(1)}^\dagger)_l + (\boldsymbol{q}_{\kappa(2)}^\dagger)_i (\boldsymbol{q}_{\kappa(2)}^\dagger)_j (\boldsymbol{q}_{\kappa(2)}^\dagger)_k (\boldsymbol{q}_{\kappa(2)}^\dagger)_l$$
$$+ \frac{1}{4} \Big((\boldsymbol{q}_{\kappa(1)}^\dagger)_i (\boldsymbol{q}_{\kappa(2)}^\dagger)_j + (\boldsymbol{q}_{\kappa(2)}^\dagger)_i (\boldsymbol{q}_{\kappa(1)}^\dagger)_j \Big) \Big((\boldsymbol{q}_{\kappa(1)}^\dagger)_k (\boldsymbol{q}_{\kappa(2)}^\dagger)_l + (\boldsymbol{q}_{\kappa(2)}^\dagger)_k (\boldsymbol{q}_{\kappa(1)}^\dagger)_l \Big) \Big], \tag{120}$$

where $(\boldsymbol{q}_{\kappa(a)}^\dagger)_i$ denotes the ith component of the 3-D vector $\boldsymbol{q}_{\kappa(a)}^\dagger$. We define the 6×6 symmetric matrix $\boldsymbol{B}$ and the 6-D vector $\boldsymbol{c}$ by

$$\boldsymbol{B} = \begin{pmatrix} B_{1111} & B_{1122} & B_{1133} & \sqrt{2}B_{1123} & \sqrt{2}B_{1131} & \sqrt{2}B_{1112} \\ B_{2211} & B_{2222} & B_{2233} & \sqrt{2}B_{2223} & \sqrt{2}B_{2231} & \sqrt{2}B_{2212} \\ B_{3311} & B_{3322} & B_{3333} & \sqrt{2}B_{3323} & \sqrt{2}B_{3331} & \sqrt{2}B_{3312} \\ \sqrt{2}B_{2311} & \sqrt{2}B_{2322} & \sqrt{2}B_{2333} & 2B_{2323} & 2B_{2331} & 2B_{2312} \\ \sqrt{2}B_{3111} & \sqrt{2}B_{3122} & \sqrt{2}B_{3133} & 2B_{3123} & 2B_{3131} & 2B_{3112} \\ \sqrt{2}B_{1211} & \sqrt{2}B_{1222} & \sqrt{2}B_{1233} & 2B_{1223} & 2B_{1231} & 2B_{1212} \end{pmatrix}, \tag{121}$$

$$\boldsymbol{c} = \begin{pmatrix} 1 & 1 & 1 & 0 & 0 & 0 \end{pmatrix}^\top, \tag{122}$$

and solve the following simultaneous linear equations for $\boldsymbol{\tau} = (\tau_i)$:

$$\boldsymbol{B}\boldsymbol{\tau} = \boldsymbol{c}. \tag{123}$$

The metric matrix $\boldsymbol{T}$ is given by

$$\boldsymbol{T} = \begin{pmatrix} \tau_1 & \tau_6/\sqrt{2} & \tau_5/\sqrt{2} \\ \tau_6/\sqrt{2} & \tau_2 & \tau_4/\sqrt{2} \\ \tau_5/\sqrt{2} & \tau_4/\sqrt{2} & \tau_3 \end{pmatrix}. \tag{124}$$

For the translation computation in Step 5, we simply let $t_{x\kappa} = \tilde{t}_{x\kappa}$ and $t_{y\kappa} = \tilde{t}_{y\kappa}$, $\kappa = 1, ..., 2M$. The third components $t_{z\kappa}$ are left indeterminate. For the rotation computation in Step 10, we compute the SVD

$$\left(\boldsymbol{m}^{\dagger}_{\kappa(1)} \ \boldsymbol{m}^{\dagger}_{\kappa(2)} \ \boldsymbol{0} \right) = \boldsymbol{V}_{\kappa} \boldsymbol{\Lambda}_{\kappa} \boldsymbol{U}_{\kappa}^{\top}. \tag{125}$$

Then, the $\boldsymbol{R}_{\kappa}$ is given by

$$\boldsymbol{R}_{\kappa} = \boldsymbol{U}_{\kappa} \mathrm{diag}(1, 1, \det(\boldsymbol{V}_{\kappa} \boldsymbol{U}_{\kappa}^{\top})) \boldsymbol{V}_{\kappa}^{\top}. \tag{126}$$

The matrix Π_{κ} in Step 11 is given by

$$\Pi_{\kappa} = \begin{pmatrix} 0 & \cdots & 0 & \overset{(2\kappa-1)}{1} & \overset{(2\kappa)}{0} & 0 & \cdots & 0 \\ 0 & \cdots & 0 & 0 & 1 & 0 & \cdots & 0 \\ 0 & \cdots & 0 & 0 & 0 & 0 & \cdots & 0 \end{pmatrix}. \tag{127}$$

and the matrix $\boldsymbol{\Omega}_{\kappa}$ in Step 13 is simply $\boldsymbol{\Omega}_{\kappa} = \mathrm{diag}(-1, -1, 1)$.

C.4. Weak Perspective Projection

If the weak perspective projection model of Eqs. (47) is assumed (Fig. 10(b)), the metric constraint of Eq. (54) takes the following form [19]:

$$(\boldsymbol{q}^{\dagger}_{\kappa(1)}, \boldsymbol{T} \boldsymbol{q}^{\dagger}_{\kappa(1)}) = (\boldsymbol{q}^{\dagger}_{\kappa(2)}, \boldsymbol{T} \boldsymbol{q}^{\dagger}_{\kappa(2)}) = \frac{f_{\kappa}^2}{t_{z\kappa}^2}, \qquad (\boldsymbol{q}^{\dagger}_{\kappa(1)}, \boldsymbol{T} \boldsymbol{q}^{\dagger}_{\kappa(2)}) = 0. \tag{128}$$

Dropping the term $f_{\kappa}^2/t_{z\kappa}^2$, we determine the metric matrix $\boldsymbol{T}$ from the resulting two equations by least squares. The computation of Step 5 goes as follows [19]. We define the $3 \times 3 \times 3 \times 3$ tensor $\mathcal{B} = (B_{ijkl})$ by

$$\begin{aligned} B_{ijkl} = \sum_{\kappa=1}^{M} \Big[& (\boldsymbol{q}^{\dagger}_{\kappa(1)})_i (\boldsymbol{q}^{\dagger}_{\kappa(1)})_j (\boldsymbol{q}^{\dagger}_{\kappa(1)})_k (\boldsymbol{q}^{\dagger}_{\kappa(1)})_l - (\boldsymbol{q}^{\dagger}_{\kappa(1)})_i (\boldsymbol{q}^{\dagger}_{\kappa(1)})_j (\boldsymbol{q}^{\dagger}_{\kappa(2)})_k (\boldsymbol{q}^{\dagger}_{\kappa(2)})_l \\ & - (\boldsymbol{q}^{\dagger}_{\kappa(2)})_i (\boldsymbol{q}^{\dagger}_{\kappa(2)})_j (\boldsymbol{q}^{\dagger}_{\kappa(1)})_k (\boldsymbol{q}^{\dagger}_{\kappa(1)})_l + (\boldsymbol{q}^{\dagger}_{\kappa(2)})_i (\boldsymbol{q}^{\dagger}_{\kappa(2)})_j (\boldsymbol{q}^{\dagger}_{\kappa(2)})_k (\boldsymbol{q}^{\dagger}_{\kappa(2)})_l \\ & + \frac{1}{4} \Big((\boldsymbol{q}^{\dagger}_{\kappa(1)})_i (\boldsymbol{q}^{\dagger}_{\kappa(2)})_j (\boldsymbol{q}^{\dagger}_{\kappa(1)})_k (\boldsymbol{q}^{\dagger}_{\kappa(2)})_l + (\boldsymbol{q}^{\dagger}_{\kappa(2)})_i (\boldsymbol{q}^{\dagger}_{\kappa(1)})_j (\boldsymbol{q}^{\dagger}_{\kappa(1)})_k (\boldsymbol{q}^{\dagger}_{\kappa(2)})_l \\ & + (\boldsymbol{q}^{\dagger}_{\kappa(1)})_i (\boldsymbol{q}^{\dagger}_{\kappa(2)})_j (\boldsymbol{q}^{\dagger}_{\kappa(2)})_k (\boldsymbol{q}^{\dagger}_{\kappa(1)})_l + (\boldsymbol{q}^{\dagger}_{\kappa(2)})_i (\boldsymbol{q}^{\dagger}_{\kappa(1)})_j (\boldsymbol{q}^{\dagger}_{\kappa(2)})_k (\boldsymbol{q}^{\dagger}_{\kappa(1)})_l \Big) \Big], \end{aligned} \tag{129}$$

and compute the 6×6 symmetric matrix B in Eq. (121). Let $\boldsymbol{\tau} = (\tau_i)$ be the 6-D unit eigenvector of B for the smallest eigenvalue. Then, the metric matrix $\boldsymbol{T}$ is given by Eq. (124) if $\det \boldsymbol{T} \geq 0$. If $\det \boldsymbol{T} < 0$, we change the sing of $\boldsymbol{T}$. For the translation computation in Step 5, we first compute

$$t_{z\kappa} = f_{\kappa} \sqrt{\frac{2}{(\boldsymbol{q}^{\dagger}_{\kappa(1)}, \boldsymbol{T} \boldsymbol{q}^{\dagger}_{\kappa(1)}) + (\boldsymbol{q}^{\dagger}_{\kappa(2)}, \boldsymbol{T} \boldsymbol{q}^{\dagger}_{\kappa(2)})}}. \tag{130}$$

Next, we let
$$t_{x\kappa} = \frac{t_{z\kappa}}{f_\kappa}\tilde{t}_{x\kappa}, \qquad t_{y\kappa} = \frac{t_{z\kappa}}{f_\kappa}\tilde{t}_{y\kappa}. \tag{131}$$

For the rotation computation in Step 10, we compute the SVD
$$\frac{t_{z\kappa}}{f_\kappa}\begin{pmatrix} \boldsymbol{m}_{\kappa(1)}^\dagger & \boldsymbol{m}_{\kappa(2)}^\dagger & \boldsymbol{0} \end{pmatrix} = \boldsymbol{V}_\kappa \boldsymbol{\Lambda}_\kappa \boldsymbol{U}_\kappa^\top, \tag{132}$$

and determine $\boldsymbol{R}_\kappa$ by Eq. (126). The matrix Π_κ in Step 11 is given by

$$\Pi_\kappa = \frac{f_\kappa}{t_{z\kappa}} \begin{pmatrix} 0 & \cdots & 0 & \overset{(2\kappa-1)}{1} & \overset{(2\kappa)}{0} & 0 & \cdots & 0 \\ 0 & \cdots & 0 & 0 & 1 & 0 & \cdots & 0 \\ 0 & \cdots & 0 & 0 & 0 & 0 & \cdots & 0 \end{pmatrix}, \tag{133}$$

and the matrix Ω_κ in Step 13 is simply $\Omega_\kappa = \mathrm{diag}(-1, -1, 1)$.

C.5. Paraperspective Projection

If the weak paraperspective projection model of Eqs. (48) is assumed (Fig. 10(c)), the metric constraint of Eq. (54) takes the following form [19]:

$$(\boldsymbol{q}_{\kappa(1)}^\dagger, \boldsymbol{T}\boldsymbol{q}_{\kappa(1)}^\dagger) = \frac{f_\kappa^2}{\alpha_\kappa t_{z\kappa}^2}, \quad (\boldsymbol{q}_{\kappa(2)}^\dagger, \boldsymbol{T}\boldsymbol{q}_{\kappa(2)}^\dagger) = \frac{f_\kappa^2}{\beta_\kappa t_{z\kappa}^2}, \quad (\boldsymbol{q}_{\kappa(1)}^\dagger, \boldsymbol{T}\boldsymbol{q}_{\kappa(2)}^\dagger) = \frac{\gamma_\kappa f_\kappa^2}{t_{z\kappa}^2}, \tag{134}$$

where
$$\alpha_\kappa = \frac{1}{1 + \tilde{t}_{x\kappa}^2/f_\kappa^2}, \qquad \beta_\kappa = \frac{1}{1 + \tilde{t}_{y\kappa}^2/f_\kappa^2}, \qquad \gamma_\kappa = \frac{\tilde{t}_{x\kappa}\tilde{t}_{y\kappa}}{f_\kappa^2}. \tag{135}$$

We eliminate $f_\kappa^2/t_{z\kappa}^2$ from Eqs. (134) and determine the metric matrix $\boldsymbol{T}$ from the resulting two equations by least squares. The computation of Step 5 goes as follows [19]. We define the $3 \times 3 \times 3 \times 3$ tensor $\mathcal{B} = (B_{ijkl})$ by

$$B_{ijkl} = \sum_{\kappa=1}^M \Big[(\gamma_\kappa^2 + 1)\alpha_\kappa^2 (\boldsymbol{q}_{\kappa(1)}^\dagger)_i (\boldsymbol{q}_{\kappa(1)}^\dagger)_j (\boldsymbol{q}_{\kappa(1)}^\dagger)_k (\boldsymbol{q}_{\kappa(1)}^\dagger)_l$$
$$+ (\gamma_\kappa^2 + 1)\beta_\kappa^2 (\boldsymbol{q}_{\kappa(2)}^\dagger)_i (\boldsymbol{q}_{\kappa(2)}^\dagger)_j (\boldsymbol{q}_{\kappa(2)}^\dagger)_k (\boldsymbol{q}_{\kappa(2)}^\dagger)_l + (\boldsymbol{q}_{\kappa(1)}^\dagger)_i (\boldsymbol{q}_{\kappa(2)}^\dagger)_j (\boldsymbol{q}_{\kappa(1)}^\dagger)_k (\boldsymbol{q}_{\kappa(2)}^\dagger)_l$$
$$+ (\boldsymbol{q}_{\kappa(1)}^\dagger)_i (\boldsymbol{q}_{\kappa(2)}^\dagger)_j (\boldsymbol{q}_{\kappa(2)}^\dagger)_k (\boldsymbol{q}_{\kappa(1)}^\dagger)_l + (\boldsymbol{q}_{\kappa(2)}^\dagger)_i (\boldsymbol{q}_{\kappa(1)}^\dagger)_j (\boldsymbol{q}_{\kappa(1)}^\dagger)_k (\boldsymbol{q}_{\kappa(2)}^\dagger)_l$$
$$+ (\boldsymbol{q}_{\kappa(2)}^\dagger)_i (\boldsymbol{q}_{\kappa(1)}^\dagger)_j (\boldsymbol{q}_{\kappa(2)}^\dagger)_k (\boldsymbol{q}_{\kappa(1)}^\dagger)_l - \alpha_\kappa \gamma_\kappa (\boldsymbol{q}_{\kappa(1)}^\dagger)_i (\boldsymbol{q}_{\kappa(1)}^\dagger)_j (\boldsymbol{q}_{\kappa(1)}^\dagger)_k (\boldsymbol{q}_{\kappa(2)}^\dagger)_l$$
$$- \alpha_\kappa \gamma_\kappa (\boldsymbol{q}_{\kappa(1)}^\dagger)_i (\boldsymbol{q}_{\kappa(1)}^\dagger)_j (\boldsymbol{q}_{\kappa(2)}^\dagger)_k (\boldsymbol{q}_{\kappa(1)}^\dagger)_l - \alpha_\kappa \gamma_\kappa (\boldsymbol{q}_{\kappa(1)}^\dagger)_i (\boldsymbol{q}_{\kappa(2)}^\dagger)_j (\boldsymbol{q}_{\kappa(1)}^\dagger)_k (\boldsymbol{q}_{\kappa(1)}^\dagger)_l$$
$$- \alpha_\kappa \gamma_\kappa (\boldsymbol{q}_{\kappa(2)}^\dagger)_i (\boldsymbol{q}_{\kappa(1)}^\dagger)_j (\boldsymbol{q}_{\kappa(1)}^\dagger)_k (\boldsymbol{q}_{\kappa(1)}^\dagger)_l - \beta_\kappa \gamma_\kappa (\boldsymbol{q}_{\kappa(2)}^\dagger)_i (\boldsymbol{q}_{\kappa(2)}^\dagger)_j (\boldsymbol{q}_{\kappa(1)}^\dagger)_k (\boldsymbol{q}_{\kappa(2)}^\dagger)_l$$
$$- \beta_\kappa \gamma_\kappa (\boldsymbol{q}_{\kappa(2)}^\dagger)_i (\boldsymbol{q}_{\kappa(2)}^\dagger)_j (\boldsymbol{q}_{\kappa(2)}^\dagger)_k (\boldsymbol{q}_{\kappa(1)}^\dagger)_l - \beta_\kappa \gamma_\kappa (\boldsymbol{q}_{\kappa(1)}^\dagger)_i (\boldsymbol{q}_{\kappa(2)}^\dagger)_j (\boldsymbol{q}_{\kappa(2)}^\dagger)_k (\boldsymbol{q}_{\kappa(2)}^\dagger)_l$$
$$- \beta_\kappa \gamma_\kappa (\boldsymbol{q}_{\kappa(2)}^\dagger)_i (\boldsymbol{q}_{\kappa(1)}^\dagger)_j (\boldsymbol{q}_{\kappa(2)}^\dagger)_k (\boldsymbol{q}_{\kappa(2)}^\dagger)_l$$
$$+ (\gamma_\kappa^2 - 1)\alpha_\kappa \gamma_\kappa (\boldsymbol{q}_{\kappa(1)}^\dagger)_i (\boldsymbol{q}_{\kappa(1)}^\dagger)_j (\boldsymbol{q}_{\kappa(2)}^\dagger)_k (\boldsymbol{q}_{\kappa(2)}^\dagger)_l$$
$$+ (\gamma_\kappa^2 - 1)\alpha_\kappa \gamma_\kappa (\boldsymbol{q}_{\kappa(2)}^\dagger)_i (\boldsymbol{q}_{\kappa(2)}^\dagger)_j (\boldsymbol{q}_{\kappa(1)}^\dagger)_k (\boldsymbol{q}_{\kappa(1)}^\dagger)_l \Big]. \tag{136}$$

Then, we compute the 6×6 symmetric matrix B in Eq. (121), and let $\tau = (\tau_i)$ be the 6-D unit eigenvector of B for the smallest eigenvalue. The metric matrix T is given by Eq. (124) if $\det T \geq 0$. If $\det T < 0$, we change the sing of T. For the translation computation in Step 5, we first compute

$$t_{z\kappa} = f_\kappa \sqrt{\frac{2}{\alpha_\kappa(q^\dagger_{\kappa(1)}, Tq^\dagger_{\kappa(1)}) + \beta_\kappa(q^\dagger_{\kappa(2)}, Tq^\dagger_{\kappa(2)})}}. \tag{137}$$

Next, we compute $t_{x\kappa}$ and $t_{y\kappa}$ by Eqs. (131). For the rotation computation in Step 10, we compute

$$r^\dagger_{\kappa(3)} = \frac{t_{z\kappa}/f_\kappa}{1 + (t_{x\kappa}/t_{z\kappa})^2 + (t_{y\kappa}/t_{z\kappa})^2} \left(\frac{t_{z\kappa}}{f_\kappa} m^\dagger_{\kappa(1)} \times m^\dagger_{\kappa(2)} - \frac{t_{x\kappa}}{t_{z\kappa}} m^\dagger_{\kappa(1)} - \frac{t_{y\kappa}}{t_{z\kappa}} m^\dagger_{\kappa(2)} \right),$$

$$r^\dagger_{\kappa(1)} = \frac{t_{z\kappa}}{f_\kappa} m^\dagger_{\kappa(1)} + \frac{t_{x\kappa}}{t_{z\kappa}} r^\dagger_{\kappa(3)}, \qquad r^\dagger_{\kappa(2)} = \frac{t_{z\kappa}}{f_\kappa} m^\dagger_{\kappa(2)} + \frac{t_{y\kappa}}{t_{z\kappa}} r^\dagger_{\kappa(3)}. \tag{138}$$

Then, we compute the SVD

$$\left(r^\dagger_{\kappa(1)} \; r^\dagger_{\kappa(2)} \; r^\dagger_{\kappa(3)} \right) = V_\kappa \Lambda_\kappa U_\kappa^\top. \tag{139}$$

The rotation matrices R_κ are given by Eq. (126). The matrix Π_κ in Step 11 is given by

$$\Pi_\kappa = \frac{f_\kappa}{t_{z\kappa}} \begin{pmatrix} 0 & \cdots & 0 & \overset{(2\kappa-1)}{1} & \overset{(2\kappa)}{0} & 0 & \cdots & 0 \\ 0 & \cdots & 0 & 0 & 1 & 0 & \cdots & 0 \\ 0 & \cdots & 0 & -t_{x\kappa}/t_{z\kappa} & -t_{y\kappa}/t_{z\kappa} & 0 & \cdots & 0 \end{pmatrix}, \tag{140}$$

and the matrix Ω_κ in Step 13 is given by

$$\Omega_\kappa = \frac{2 t_\kappa t_\kappa^\top}{\|t_\kappa\|^2} - I. \tag{141}$$

C.6. Generic Model

If the generic model of Eqs. (45) is assumed, the metric constraint of Eq. (54) takes the following form [20]:

$$(q^\dagger_{\kappa(1)}, Tq^\dagger_{\kappa(1)}) = \frac{1}{\zeta_\kappa^2} + \beta_\kappa^2 \tilde{t}_{x\kappa}^2, \qquad (q^\dagger_{\kappa(2)}, Tq^\dagger_{\kappa(2)}) = \frac{1}{\zeta_\kappa^2} + \beta_\kappa^2 \tilde{t}_{y\kappa}^2,$$

$$(q^\dagger_{\kappa(1)}, Tq^\dagger_{\kappa(2)}) = \beta_\kappa^2 \tilde{t}_{x\kappa} \tilde{t}_{y\kappa}, \tag{142}$$

We eliminate $1/\zeta_\kappa^2$ and β_κ^2 from Eqs. (142) and determine the metric matrix T from the resulting two equations by least squares. The computation of Step 5 goes as follows [20]. We let

$$A_\kappa = \tilde{t}_{x\kappa} \tilde{t}_{y\kappa}, \qquad C_\kappa = \tilde{t}_{x\kappa}^2 - \tilde{t}_{y\kappa}^2, \tag{143}$$

and define the $3 \times 3 \times 3 \times 3$ tensor $\mathcal{B} = (B_{ijkl})$ by

$$B_{ijkl} = \sum_{\kappa=1}^{M} \Big[A_\kappa^2 \big((q_{\kappa(1)}^\dagger)_i (q_{\kappa(1)}^\dagger)_j (q_{\kappa(1)}^\dagger)_k (q_{\kappa(1)}^\dagger)_l + (q_{\kappa(2)}^\dagger)_i (q_{\kappa(2)}^\dagger)_j (q_{\kappa(2)}^\dagger)_k (q_{\kappa(2)}^\dagger)_l$$
$$- (q_{\kappa(1)}^\dagger)_i (q_{\kappa(1)}^\dagger)_j (q_{\kappa(2)}^\dagger)_k (q_{\kappa(2)}^\dagger)_l - (q_{\kappa(2)}^\dagger)_i (q_{\kappa(2)}^\dagger)_j (q_{\kappa(1)}^\dagger)_k (q_{\kappa(1)}^\dagger)_l \big)$$
$$+ \frac{1}{4} C_\kappa^2 \big((q_{\kappa(1)}^\dagger)_i (q_{\kappa(2)}^\dagger)_j (q_{\kappa(1)}^\dagger)_k (q_{\kappa(2)}^\dagger)_l + (q_{\kappa(2)}^\dagger)_i (q_{\kappa(1)}^\dagger)_j (q_{\kappa(1)}^\dagger)_k (q_{\kappa(2)}^\dagger)_l$$
$$+ (q_{\kappa(1)}^\dagger)_i (q_{\kappa(2)}^\dagger)_j (q_{\kappa(2)}^\dagger)_k (q_{\kappa(1)}^\dagger)_l + (q_{\kappa(2)}^\dagger)_i (q_{\kappa(1)}^\dagger)_j (q_{\kappa(2)}^\dagger)_k (q_{\kappa(1)}^\dagger)_l \big)$$
$$- \frac{1}{2} A_\kappa C_\kappa \big((q_{\kappa(1)}^\dagger)_i (q_{\kappa(1)}^\dagger)_j (q_{\kappa(1)}^\dagger)_k (q_{\kappa(2)}^\dagger)_l + (q_{\kappa(1)}^\dagger)_i (q_{\kappa(1)}^\dagger)_j (q_{\kappa(2)}^\dagger)_k (q_{\kappa(1)}^\dagger)_l$$
$$+ (q_{\kappa(1)}^\dagger)_i (q_{\kappa(2)}^\dagger)_j (q_{\kappa(1)}^\dagger)_k (q_{\kappa(1)}^\dagger)_l + (q_{\kappa(2)}^\dagger)_i (q_{\kappa(1)}^\dagger)_j (q_{\kappa(1)}^\dagger)_k (q_{\kappa(1)}^\dagger)_l$$
$$- (q_{\kappa(1)}^\dagger)_i (q_{\kappa(2)}^\dagger)_j (q_{\kappa(2)}^\dagger)_k (q_{\kappa(2)}^\dagger)_l - (q_{\kappa(2)}^\dagger)_i (q_{\kappa(1)}^\dagger)_j (q_{\kappa(2)}^\dagger)_k (q_{\kappa(2)}^\dagger)_l$$
$$- (q_{\kappa(2)}^\dagger)_i (q_{\kappa(2)}^\dagger)_j (q_{\kappa(1)}^\dagger)_k (q_{\kappa(2)}^\dagger)_l - (q_{\kappa(2)}^\dagger)_i (q_{\kappa(2)}^\dagger)_j (q_{\kappa(2)}^\dagger)_k (q_{\kappa(1)}^\dagger)_l \big) \Big]. \qquad (144)$$

Then, we compute the 6×6 symmetric matrix B in Eq. (121), and let $\tau = (\tau_i)$ be the 6-D unit eigenvector of B for the smallest eigenvalue. The metric matrix T is given by Eq. (124) if $\det T \geq 0$. If $\det T < 0$, we change the sing of T. For the translation computation in Step 5, we solve the following simultaneous linear equations for $1/\zeta_\kappa$ and β_κ^2:

$$\begin{pmatrix} 2 & \tilde{t}_{x\kappa}^2 + \tilde{t}_{y\kappa}^2 \\ \tilde{t}_{x\kappa}^2 + \tilde{t}_{y\kappa}^2 & \tilde{t}_{x\kappa}^4 + \tilde{t}_{y\kappa}^4 + \tilde{t}_{x\kappa}^2 \tilde{t}_{y\kappa}^2 \end{pmatrix} \begin{pmatrix} 1/\zeta_\kappa^2 \\ \beta_\kappa^2 \end{pmatrix}$$
$$= \begin{pmatrix} (q_{\kappa(1)}^\dagger, T q_{\kappa(1)}^\dagger) + (q_{\kappa(2)}^\dagger, T q_{\kappa(2)}^\dagger) \\ \tilde{t}_{x\kappa}^2 (q_{\kappa(1)}^\dagger, T q_{\kappa(1)}^\dagger) + \tilde{t}_{y\kappa}^2 (q_{\kappa(2)}^\dagger, T q_{\kappa(2)}^\dagger) + \tilde{t}_{x\kappa} \tilde{t}_{y\kappa} (q_{\kappa(1)}^\dagger, T q_{\kappa(2)}^\dagger) \end{pmatrix}. \qquad (145)$$

Next, we let

$$\begin{pmatrix} t_{x\kappa} \\ t_{y\kappa} \end{pmatrix} = \zeta_\kappa \begin{pmatrix} \tilde{t}_{x\kappa} \\ \tilde{t}_{y\kappa} \end{pmatrix}. \qquad (146)$$

The third components $t_{z\kappa}$ are left indeterminate. For the rotation computation in Step 10, we compute

$$r_{\kappa(3)}^\dagger = \zeta_\kappa \Big(\frac{\zeta_\kappa m_{\kappa(1)}^\dagger \times m_{\kappa(2)}^\dagger - \beta_\kappa (t_{x\kappa} m_{\kappa(1)}^\dagger + t_{y\kappa} m_{\kappa(2)}^\dagger)}{1 + \beta_\kappa^2 (t_{x\kappa}^2 + t_{y\kappa}^2)} \Big),$$

$$r_{\kappa(1)}^\dagger = \zeta_\kappa m_{\kappa(1)}^\dagger + \beta_\kappa t_{x\kappa} r_{\kappa(3)}^\dagger, \qquad r_{\kappa(2)}^\dagger = \zeta_\kappa m_{\kappa(2)}^\dagger + \beta_\kappa t_{y\kappa} r_{\kappa(3)}^\dagger. \qquad (147)$$

Then, we compute the SVD of Eq. (139), and R_κ are given by Eq. (126). The matrix Π_κ in Step 11 is given by

$$\Pi_\kappa = \begin{pmatrix} 0 & \cdots & 0 & \overset{(2\kappa-1)}{1/\zeta_\kappa} & \overset{(2\kappa)}{0} & 0 & \cdots & 0 \\ 0 & \cdots & 0 & 0 & 1/\zeta_\kappa & 0 & \cdots & 0 \\ 0 & \cdots & 0 & -\beta_\kappa t_{x\kappa}/\zeta_\kappa & -\beta_\kappa t_{y\kappa}/\zeta_\kappa & 0 & \cdots & 0 \end{pmatrix}, \qquad (148)$$

and the matrix Ω_κ in Step 13 is given by

$$\Omega_\kappa = \frac{2n_\kappa n_\kappa^\top}{\|n_\kappa\|^2} - I, \qquad n_\kappa = \begin{pmatrix} 1 \\ 0 \\ -\beta_\kappa t_{x\kappa} \end{pmatrix} \times \begin{pmatrix} 0 \\ 1 \\ -\beta_\kappa t_{y\kappa} \end{pmatrix}. \qquad (149)$$

References

[1] Basri, R. *Int. J. Comput. Vision* 1996, 19, 169–179.

[2] Deguchi, K.; Sasano, T.; Arai, H.; Yoshikawa, H. *IEICE Trans. Inf. & Syst.* 1996, E79-D, 1329–1336.

[3] Dempster, A. P.; Laird, N. M.; Rubin, D. B. *J. Roy. Statist. Soc.* 1977, B-39, 1–38.

[4] Faugeras, O. *Three-Dimensional Computer Vision: A Geometric Viewpoint*; MIT Press, Cambridge, MA, 1993.

[5] Faugeras, O; Luong, Q.-T. *The Geometry of Multiple Images*; MIT Press, Cambridge, MA, 2001.

[6] Forsyth, D. A.; Ponce, J. *Computer Vision: A Modern Approach*; Prentice Hall, Upper Saddle River, NJ, 2003.

[7] Hartley, R. I. *Int. J. Comput. Vision.* 1998, 21-6, 41–61.

[8] Hartley, R.; Zisserman, A. *Multiple View Geometry in Computer Vision*; Cambridge University Press, Cambridge, 2000.

[9] Heyden, A.; Åström, K. In *Proc. IEEE Conf. Comput. Vision Patt. Recog.*, June 1997, Puerto Rico, pp. 438–443.

[10] Heyden A.; Åström, K. In *Proc. 7th Int. Conf. Computer. Vision*; September 1999, Kerkyra, Greece, Vol. 1, pp. 350–355.

[11] Heyden, A.; Berthilsson, R.; Sparr, G. *Image Vision Comput.* 1999, 17, 981–991.

[12] Heyden A.; Pollefeys, M. In *Emerging Topics in Computer Vision*; G. Medioni, G.; Kang, S. B.; Eds.; Prentice Hall, Upper Saddle River, NJ, pp. 44–106.

[13] Kanatani, K. *Group-Theoretical Methods in Image Understanding*; Springer, Berlin, 1990.

[14] Kanatani, K. *Geometric Computation for Machine Vision*; Oxford University Press, Oxford, 1993.

[15] Kanatani, K. *Statistical Optimization for Geometric Computation: Theory and Practice*; Elsevier Science, Amsterdam, 1996; reprinted by Dover Publications, New York, NY, 2005.

[16] Kanatani, K. In *Proc. 8th Int. Conf. Comput. Vision*; July 2001, Vancouver, Canada, Vol. 2, pp. 301–306.

[17] Kanatani, K. *Int. J. Image Graphics* 2002, 2, 179–197.

[18] Kanatani, K. In *Proc. 7th Euro. Conf. Comput. Vision*; May 2002, Copenhagen, Denmark, Vol. 3, pp. 335–349.

[19] Kanatani, K.; Sugaya, Y. *Mem. Fac. Eng. Okayama Univ.* 2004, 38, 61–72.

[20] Kanatani, K.; Sugaya, Y.; Ackermann, H. *IEICE Trans. Inf. & Syst.* 2007, E90-D, 851–858.

[21] Ma, Y.; Soatto, S.; Košecká, J.; Sastry, S. S. *An Invitation to 3-D Vision: From Images to Geometric Models*; Springer, New York, NY, 2004.

[22] Marr, D. *Vision: A Computational Investigation into the Human Representation and Processing of Visual Information*; W. H. Freeman, San Francisco, CA, 1982.

[23] Medioni, G.; Kang, S. B.; Eds. *Emerging Topics in Computer Vision*, Prentice Hall, Upper Saddle River, NJ, 2004.

[24] Morita, M.;. Kanade, T. *IEEE Trans. Patt. Anal. Mach. Intell.* 1997, 19-8, 858–867.

[25] Mahamud, S.; Hebert, M.; In *Proc. IEEE Conf. Comput. Vision Patt. Recog.*; June 2000, Hilton Head Island, SC, Vol. 2, pp. 430–437.

[26] Mahamud, S.; Hebert, H.; Omori, Y.; Ponce, J. In *Proc. IEEE Conf. Comput. Vision Patt. Recog.*; December 2001, Kauai, HI, Vol. 1, pp. 1018–1025.

[27] Poelman, C. J.; and Kanade, T. *IEEE Trans. Patt. Anal. Mach. Intell.* 1997, 19, 206–218.

[28] Pollefeys, M.; Koch, R.; Van Gool, L. *Int. J. Comput. Vision* 1999, 32, 145–150.

[29] Quan, L. *Int. J. Comput. Vision* 1996, 19, 93–105.

[30] Seo, Y.; Heyden, A. In *Proc. 15th Int. Conf. Patt. Recog.*; September 2000, Barcelona, Spain, Vol. 1, pp. 69–71.

[31] Seo Y.; Heyden, H. *Image and Vision Computing* 2004, 22, 919–926.

[32] Seo, Y.; K.-S. Hong, K.-S. *IEICE Trans. Inf. & Syst.* 2001, E84-D, 1626–1632.

[33] Shapiro, L. S.; Zisserman, A.; Brady, M. *Int. J. Comput. Vision* 1995, 16, 147–182.

[34] Sturm, P.; Triggs, B. In *Proc. 4th Euro. Conf. Comput. Vision*; April 1996, Cambridge, Vol. 2, pp. 709–720.

[35] Sugaya Y.; Kanatani, K. *IEICE Trans. Inf. & Syst.* 2003, E86-D, 1095–1102.

[36] Sugaya, Y.; Kanatani, K. *IEICE Trans. Inf. & Syst.* 2004, E87-D, 1031–1038.

[37] Sugaya, Y.; Kanatani, K. *IEICE Trans. Inf. & Syst.* 2004, E87-D, 1935–1942.

[38] Sugaya, Y.; Kanatani, K. *Mem. Fac. Eng. Okayama Univ.* 2005, 39, 56–62.

[39] Sugimoto, A. *Int. J. Comput. Vision* 1996, 19, 181–201.

[40] Tomasi, C.; Kanade, T. *Int. J. Comput. Vision* 1992, 9, 137–154.

[41] Triggs, B. In *Proc. IEEE Conf. Comput. Vision Patt. Recog.*; June 1996, San Francisco, CA, pp. 845–851.

[42] Triggs, B. In *Proc. IEEE Conf. Compt. Vision Patt. Recog.*; June 1997, San Juan, Puerto Rico, pp. 609–614.

[43] Ueshiba, T.; Tomita, F. *Proc. 5th Euro. Conf. Comput. Vision*; June 1998, Freiburg, Germany, Vol. 1, pp. 296–310.

[44] Xu, B.; Zhang, Z. *Epipolar Geometry in Stereo, Motion and Object Recognition*; Kluwer, Dordrecht, The Netherlands, 1996.

[45] Zhang, Z. In *Emerging Topics in Computer Vision*, Medioni, G.; Kang, S. B.; Eds.; Prentice Hall, Upper Saddle River, NJ, pp. 4–40.

Chapter 2

PROJECT DIVA: GUIDANCE AND VISION SURVEILLANCE TECHNIQUES FOR AN AUTONOMOUS AIRSHIP

Alexandra Moutinho[1*], *Luiz Mirisola*[2†], *José Azinheira*[1‡] *and Jorge Dias*[2§]
[1]IDMEC/Instituto Superior Técnico, TULisbon
Av. Rovisco Pais, 1049-001 Lisbon, Portugal
[2]ISR - Institute of Systems and Robotics, University of Coimbra, Portugal

Abstract

Unmanned Aerial Vehicles have a wide spectrum of potential civilian applications as observation and data acquisition platforms. Most of the aerial surveillance applications require low altitude, low speed platforms. The vehicle should ideally be able to hover above an area, allow long duration studies, take-off and land vertically without the need for runways. For such a scenario, lighter-than-air (LTA) vehicles are often better suited than airplanes and helicopters. This chapter introduces the Portuguese airship project named DIVA, and the approaches currently under development and that are being implemented in the areas of robotic integration, navigation and guidance, and vision based surveillance. The airship platform is described, along with an overview of the architecture developed, to document the practical experience gained in this field. Looking at the typical autonomous mission objective, we present a control approach for airship path-tracking, covering the whole flight envelope from hover to the normal cruise flight. An asymptotically stable backstepping controller is designed from the airship nonlinear dynamics and kinematics. Some practical issues are then considered and the control law is improved to take into account input saturations and wind disturbances, maintaining its asymptotic stability for a bounded wind estimation error. The presented simulation results illustrate the controller performance during a full realistic mission that covers all the usual tasks: vertical take-off and landing, stabilization and route path-tracking. Wind disturbances are also included. In vision systems used in aerial robotics, inertial and earth field magnetic sensors can provide valuable data

---

[*]E-mail address: moutinho@dem.ist.utl.pt
[†]E-mail address: lgm@isr.uc.pt
[‡]E-mail address: jraz@dem.ist.utl.pt
[§]E-mail address: jorge@isr.uc.pt

about the observer ego-motion, as well as an absolute orientation reference. Here, the inertial orientation measurements are used to compensate the rotational degrees of freedom in two different computer vision tasks: first, inertial data is used to project images on a leveled plane, relaxing the demands on interest point matching algorithms when performing image mosaicing; second, in the rotation-compensated, pure translation case, full homographies are reduced to planar homologies, and the heights over the ground plane on two views are calculated more accurately. Visual odometry for the airship 3D trajectory is performed by calculating the focus of expansion during the motion. These results are important in the context of vision based aerial mapping and aerial vision surveillance. A technique is also presented that allows aerial vision tracking of dynamic objects moving in the ground. Experimental results are presented, illustrating the approach and some achievements.

1. Introduction

Besides their use as military surveillance platforms, Unmanned Aerial Vehicles (UAVs) have a wide spectrum of potential civilian applications as observation and data acquisition platforms. They can be utilized in several environmental monitoring applications related to biodiversity, ecological and climate research and monitoring. Inspection oriented applications cover different areas such as mineral and archaeological prospecting, agricultural and livestock studies, crop field prediction, land use surveys in rural and urban regions, and also inspection of manmade structures such as pipelines, power transmission lines, dams and roads. UAV gathered data can also be used in a complementary way concerning information obtained by satellites, balloons, manned aircraft or on ground. Detailed UAVs applications scenarios and road maps are presented in [1] and [2] from civil and military perspectives, respectively.

Most of the applications before cited have profiles that require maneuverable low altitude, low speed airborne data gathering platforms. The vehicle should ideally be able to hover above an area, present extended airborne capabilities for long duration studies, take-off and land vertically without the need of runway infrastructures, have a large payload to weight ratio, among other requirements. For this scenario, lighter-than-air (LTA) vehicles are better suited than airplanes and helicopters [3], mainly because: they derive the largest part of their lift from aerostatic, rather than aerodynamic forces; they are safer and, in case of failure, present a graceful degradation; they are intrinsically of higher stability than other platforms.

In this context, the Portuguese Project DIVA - *Dirigível Instrumentado para Vigilância Aérea* - was proposed. DIVA focuses on the establishment of the technologies required to substantiate autonomous operation of unmanned robotic airships for environmental monitoring and aerial inspection missions. This includes sensing and processing infrastructures, control and guidance capabilities, and the ability to perform mission, navigation, and sensor deployment planning and execution.

Other important researches related to outdoor autonomous airships in the world at this moment are the AURORA Project [4] in Brazil, sharing a partnership with the DIVA Project, the Lotte Project [5] at Germany, the French projects at LAAS-CNRS [6, 7], and LSCUniversit d'Evry [8]. In the USA there is a partnership between the projects of STWing-SEAS [9] of University of Pennsylvania and the EnviroBLIMP at CMU.

Aiming at the autonomous airship goal, aerial platform positioning and path-tracking should be assured by a control and navigation system. Such a system needs to cope with the highly nonlinear and underactuated airship dynamics, ranging from hovering flight (HF) to cruise or aerodynamic flight (AF). Hovering flight is defined here as a flight in low airspeed condition. In addition, the abrupt and continuous transition between the HF and AF in the dynamics, and the different use of actuators necessary within each region, makes that a very difficult issue to be dealt with by the control scheme.

Basically, two main approaches can be considered for the automatic control and navigation system of an airship. The first one relies on the linear control theory to design individual compensators to satisfy closed-loop specifications, based on linearized models of the airship dynamics. One important result of the linearization approach is the separation of two independent (decoupled) motions: the motion in the vertical plane, named longitudinal, and the motion in the horizontal plane, named lateral. Following this approach, experimental results were obtained for the AURORA airship for path following through a set of pre-defined points in latitude/longitude, along with an automatic altitude control [10].

Also based on a linearized airship model, Wimmer et al. [5] introduced a robust controller design method to compensate for the lack of knowledge about the Lotte airship dynamic behavior and model parameters. The decoupled longitudinal and lateral control systems both consist of an inner H_∞-controller for the dynamics and an outer SISO P- or PI-controller for the remaining states. Experimental results are shown therein for the pitch and velocity control. We remark that, as far as the authors are aware, both experimental results (from Lotte and AURORA Projects) on automatic control for outdoor airships are the only ones reported in the literature at this moment. For the lateral control problem, an alternative H_∞ approach for the airship heading control is proposed in [11], and a H_2/H_∞ approach for the design of a lateral PD-PI controller for the AURORA airship is proposed in [12]. Other works in the AURORA Project following this methodology of linear based controllers can be found in [3, 10, 13].

The second approach for the airship automatic control system consists on the search for a single global control scheme covering all the aerodynamic range, such that the different flight regions, from HF to AF, are considered inside a sole formulation. For security reasons, as well as simplicity and flexibility, a global nonlinear control is more interesting than a linearized and decoupled one. At present, Backstepping is the main nonlinear approach under investigation for the AURORA airship [14]. Other important nonlinear approaches for the UAV control are the Sliding Mode technique [15, 16] and Dynamic Inversion (or Feedback Linearization) [17, 18].

In the Backstepping approach [19], a Lyapunov-based technique, by formulating a scalar positive function of the system states and then choosing a control law to make this function decrease, we have the guarantee that the nonlinear control system thus designed will be asymptotically stable, and still robust to some unmatched uncertainties. In a previous work [14], a Backstepping control strategy for the stabilization of the AURORA airship has been proposed. It introduced a synthetic modeling of the airship dynamics, resulting in an original formulation of the system kinematics and dynamics with an appropriate change of variables allowing the application of Backstepping techniques for the design of the UAV stabilization control. The saturation of the control signals was also considered in the design, since at low airspeeds, or when in hovering state, the airship is usually underactuated. In order to cope with limitations due to reduced actuation, the idea of Teel [20] is followed. It uses a nonlinear combination of saturation functions of linear feedbacks that globally stabilizes a chain of integrators. The problem of actuator limits is of fundamental interest for this kind of underactuated systems, making this a theoretical study closely related to practical applications. In addition, a guidance strategy was proposed to deal with the airship lateral underactuation in face of wind disturbances.

Based on this work [14], we now present a Backstepping controller applied to the DIVA airship model, and where the kinematics are described by the Euler angles in place of the quaternions used previously. Also, the control law applicability range is broaden so it is valid over the entire flight envelope and not only for stabilization purposes.

Other successful applications of the Backstepping approach for UAV control can be found in [6, 21, 22]. A Backstepping technique has been proposed by the LAAS/CNRS autonomous blimp project [6, 7]. The global control strategy studied is obtained by switching between four sub-controllers, one for each of the flight phases considered. Each controller is however still based on linearized models of the airship, what leads once again to the separate control of the longitudinal and lateral motions. Also considering the nonlinear control techniques, Beji et al. [23] introduced a Backstepping tracking feedback control for ascent and descent flight maneuvers, where the objective is to stabilize the airship engine around trimmed flight trajectories. The treatment of the actuators saturation is also considered in the Backstepping design of Freeman and Praly [24], so that the boundedness of the control signals and its derivative is propagated through each step of the recursive design. Finally, Metni et al. [25] follow the same idea for an UAV with orientation limits.

The trajectory of a mobile observer can be recovered from images of a planar surface using interest point matching and the well known planar homography model. The recovered homography matrix is then decomposed into the rotational and translational motion parameters, yielding four possible solutions among which two can not be immediately discarded. Various geometric constraints have been proposed to recover the right motion. This was already performed for an airship using clustering and blob-based interesting point matching algorithms and building a image mosaic to improve the trajectory estimate [26].

The trajectory of an UAV can also be recovered by tracking known fixed targets on the ground, what requires modifying the environment [27]. With an on-board stereo camera, the height over the ground plane can be recovered with efficient sparse stereo techniques, although subject to limitations in range related to the stereo baseline size [28].

Aerial vehicles have been utilized to produce 3D maps of the ground using a variety of different sensors. For example, stereo images taken by a remotely controlled blimp were used to build a dense 3D map of the ground surface, in the form of a DEM (Digital Elevation Map), and also performing localization by tracking the position of automatically detected landmarks on the ground [29]. Stereo imagery has also been combined with other vision techniques such as color segmentation [30]. Airborne range sensing devices such as laser range finders or radars have also been extensively used to build 3D maps actually exploited in domains such as geology [31]. Statistical techniques such as Markov Random Fields have been applied on such DEMs to reduce noise, in various levels including pixel-to-pixel graphs and larger segmented regions [32].

This chapter introduces the Portuguese DIVA airship project, and the approaches currently under development and that are being implemented in the areas of robotic integration, navigation and guidance, and vision based surveillance.

Looking at the typical autonomous mission objective, and after this introductory section, the remaining parts of this chapter are organized as follows.

Section 2. presents the DIVA airship, and lists the requirements for the on-board and ground station systems, providing a development road-map for the experimental platform utilized to obtain the current results. An overview of the system developed so far is also described.

Section 3. is dedicated to the guidance and control objectives. It describes the airship dynamics and kinematics under constant wind and presents a control approach for airship path-tracking, covering the whole flight envelope from hover to the normal cruise flight. An asymptotically stable backstepping controller is designed from the airship nonlinear equations. Some practical issues are then considered and the control law is improved to take into account input saturations and wind disturbances, maintaining its asymptotic stability for a bounded wind estimation error. The control allocation problem is focused as well as the reference shaping to deal with the airship underactuation. It presents simulation results illustrating the controller performance during a full realistic mission that covers all the usual tasks: vertical take-off and landing, stabilization and route path-tracking. Wind disturbances are also included.

Section 4. is dedicated to visual perception. Trajectory recovery from images is discussed, exploiting the inertial orientation measurements to separate rotational and translational components, and using a pure translation movement model. The approach followed uses a monocular camera and does not require artificial targets on the ground. Next, rotation compensated images projected on the horizontal ground plane are further exploited to build a coarse Digital Elevation Map (DEM), performing 3D mapping from monocular aerial images and with a very fast process after pixel correspondences are found. Finally, section 5. stresses the concluding remarks.

2. DIVA Prototype

The DIVA is a nonrigid airship (see Fig. 1(a)) with $9.4m$ long, $1.9m$ diameter and with a volume of $18m^3$. Its payload capacity is approximately $14kg$, and the maximum speed attained is around $70km/h$.

The airship actuators input is given by $U = [T_X, \delta_v, T_D, T_Y, \delta_a, \delta_e, \delta_r]^T$ (see Fig.1(b)), where T_X is the total main propellers thrust, δ_v is the vectoring angle, T_D is the difference between right and left thrust, T_Y is the stern propeller lateral thrust, and $\delta_a, \delta_e, \delta_r$ are the tail surfaces deflections, corresponding to aileron, elevator and rudder, respectively. The aileron input δ_a is generated through the opposite deflection of each of the fins yielding a rolling moment.

2.1. Technical Characteristics of the Airship

This section details the technical characteristics of the system which collected the data used on the experiments shown in section 4.. The system is composed by an embedded system and a ground station, which communicate through a wireless link.

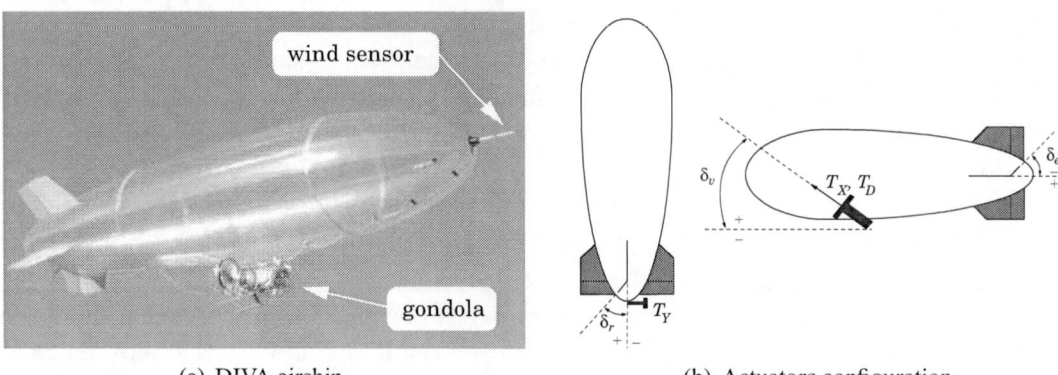

(a) DIVA airship. (b) Actuators configuration.

Figure 1. DIVA prototype.

2.1.1. Embedded System

Operational Requisites

- Periodically, the embedded system must read data from all sensors and transmit the data to the ground station, with a period compatible with the vehicle dynamics.

- The embedded system must integrate at least one camera, capturing images with sufficient resolution and frame rate. A modern camera with automatic gain adjustment is necessary as illumination conditions change often during the flight.

- The embedded system must store all telemetry data, besides transmitting them to the ground station, to avoid losing data if the data link is lost. Camera images also are stored (the heaviest burden to the CPU and storage system).

- If DGPS (Differential GPS) corrections are needed, the embedded system must receive DGPS data from the ground station via a reliable data link. GPS readings when DGPS data is intermittent can often "jump" between corrected and uncorrected states, which may be worse than having no DGPS correction at all. A separate low speed link for DGPS data may be advisable.

- Payload weight is a severe limitation for airships. Thus, energy consumption must be minimized to decrease battery weight.

Safety Characteristics

- The embedded system must have a Remote Control (RC) receiver and allow a human pilot to manually pilot the airship.

- There must be a device electrically independent of the CPU to read the servo commands from the RC receiver, able to continue working in case of CPU malfunctioning. It receives commands from the RC receiver, and relays them to the servo motors and to the CPU to be read and stored.

- The embedded system must be sufficiently resistant to vibration and tilting to resist the flight and motor induced vibration. Vibration isolation (lightweight) may be necessary. Vibration-resistant data storage is needed to store imagery.

2.1.2. Ground Station

This section presents the technical characteristics of the ground station.

Data Collection And Storage

- Although the embedded system stores all state variables, the data may not be recoverable due to accidents or malfunctioning. Therefore, the ground station should receive and store data from the embedded system.

- The ground station must be easily reconfigurable if there is a change in the data format sent by the embedded system (e.g., if a new sensor is added).

- The stored data must be easily converted to a format readable by commercial mathematical software such as MATLAB® (e.g., an ASCII format).

User Interface: Vehicle Safety And Monitoring The ground station must monitor the vehicle state not only to detect hazardous situations (safety issues), but also to avoid useless flights and waste of time in case of malfunctioning. The human pilot and the algorithm developers should determine which variables should be monitored.

- Monitor all critical state variables, i.e., the ones which are essential to the flight safety or data recording, like tachometers, GPS, camera status, etc. The user interface must show the state of the most critical variables clearly and continuously (but not necessarily the actual numeric value of all of them), indicating critical failures with alarms (red lights and/or sounds).

- This monitoring must be active before take-off, so that the ground station operator can abort the flight if a critical system is not operating.

2.2. Overview of Architecture

This section provides an overview of the system architecture developed and utilized to obtain the datasets used on this chapter, as well as parts still in development. Remotely piloted flights were performed with telemetry and image recording with the on-board hardware architecture shown in Fig. 2. The images and sensor data were time-stamped immediately after reading with the same CPU clock.

The C++ embedded software includes a main loop to read, store and transmit sensor data, and threads to capture and store images. CORBA middleware [33] is used for the data transmission, avoiding manual data marshaling and allowing fast reconfiguration when the telemetry format is changed, as when a new sensor is added.

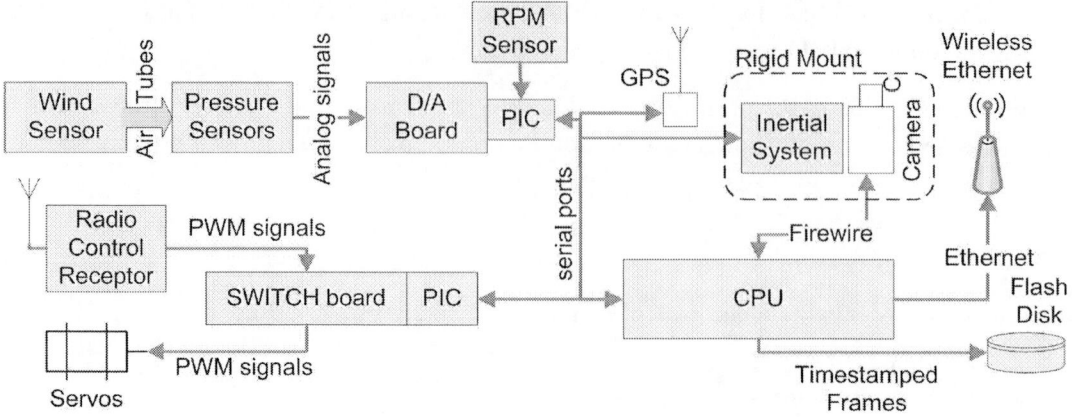

Figure 2. The embedded system of the DIVA project.

Figure 3 shows hardware components used in the DIVA airship, some of which are described below:

(a) Xsens MTB-9 IMU  (b) Garmin GPS35  (c) Wind sensor on the airship nose

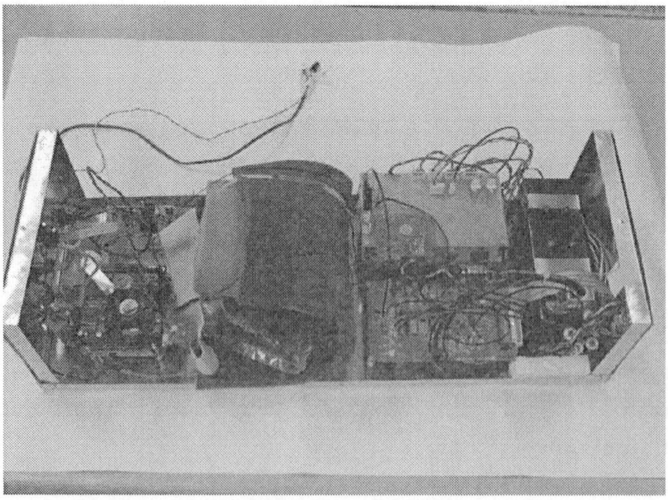

(d) The embedded system.

Figure 3. Hardware components of the DIVA prototype.

CPU The data sets used on this chapter were collected with a M570-BAB Board; VIA EDEN $600MHz$ processor; PC/104+; $512MB$ RAM. Nevertheless, an upgrade is necessary to be able to store images at a higher frame rate and to transmit images to the ground.

Inertial System (Xsens MTB-9 system) Outputs direct readings from its internal sensors (magnetometer, accelerometer and inclinometer), and also filtered, driftless absolute orientation. Low weight ($35g$).

GPS Receptor (Garmin GPS35 12 channel GPS unit) Low weight GPS receptor and antenna integrated in a single mouse-sized package.

Wind Sensor It is mounted on the airship nose and measures barometric altitude, the angle of attack, sideslip angle and wind speed. It is not yet calibrated.

RPM Sensor Measures the rotation speed of the motors, by counting the number of interruptions on an Infra Red light signal that is cut by the propeller movement.

Switch Board The servo-motors on the flaps and motors accelerators are commanded by Pulse Width Modulation (PWM) signals sent by the human pilot via RC. The Switch Board reads the signals from the Radio Control Receiver, transmits their values in digital form to the CPU, for recording, and retransmits PWM signals to the servo motors. In future automatic flights, the same board may switch to transmit commands sent from the CPU to the servo motors.

Besides the on-board system, there is also support equipment on the ground (see Fig. 4). The human pilot commands the airship with a standard aero model Remote Control Unit, sending PWM signals to command the servo-motors on the flaps and motor accelerators. Also, a laptop connected to a Wireless Access Point receives, stores and displays telemetry data from the airship.

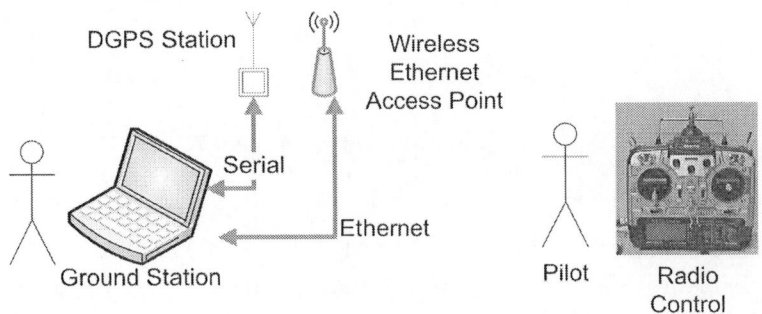

Figure 4. The Ground Station.

3. Guidance and Control

This section is dedicated to the guidance and control problem of the DIVA airship. After presenting its dynamic model, a backstepping path-tracking controller is designed. Control implementation issues are addressed prior to the presentation of the simulation results obtained for a realistic complete mission.

3.1. Airship Dynamic Model

This section describes the dynamic and cinematic equations of the DIVA airship. It also presents an estimator for the unknown wind disturbances.

3.1.1. Airship Dynamics

For the derivation of the mathematical model of the DIVA airship flight dynamics, based on the AURORA airship model, the following aspects were taken into account [34, 35]: (i) the model considers the airship virtual masses and inertias due to the large volume of air displaced by the airship; (ii) the airship motion is referenced to a system of orthogonal axes fixed to the vehicle (Fig. 5) whose origin is the Center of Volume (CV), assumed to coincide with the gross Center of Buoyancy (CB); (iii) the airship is assumed to be a rigid body, so that aeroelastic effects are ignored.

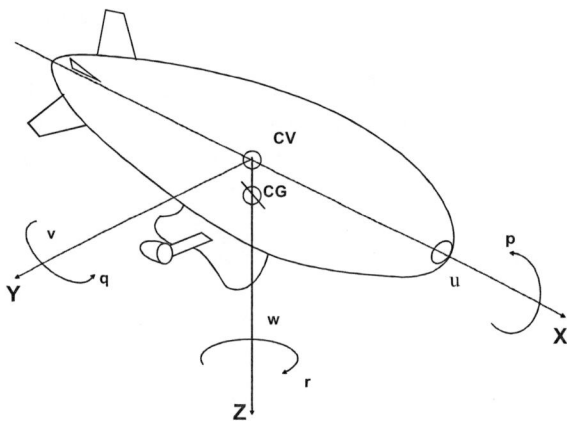

Figure 5. Airship local reference frame.

Let $\{i\}$ represent the inertial frame (which, for simplicity, is considered coincident with the geographical North-East-Down (NED) frame), $\{l\}$ be the body-fixed coordinate frame, and $S \in SO(3) := \{S \in \mathbb{R}^{3\times 3} : SS^T = I, \det(S) = +1\}$ be the rotation matrix from $\{i\}$ to $\{l\}$ frame. In the $\{l\}$ frame, the airship linear and angular velocities are given by $v \in \mathbb{R}^3$ and $\omega \in \mathbb{R}^3$, respectively. Let us recall the dynamic equation as it was deduced in [35], using the formulation of Newton second law expressed in the local frame $\{l\}$:

$$M\dot{V} = F_k + F_w + F_g + F_p + F_a \quad (1)$$

where $M \in \mathbb{R}^{6 \times 6}$ is the generalized mass and inertia matrix, $V = [v^T, \omega^T]^T \in \mathbb{R}^6$ is the inertial velocity in the local frame, and the generalized forces appearing in the right hand side are respectively the kinematic, wind induced, gravity, propulsion and aerodynamic forces.

If we assume the wind as constant in the earth frame, with linear inertial velocity w and without angular component, the local linear velocity v may be written as the sum of the air velocity v_a, which is the relative velocity of the vehicle in the air flow (associated with the airspeed), and the wind velocity in local frame v_w:

$$v = v_a + v_w = v_a + Sw \qquad (2)$$

Defining the air velocity state as $x = [v_a^T, \omega^T]^T \in \mathbb{R}^6$, the airship dynamic equation may finally be written in a compact form as a function of the air velocity [36]:

$$M\dot{x} = -\Omega_6 M x + E_g S g + F + f \qquad (3)$$

where $\Omega_6 \triangleq \mathrm{diag}\{\Omega_3, \Omega_3\} \in \mathbb{R}^{6\times 6}$, and the synthetic matrix notation is used for the cross-product $\Omega_3 = \omega \times \in \mathbb{R}^{3\times 3}$.

Also, g is the gravity vector in the inertial frame, and $E_g \triangleq \begin{bmatrix} m_w I_3 \\ mC_3 \end{bmatrix}$, where m is the airship scalar mass and m_w is its weighting mass. The forces vector corresponds to $F + f = \begin{bmatrix} F_p + F_a \\ T_p + T_a \end{bmatrix}$ separating $F = F(x)$, for the state only depending part, and f for the actuation or control force input, with an aerodynamic part in (F_a, T_a), and a propulsion part in (F_p, T_p).

3.1.2. Airship Kinematics

Let us define the airship position vector $\eta = [p^T, \phi^T]^T \in \mathbb{R}^6$ as being composed by its Cartesian coordinates $p \in \mathbb{R}^3$ in the $\{i\}$ frame, and the angular Euler attitude $\phi \in \mathbb{R}^3$. The kinematics involves the transformation between velocity and position [37]:

$$\dot{p} = S^T v \qquad (4a)$$
$$\dot{\phi} = R\omega \qquad (4b)$$

with the coefficient matrix $R \in \mathbb{R}^{3\times 3}$ relating the Euler angles with their derivatives and the angular rates.

The position derivative is related to the airship air velocity as:

$$\dot{\eta} = \begin{bmatrix} \dot{p} \\ \dot{\phi} \end{bmatrix} = \begin{bmatrix} S^T & 0 \\ 0 & R \end{bmatrix} \begin{bmatrix} v \\ \omega \end{bmatrix} = \begin{bmatrix} S^T & 0 \\ 0 & R \end{bmatrix} \begin{bmatrix} v_a + Sw \\ \omega \end{bmatrix} \qquad (5)$$

We may write (5) as:

$$\dot{\eta} = \begin{bmatrix} S^T & 0 \\ 0 & R \end{bmatrix} \begin{bmatrix} v_a \\ \omega \end{bmatrix} + \begin{bmatrix} I \\ 0 \end{bmatrix} w \qquad (6)$$

with I the identity matrix, or:

$$\dot{\eta} = Jx + Bw \qquad (7)$$

where $J \triangleq \begin{bmatrix} S^T & 0 \\ 0 & R \end{bmatrix} \in \mathbb{R}^{6\times 6}$ and $B \triangleq [I, 0]^T \in \mathbb{R}^{6\times 3}$.

The system may then be expressed by the following equations describing the dynamics, kinematics and constant wind:

$$\dot{x} = Kx + M^{-1}(E_g S g + F + f) \tag{8a}$$
$$\dot{\eta} = Jx + Bw \tag{8b}$$
$$\dot{w} = 0 \tag{8c}$$

where $K \triangleq -M^{-1}\Omega_6 M \in \mathbb{R}^{6\times 6}$ is linearly dependent of the angular velocity ω, whereas M is constant or slowly varying with altitude (since the inertia terms depend on the air density).

The kinematic derivative relations satisfy the equations:

$$\dot{S} = -\Omega_3 S \tag{9a}$$
$$\dot{R} = -R\dot{R}^{-1}R \tag{9b}$$
$$\dot{J} = JH \tag{9c}$$

with $H \triangleq \begin{bmatrix} \Omega_3 & 0 \\ 0 & -\dot{R}^{-1}R \end{bmatrix} \in \mathbb{R}^{6\times 6}$.

3.1.3. Wind Estimator

Since the wind disturbance is unknown, an estimator may be built based on (8b)-(8c). As assumed earlier, the wind input is not affecting the angular position part in (6), and therefore only the cartesian position p of the airship should be considered:

$$\dot{p} = S^T v_a + w \tag{10}$$

The estimator states may then be $(\hat{p}, \hat{w})$, and its dynamics may be chosen as:

$$\frac{d}{dt}\begin{bmatrix} \hat{p} \\ \hat{w} \end{bmatrix} = \begin{bmatrix} S^T v_a \\ 0 \end{bmatrix} + \begin{bmatrix} L_p & I_3 \\ L_w & 0 \end{bmatrix}\begin{bmatrix} p - \hat{p} \\ \hat{w} \end{bmatrix} \tag{11}$$

leading to a dynamics of the estimation error ϵ obtained from (10)-(11) and given by:

$$\dot{\epsilon} = \frac{d}{dt}\begin{bmatrix} p - \hat{p} \\ w - \hat{w} \end{bmatrix} = \begin{bmatrix} -L_p & I_3 \\ -L_w & 0 \end{bmatrix}\epsilon = A_\epsilon \epsilon \tag{12}$$

where the two constant matrices (L_p, L_w) are chosen so that A_ϵ be Hurwitz.[1]

Then there exists a positive definite symmetric matrix $P_\epsilon > 0$ such that:

$$\frac{d}{dt}(\epsilon^T P_\epsilon \epsilon) = -\epsilon^T Q_\epsilon \epsilon \tag{13}$$

---

[1] For instance, taking $L_p = \alpha I_3$ and $L_w = \frac{\alpha^2}{\beta} I_3$ leads to three pairs of poles at $-\frac{\alpha}{2}(1 \pm \sigma i)$ with $\sigma = \sqrt{\frac{4-\beta^2}{\beta}}$, i.e. with a damping factor $\xi = \frac{1}{2}\sqrt{\beta}$.

where the matrix $Q_\epsilon > 0$ is symmetric positive definite, and chosen in a block diagonal form. The matrix P_ϵ is the solution of the Riccati equation:

$$A_\epsilon^T P_\epsilon + P_\epsilon A_\epsilon = -Q_\epsilon = -\begin{bmatrix} Q_p & 0 \\ 0 & Q_w \end{bmatrix} \quad (14)$$

with Q_p and Q_w diagonal matrices, so that the Lyapunov function of the estimator and its derivative are:

$$W_e = \epsilon^T P_\epsilon \epsilon \quad (15)$$
$$\dot{W}_e = -\epsilon^T Q_\epsilon \epsilon = -\tilde{p}^T Q_p \tilde{p} - \tilde{w}^T Q_w \tilde{w} \quad (16)$$

where the estimation errors are:

$$\tilde{p} = p - \hat{p} \quad (17)$$
$$\tilde{w} = w - \hat{w} \quad (18)$$

The asymptotically stable wind estimator ($\dot{W}_e < 0$) will then assure that the estimation errors (17)-(18) will converge to zero.

3.2. Path-Tracking Controller Design

3.2.1. Backstepping Design Approach

Let us consider a generic control problem with output y. We first define two auxiliary outputs involving the output y and its derivative $\dot{y}$:

$$\begin{cases} y_1 = ay + \dot{y} \\ y_2 = \dot{y} \end{cases} \Rightarrow \begin{cases} \dot{y}_1 = a\dot{y} + \ddot{y} \\ \dot{y}_2 = \ddot{y} \end{cases} \quad (19)$$

where a is a positive scalar to be used as design parameter. It is easily seen that when both auxiliary outputs are taken to the origin, the regulation of the main output y is then achieved.

A tentative Lyapunov function may be:

$$W_0 = \frac{1}{2} y_1^T y_1 + \frac{1}{2} y_2^T y_2 \quad (20)$$

Its derivative is:

$$\dot{W}_0 = y_1^T \dot{y}_1 + y_2^T \dot{y}_2 = (ay + \dot{y})^T (a\dot{y} + \ddot{y}) + \dot{y}^T \ddot{y} = (ay + 2\dot{y})^T (a\dot{y} + \ddot{y}) - a\dot{y}^T \dot{y} \quad (21)$$

If the control is chosen in order to give:

$$a\dot{y} + \ddot{y} = -\Lambda (ay + 2\dot{y}) \quad (22)$$

where $\Lambda = \Lambda^T$ is a positive definite matrix, then the derivative:

$$\dot{W}_0 = -(ay + 2\dot{y})^T \Lambda (ay + 2\dot{y}) - a\dot{y}^T \dot{y} \quad (23)$$

will clearly be negative definite and the system will be globally asymptotically stable.

3.2.2. Backstepping Design Applied to Path-Tracking

We shall now proceed applying the control design described in the previous section to the path-tracking problem.

Let us assume a point p_r with a constant ground velocity v_r is to be tracked with constant attitude along a rectilinear path AB (see Fig. 6):

$$\dot{p}_r = S_r^T v_r \qquad (24)$$

where $S_r \in \mathbb{R}^{3\times 3}$ is the constant transformation matrix from the inertial frame to the desired path.

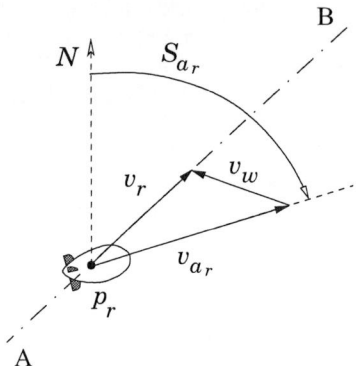

Figure 6. Air velocity reference estimation (2D).

As the wind velocity v_w is considered, the desired air velocity v_{a_r} may be deduced from the desired ground velocity v_r. Moreover, since the airship is to align itself with this air velocity, we have a reference for the attitude given by the transformation S_{a_r} from the inertial frame to the air velocity v_{a_r}, which is described by the desired attitude vector ϕ_{a_r}. This leads to the reference position $\eta_r = [p_r^T, \phi_{a_r}^T]^T$.

The derivative of this reference position is:

$$\dot{\eta}_r = \begin{bmatrix} S_r^T v_r \\ 0 \end{bmatrix} = J_r x_r \qquad (25)$$

where the reference velocity state is $x_r = \begin{bmatrix} v_r \\ 0 \end{bmatrix} \in \mathbb{R}^6$ and $J_r = \begin{bmatrix} S_r^T & 0 \\ 0 & 0 \end{bmatrix} \in \mathbb{R}^{6\times 6}$.

Note that, although we have assumed a rectilinear reference path, the approach may also be extended to the cases where the reference path varies slowly, with negligible derivatives when compared to the state derivative.

Let us now consider a candidate Lyapunov function similar to (20):

$$W_t = \frac{1}{2} y_1^T y_1 + \frac{1}{2} y_2^T y_2 \qquad (26)$$

where the output auxiliary variables y_1 and y_2 are again derived from the output y and its derivative $\dot{y}$, but where $y = \eta - \eta_r$ is the position tracking error:

$$\begin{cases} y_1 = ay + \dot{y} = a(\eta - \eta_r) + \dot{\eta} - \dot{\eta}_r = a(\eta - \eta_r) + Jx + Bw - J_r x_r \\ y_2 = \dot{y} = \dot{\eta} - \dot{\eta}_r = Jx + Bw - J_r x_r \end{cases} \qquad (27)$$

and where equations (8b) and (25) were used.

The derivative of the tentative Lyapunov function (26) is then:

$$\dot{W}_t = y_1^T \dot{y}_1 + y_2^T \dot{y}_2 = (ay + \dot{y})^T (a\dot{y} + \ddot{y}) + \dot{y}^T \ddot{y} = (ay + 2\dot{y})^T (a\dot{y} + \ddot{y}) - a\dot{y}^T \dot{y} \quad (28)$$

If the control is chosen such that:

$$a\dot{y} + \ddot{y} = -\Lambda(ay + 2\dot{y}) \quad (29)$$

or:

$$a(Jx + Bw - J_r x_r) + J\dot{x} + JHx = -\Lambda \left(a(\eta - \eta_r) + 2(Jx + Bw - J_r x_r) \right) \quad (30)$$

where we used:

$$\dot{y}_2 = \ddot{y} = J\dot{x} + JHx \quad (31)$$

this leads to the control law:

$$\begin{aligned} J\dot{x} &= -\Lambda \left(a(\eta - \eta_r) + 2(Jx + Bw - J_r x_r) \right) - JHx - a(Jx + Bw - J_r x_r) \\ &= -a\Lambda(\eta - \eta_r) - JHx - (aI + 2\Lambda)(Jx + Bw - J_r x_r) \end{aligned} \quad (32)$$

As wind is estimated, the suggested control law is:

$$J\dot{x} = -a\Lambda(\eta - \eta_r) - JHx - (aI + 2\Lambda)(Jx + B\hat{w} - J_r x_r) \quad (33)$$

and (29) should be rewritten as:

$$a\dot{y} + \ddot{y} = -\Lambda(ay + 2\dot{y}) + (aI + 2\Lambda)B\tilde{w} \quad (34)$$

Introducing $y_0 \in \mathbb{R}^6$ such that $y_0 \triangleq ay + 2\dot{y}$, and defining $G \in \mathbb{R}^{6\times 3}$ such that $G \triangleq (\frac{a}{2}\Lambda^{-1} + I)B$, the tentative Lyapunov derivative appears as:

$$\dot{W}_t = -\Lambda y_0^T (y_0 - 2G\tilde{w}) - ay_2^T y_2 \quad (35)$$

or, completing the squares[2]:

$$\dot{W}_t = -\Lambda(y_0 - G\tilde{w})^T (y_0 - G\tilde{w}) + \Lambda \tilde{w}^T G^T G \tilde{w} - ay_2^T y_2 \quad (36)$$

If we now consider a corrected tentative Lyapunov function with the wind estimator:

$$W = W_t + W_e \quad (37)$$

the derivative $\dot{W}$ may be written using (36) and (16) as:

$$\dot{W} = -\Lambda(y_0 - G\tilde{w})^T (y_0 - G\tilde{w}) - ay_2^T y_2 - \tilde{p}^T Q_p \tilde{p} - \tilde{w}^T (Q_w - \Lambda G^T G) \tilde{w} \quad (38)$$

---

[2]From the expansion of a square:

$$(y_0 - G\tilde{w})^T (y_0 - G\tilde{w}) = y_0^T (y_0 - G\tilde{w}) - (G\tilde{w})^T (y_0 - G\tilde{w}) = y_0^T (y_0 - 2G\tilde{w}) + (G\tilde{w})^T G\tilde{w}$$

it is easily deduced that: $y_0^T (y_0 - 2G\tilde{w}) = (y_0 - G\tilde{w})^T (y_0 - G\tilde{w}) - \tilde{w}^T G^T G\tilde{w}$

which is definite negative if:
$$Q_w - \Lambda G^T G > 0 \qquad (39)$$

The control law may be deduced from equations (8a) and (33), leading to:
$$f = M(\dot{x} - Kx) - E_g S g - F \qquad (40)$$
$$\dot{x} = -aJ^{-1}\Lambda(\eta - \eta_r) - Hx - J^{-1}\Lambda_2^2(Jx + B\hat{w} - J_r x_r) \qquad (41)$$

where $\Lambda_2^2 = (aI + 2\Lambda)$. The force control input is then given by:
$$f = -M\left[A_1\left(Jx + B\hat{w} - J_r x_r\right) + B_1\left(\eta - \eta_r\right) + \Gamma_1 x\right] - E_g S g - F \qquad (42)$$

with $A_1 = J^{-1}\Lambda_2^2$, $B_1 = aJ^{-1}\Lambda$ and $\Gamma_1 = H + K$, resulting in an asymptotically stable closed-loop system.

However, the force control input, as it is, may result in excessively high demands for a real system subject to input constraints. In the next section the control solution (42) will be adapted to deal with this matter.

3.2.3. Control Design with Saturation Constraints

In order to include saturation limits into the control design, let us rewrite equation (29), corresponding to a second derivative demand:
$$\ddot{y} = -a\dot{y} - \Lambda\left(ay + 2\dot{y}\right) = -\left(aI + 2\Lambda\right)\dot{y} - a\Lambda y = -\Lambda_2^2 \dot{y} - \Lambda_1 y \qquad (43)$$

with $\Lambda_1 = a\Lambda$ and Λ_2^2 as defined in the previous section.

Defining the second Lyapunov function as:
$$W_2 = \frac{1}{2} y_2^T y_2 \qquad (44)$$

with, as before, $y_2 = \dot{y}$, its derivative may be expressed as:
$$\dot{W}_2 = \dot{y}^T \ddot{y} = -\dot{y}^T \left(\Lambda_2^2 \dot{y} + \Lambda_1 y\right) = -z_2^T \left(z_2 + z_1\right) \qquad (45)$$

where $z_1 = \Lambda_2^{-1}\Lambda_1 y$ and $z_2 = \Lambda_2 \dot{y}$. Writing (43) as function of z_1 and z_2 yields:
$$\ddot{y} = -\Lambda_2(z_2 + z_1) \qquad (46)$$

Before proceeding, we will now define *linear saturation* as well as its properties, and provide an important theorem used in the proof of stability of the saturated control.

Definition 1. *As a particular case and extension of the linear saturation definition proposed by Teel [20], let us introduce the element-wise nondecreasing saturation function $\sigma : \mathbb{R}^n \to \mathbb{R}^n$, defined by a vector m of n positive values m_i, with $m_i > r > 0$, and such that:*
$$\forall z \in \mathbb{R}^n,\ \sigma[z] = \Sigma z \qquad (47)$$

where the diagonal matrix Σ is defined by:
$$\begin{aligned} |z_i| < m_i &\Rightarrow \Sigma_i = 1 \\ |z_i| \geq m_i &\Rightarrow \Sigma_i = \frac{m_i}{|z_i|} \end{aligned} \qquad (48)$$

Properties. *It may easily be verified that the definition yields the following properties [20]:*

$$\begin{cases} \forall z \in \mathbb{R}^n \,;\; z^T \sigma[z] > 0 \\ \forall z \in \mathbb{R}^n \,;\; |\sigma[z]| \leq R \\ |z| < r \Rightarrow \sigma[z] = z \end{cases} \quad (49)$$

where $|z| = \sqrt{z^T z}$ is the norm of vector z as defined in $\mathbb{R}^n$ and $R^2 = \sum_{i=1}^{n} m_i^2$.

Theorem 1. *If two saturations σ_1 and σ_2 are defined, such that $R_1 < \frac{1}{2} r_2$, then:*

$$\forall (z_1, z_2) \in \mathbb{R}^n,\; |z_2| > \frac{1}{2} r_2 \Rightarrow z_2^T \sigma_2 [z_2 + \sigma_1 [z_1]] > 0 \quad (50)$$

Proof. Since $|z_2| > \frac{1}{2} r_2$ and $|\sigma_1[z_1]| \leq R_1 < \frac{1}{2} r_2$, one can write the orthogonal projection of the saturated vector $\sigma_1[z_1]$ on z_2 as:

$$\sigma_1 [z_1] = \lambda_1 z_2 + v_1 \quad (51)$$

where $|\lambda_1| < 1$, $z_2^T v_1 = 0$, and $|\lambda_1 z_2 + v_1| < \frac{1}{2} r_2$.
Then:

$$\begin{aligned} z_2^T \sigma_2 [z_2 + \sigma_1 [z_1]] &= z_2^T \sigma_2 \left[(1 + \lambda_1) z_2 + v_1 \right] \\ &= z_2^T \Sigma_2 ((1 + \lambda_1) z_2 + v_1) \\ &= (1 + \lambda_1) z_2^T \Sigma_2 z_2 > 0 \end{aligned} \quad (52)$$

We can now proceed and introduce the second derivative (46) saturated demand:

$$\ddot{y}_s = -\Lambda_2 \sigma_2 [z_2 + \sigma_1 [z_1]] \quad (53)$$

From Theorem 1, if $|z_2| > \frac{1}{2} r_2$, then $\dot{W}_2 = -z_2^T \sigma_2 [z_2 + \sigma_1 [z_1]]$ will be negative definite for saturations σ_1 such that $|\sigma_1 [z_1]| \leq R_1 < \frac{1}{2} r_2$.

Since the saturated system is asymptotically stable, after a time T_2 the variable z_2 will enter the linear zone of its saturation and remain inside of it, namely with $|z_2| < \frac{1}{2} r_2$.

After time T_2 the saturated demand will be equal to:

$$\ddot{y}_s = -\Lambda_2 (z_2 + \sigma_1 [z_1]) \quad (54)$$

Introducing (54) into (28) yields:

$$\begin{aligned} \dot{W}_{ts} &= (ay + 2\dot{y})^T (a\dot{y} + \ddot{y}_s) - a\dot{y}^T \dot{y} \\ &= (ay + 2\dot{y})^T (a\dot{y} - \Lambda_2 (z_2 + \sigma_1 [z_1])) - a\dot{y}^T \dot{y} \\ &= (ay + 2\dot{y})^T (a\dot{y} - \Lambda_2 z_2 - \Lambda_2 \sigma_1 [z_1]) - a\dot{y}^T \dot{y} \\ &= (ay + 2\dot{y})^T (a\dot{y} - (aI_7 + 2\Lambda) \dot{y} - \Lambda_2 \sigma_1 [z_1]) - a\dot{y}^T \dot{y} \\ &= -(ay + 2\dot{y})^T \Lambda (2\dot{y} + \Lambda^{-1} \Lambda_2 \sigma_1 [z_1]) - a\dot{y}^T \dot{y} \end{aligned} \quad (55)$$

Using the definition of the saturation $\sigma_1[z_1] = \Sigma_1 z_1 = \Sigma_1 \Lambda_2^{-1} a \Lambda y$, from (55) we get:

$$\dot{W}_{ts} = -(2\dot{y} + ay)^T \Lambda \left(2\dot{y} + \Lambda^{-1} \Lambda_2 \Sigma_1 \Lambda_2^{-1} a \Lambda y\right) - a\dot{y}^T \dot{y} \tag{56}$$

Two scenarios are now possible: (i) z_1 is not saturated, in which case $\Sigma_1 = I$, resulting in $\dot{W}_{ts} < 0$; or (ii) z_1 is saturated and $\Sigma_{1_i} = \frac{m_{1_i}}{|z_{1_i}|} \leq 1$. Let us further analyze this case.

Taking $z_0 = 2\Lambda^{1/2}\dot{y}$, $s = a\Lambda^{1/2}\Sigma_1 y$, and $Z = \Sigma_1^{-1}$, we have:

$$\dot{W}_{ts} = -(z_0 + Zs)^T (z_0 + s) - a\dot{y}^T \dot{y} \tag{57}$$

If we consider the decomposition of the vectors in their components, $z_0 = [z_i]$, $s = [s_i]$ and also the diagonal matrix $Z = [\lambda_i]$ with elements $\lambda_i = \frac{|z_{1_i}|}{m_{1_i}} \geq 1$, then:

$$(z_0 + s)^T (z_0 + Zs) = \sum_i (z_i + s_i)(z_i + \lambda_i s_i) \tag{58}$$

Noting that s and z_0 have behaviors similar to, respectively, z_1 and z_2, and that z_2 is in its linear zone and converging, we have that after some time $|z_i| < |s_i|, \forall i$ and then $z_i = \mu_i s_i$ with $|\mu_i| < 1$, so that:

$$(z_0 + s)^T (z_0 + Zs) = \sum_i (\mu_i s_i + s_i)(\mu_i s_i + \lambda_i s_i) \tag{59}$$

$$= \sum_i (s_i)^2 (\mu_i + 1)(\mu_i + \lambda_i) \tag{60}$$

which shows that each term is positive, making the result of the sum also positive. Therefore, $\dot{W}_{ts}$ is also negative definite.

To include the input forces limitations into the control law design, let us consider the desired demand is a saturated one, $\ddot{y} = \ddot{y}_s$. From (31) and (53) we obtain:

$$J\dot{x} + JHx = -\Lambda_2 \sigma_2 \left[\Lambda_2(Jx + Bw - J_r x_r) + \sigma_1[\Lambda_2^{-1}\Lambda_1(\eta - \eta_r)]\right] \tag{61}$$

or, solving for $\dot{x}$:

$$\dot{x} = -J^{-1}\Lambda_2 \sigma_2 \left[\Lambda_2(Jx + Bw - J_r x_r) + \sigma_1[\Lambda_2^{-1}\Lambda_1(\eta - \eta_r)]\right] - Hx \tag{62}$$

Substituting now (62) into (40) leads to the control law:

$$f_s = -M\left(J^{-1}\Lambda_2 \sigma_2 \left[\Lambda_2(Jx + Bw - J_r x_r) + \sigma_1[\Lambda_2^{-1}\Lambda_1(\eta - \eta_r)]\right] + \Gamma_1 x\right) - E_g Sg - F \tag{63}$$

Again, as the wind is estimated, the control law that considers the force input saturations is finally given by:

$$f_s = -M\left(J^{-1}\Lambda_2 \sigma_2 \left[\Lambda_2(Jx + B\hat{w} - J_r x_r) + \sigma_1[\Lambda_2^{-1}\Lambda_1(\eta - \eta_r)]\right] + \Gamma_1 x\right) - E_g Sg - F \tag{64}$$

where σ_1 and σ_2 are the velocity saturation matrices obtained from (64) with f_s corresponding to the input force maximum values related to the actuators limits (see section 3.3.1.), and that satisfy the condition $R_1 < \frac{1}{2}r_2 < |z_2|$. This control law will lead to an asymptotically stable closed-loop system as long as the estimation error is bounded according to (39).

With respect to the above control law, it is important to remark that although the reference velocity x_r is a groundspeed, the feedback velocity x is an airspeed (calculated from the actual groundspeed and the wind estimation).

3.3. Control Implementation

The control law (64) solves the airship path-tracking problem in the presence of constant translational wind while taking into account the limitations of the demanded forces input. However, this control law cannot be directly fed into the system, and needs to be adapted, as: (i) the airship is an underactuated vehicle, as detailed in the following; (ii) the position and velocity references may be shaped to reduce the consequences of saturations; (iii) the airship actuators are not directly usable as force inputs and a conversion or control allocation is to be applied in order to compute the real actuators inputs.

The airship actuation system may be split into two sets:

- force inputs that are available from two stroke engines, on each side of the gondola, with vectoring capability ranging from -30° to $+120^o$. The propellers provide a complementary lift to oppose the weighting mass, as well as a forward thrust controlling the longitudinal speed; when a differential input is added between the two propellers (meaning different rotations for the left and right engines), they also provide torque to control the rolling motion near hover; finally, a stern lateral thruster may be necessary to provide yaw control at low airspeeds, although it has not been used in the DIVA airship standard configuration;

- surface deflections of the tail (in the range of -25° to $+25^o$), which in the presence of a minimum airspeed provide torque inputs mostly for the control of the pitching and yawing motions. However, when the air is perfectly still and no wind is available, the hover control is reduced to the use of the force inputs only.

Thus, the airship real actuators input corresponds to $U = [T_X, \delta_v, T_D, T_Y, \delta_a, \delta_e, \delta_r]^T$, as described in section 2..

Before presenting the proposed solution to the control allocation problem, it is important to remark some features of the airship dynamics and actuation system. The airship dynamics is highly nonlinear and underactuated, with a very different behavior as the airspeed varies from the hovering or low airspeed flight (HF) to the cruise or aerodynamic flight (AF) [10, 34, 38]. The abrupt and continuous transition between the HF and AF in the dynamics implies a different use of actuators for each situation. For AF, the most important actuators are the propellers thrust and the aerodynamic elevator/rudder control surfaces, whereas for HF the effective actuators are the propellers total thrust and vectoring, differential propulsion, and the stern thrust when available.

3.3.1. Control Allocation

Force Inputs As it was stressed above, the relation between actuators and control inputs depends on the flight region:

- In the low airspeed region, the tail surfaces have reduced authority since the action from the surface deflections is a function of the dynamic pressure and varies as the square of the airspeed V_t, according to the aerodynamic characteristics of the airship [37]. This leaves the airship to be controlled by the force inputs only. The two main propellers correspond to 3 inputs (T_X, δ_v, T_D) - total thrust, vectoring angle, and differential thrust - providing longitudinal and vertical force, pitching and rolling torques. If available, the tail lateral thruster adds one input (T_Y), providing a side force and a yawing torque. These force actuators are slightly influenced by the airspeed but may be considered as independent in a first step.

- In aerodynamic flight, the vectoring angle is no longer necessary, leaving the airship with a reduced vertical force. The maneuvering is mostly accomplished by the tail fins. The surface deflections correspond to the three standard inputs of aileron, elevator and rudder deflections $(\delta_a, \delta_e, \delta_r)$, which mostly correspond to torque inputs, keeping the airship with reduced lateral force input.

As stated above, although it may have up to 7 actuator inputs to control 6 forces (3 forces and 3 torques), the airship is indeed an underactuated system (particularly in hovering) due to the limitations in the controllability. In order to reduce the influence of these limitations, a solution adopted in the control design was to add the tuning parameters (a, Λ), so as to decrease the closed-loop frequency, searching for a slower solution, that would be more robust to the unmodelled and approximate dynamics, as well as input saturations.

Conversion From Forces to Airship Inputs The relation from actuators to force inputs may then be established in an approximated approach neglecting the actuators dynamics, using the airspeed measurement and resolving the possible redundancies according to the usual operation of the airship [10] (the airship aerodynamic angles also have their effect, but they may be neglected in a first step, assuming small angles):

$$U = U(f, V_t) \tag{65}$$

where $U = [T_X, \delta_v, T_D, \delta_a, \delta_e, \delta_r]^T \in \mathbb{R}^6$ is the real actuators input, the force vector is represented by $f = [f_u, f_v, f_w, f_p, f_q, f_r]^T \in \mathbb{R}^6$, and V_t is the true airspeed. In the present case, the input U is computed as solution of the system composed by the 6 equations below, in agreement with the DIVA airship model:

$$\begin{aligned} f_u &= T_X \cos(\delta_v) + k_1 \delta_e & f_p &= k_2 l_4 \delta_a + b_4 \sin(\delta_v) T_D \\ f_v &= -k_2 \delta_r & f_q &= T_X b_3 \cos(\delta_v) + k_5 \delta_e \\ f_w &= -T_X \sin(\delta_v) + k_3 \delta_e & f_r &= k_2 l_6 \delta_r + b_4 \cos(\delta_v) T_D \end{aligned} \tag{66}$$

where (b_j, l_j) are geometrical constants of the airship, and $k_j(V_t)$ are second order polynomials expressing the airspeed depending authority of the tail deflections.

3.3.2. Adapted Control Law to Deal with Underactuation

As referred, the present configuration of the DIVA airship actuators results in an underactuated system. At very low airspeeds we reach the worst-case scenario, with the airship being uncontrollable due to the lack of authority from the control surfaces. The implementation of the proposed control law assumes this situation is not reached, therefore requiring that the true airspeed does not drop below a minimum, $V_t > V_{t_{min}} = 2m/s$ (note that it is quite realistic in outdoor conditions to assume a wind intensity above this level).

Even if this limit is respected, the airship may still be underactuated as the transversal forces available are too small. This means that a straightforward correction of eventual lateral and vertical position errors might lead to a saturated inputs request. In the following, we adapt the control law (64) to deal with this scenario, obtaining a faster error correction with smoother input requests.

Consider the approximated kinematic relations:

$$\dot{E} \simeq V_t \psi \qquad (67)$$
$$\dot{D} \simeq -V_t \theta \qquad (68)$$

where ψ and θ are the pitch and yaw Euler angles that describe the airship orientation, used here in place of the quaternions for simplicity. Equations (67)-(68) allow us to relate the airship orientation with its lateral and vertical positions.

Consider now the airship is to track a rectilinear path with orientation (ψ_r, θ_r) and has lateral and vertical errors respectively y and z. The angular errors are defined as:

$$\Delta \psi = \psi - \psi_r \qquad (69)$$
$$\Delta \theta = \theta - \theta_r \qquad (70)$$

Due to the airship underactuation, if we try to independently correct the position and angular errors, depending on their magnitude, we will probably have input saturation. However, if we consider the relation between position and attitude, we may consider instead the following expressions:

$$\Delta \psi' = \psi - \psi_r - k_y y \qquad (71)$$
$$\Delta \theta' = \theta - \theta_r - k_z z \qquad (72)$$

where the constants k_y and k_z are dependent of the airspeed V_t. This means we will postpone the angular corrections and use them to annulate the position errors first. The angular references (converted first to quaternions, with the roll reference $\phi_r = 0$) used in η_r in the control law (64) will then be:

$$\psi'_r = \psi_r + k_y y \qquad (73)$$
$$\theta'_r = \theta_r + k_z z \qquad (74)$$

3.4. Simulation Results

In order to test the nonlinear behavior of the presented control law, representative simulation tests were performed using the fully 6-DOF nonlinear model simulation environment developed in the DIVA Project. This simulator was built based on the AURORA experience and results [12]. The simulation case presented here concerns a complete airship mission to be implemented, starting with a vertical take-off, a path-tracking with two semicircles of 200m diameter, airship stabilization for ground hover, and finally a vertical landing (see Fig. 7).

The airship starts in position $(N_i, E_i, h_i) = (-30, -20, 0)m$ and is to go up $5m$ to the initial reference point $p_{r_0} = (N_{r_0}, E_{r_0}, h_{r_0}) = (-30, -20, 5)m$ so as to be stable

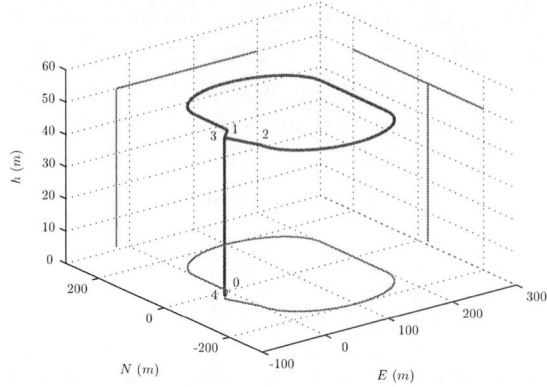

Figure 7. North N, east E and altitude h mission reference (bold) and projections (light).

and ready to start the mission. From this point, the vertical take-off begins, finishing with an approach to the first point of the horizontal path-tracking at $p_{r_1} = (N_{r_1}, E_{r_1}, h_{r_1}) = (0, 0, 50)m$. A square reference with a $200m$ side was adapted with two half-circles so as to provide a smooth reference with a continuous derivative. Although with this solution we do not always have a straight line reference, with a groundspeed reference of $7m/s$ and a $200m$ circle radius, the approximation is quite acceptable since the yaw rate is fairly small. Obviously, the angular velocity reference must be adapted to the case. When reaching the point $p_{r_2} = (N_{r_2}, E_{r_2}, h_{r_2}) = (-100, 0, 50)m$, the path-tracking gives place to the airship stabilization at the coordinates $p_{r_3} = (N_{r_3}, E_{r_3}, h_{r_3}) = (-30, -20, 50)m$, preparing it for vertical landing at $p_{r_4} = (N_{r_4}, E_{r_4}, h_{r_4}) = (-30, -20, 1)m$.

For the take-off and landing segments, some constraints are also given along with the position reference coordinates. Regarding the groundspeed, the airship is to ascend vertically from p_{r_0} at $0m/s$ and finish the approximation to p_{r_1} at $7m/s$, having a rate limit of $0.5m/s^2$. The stabilization process will start at p_{r_2} at $7m/s$ and finish at p_{r_3} at $0m/s$. The landing at p_{r_4} is required also at $0m/s$ (vertical descent). Concerning the altitude, the airship has a ascent/descent rate limit of, respectively, 1.5 and $-0.5m/s$.

In order to test the proposed controller robustness to unmodelled wind disturbances, a 3D $2m/s$ turbulence was added to the constant $4m/s$ wind blowing from northwest at $20°$.

The airship position coordinates (north N, east E and altitude h) are represented in Fig. 8. The vertical take-off and landing are well perceived in Fig. 8(a), as well as the path-tracking performance. Figure 8(b) compares the real coordinates with the references provided, allowing to identify the more problematic mission point, namely during the airship descent (see Fig. 9). The remaining noticeable errors correspond to instantaneous references changes before stabilization, which the airship smoothly corrects.

Figure 9 describes the airship north-east coordinates and heading during the mission. The preferential alignment with the wind during take-off and landing is well recognized, whereas along the tail wind segment the airship appears as slightly *crabbing*.

The airship velocities are depicted in Fig. 10. The ground velocity components are represented in Fig. 10(a). The longitudinal groundspeed u mostly follows the reference that varies between $0m/s$ for take-off and landing, and $8m/s$ during the path-tracking. Along

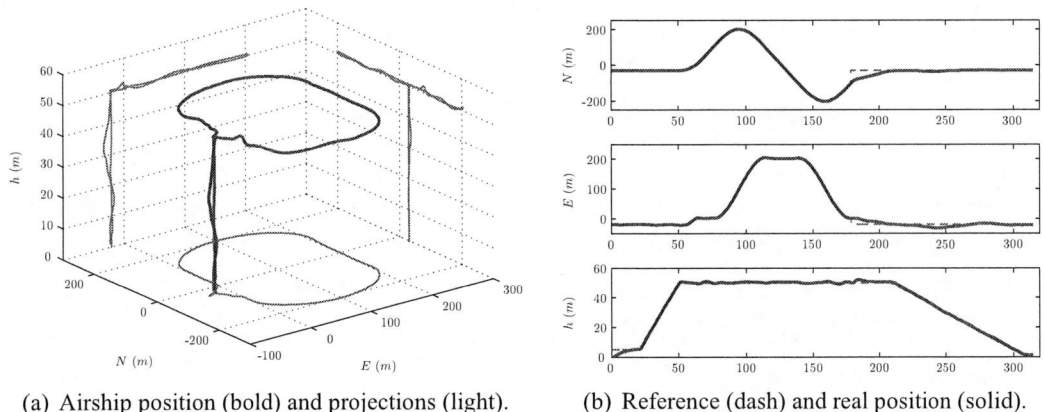

(a) Airship position (bold) and projections (light).

(b) Reference (dash) and real position (solid).

Figure 8. Airship north N, east E and altitude h coordinates.

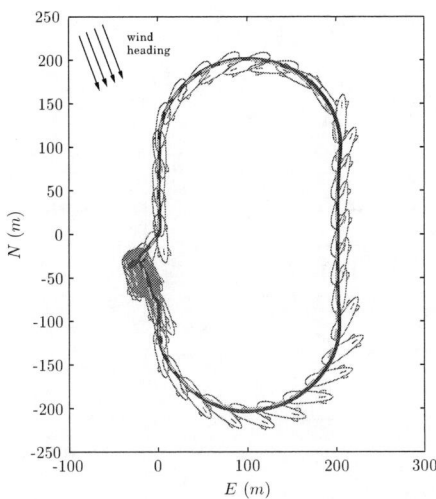

Figure 9. North-east position with airship heading (reference - dash line, real position - solid).

the circular segments, the errors are more noticeable due to the change of the wind incidence angle while the airship is turning. The lateral velocity v is also mostly influenced by the circular segments and during the tail wind segment. For the vertical velocity w, the two steps of, respectively, -1.5 and $0.5 m/s$ corresponding to the take-off and landing vertical motion are also easy to recognize.

The airspeed and aerodynamic angles can be seen in Fig. 10(b). During the whole mission, the airspeed V_t varies significantly, from values around $4m/s$ (above the set limit of $2m/s$ - see section 3.3.2.) up to $12m/s$. The airship covers a wide flight envelope, from hover to the aerodynamic flight, crossing the troublesome transition region between the two. The sideslip angle β and the angle of attack α vary between $\pm 10^o$ and, as expected, their behavior is correlated with v and w respectively.

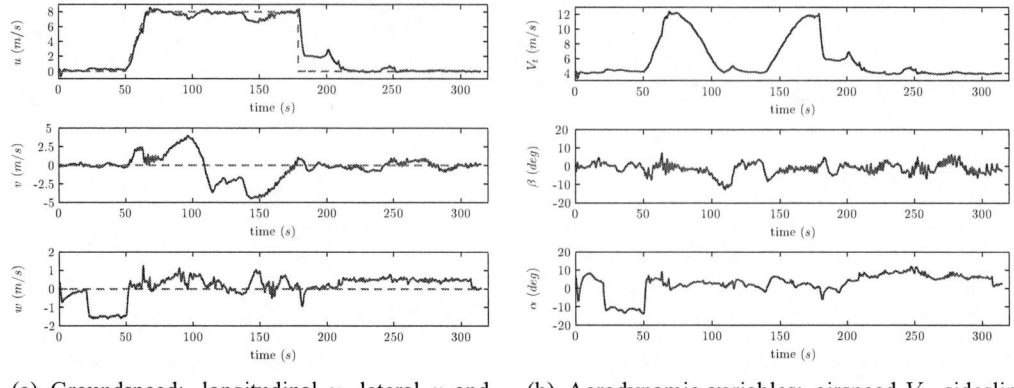

(a) Groundspeed: longitudinal u, lateral v and vertical w (reference - dash, real values - solid).

(b) Aerodynamic variables: airspeed V_t, sideslip angle β and angle of attack α.

Figure 10. Airship ground and air velocities

The saturated forces computed by (64) may be seen in Fig. 11, with the forces f_u, f_v and f_w represented in Fig. 11(a) and the moments f_p, f_q and f_r in Fig. 11(b).

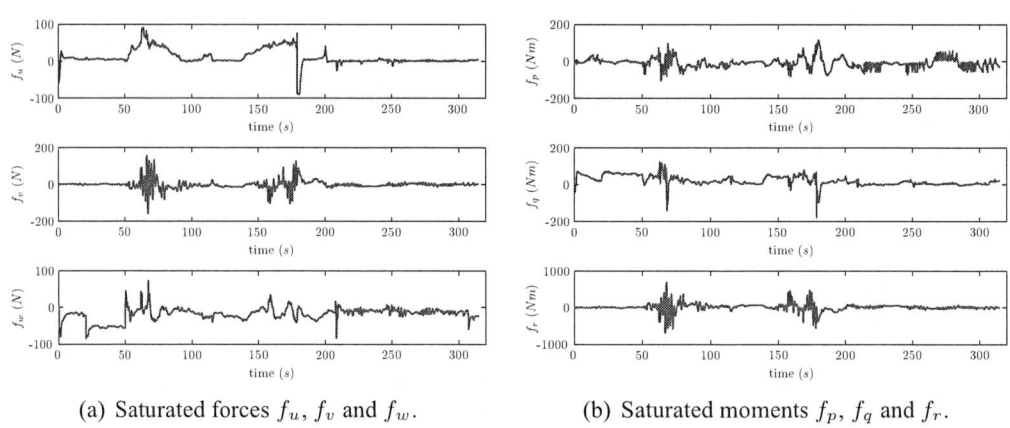

(a) Saturated forces f_u, f_v and f_w.

(b) Saturated moments f_p, f_q and f_r.

Figure 11. Saturated forces and moments request.

As justified before, these forces, which in this example show no saturation, have to be converted into airship actuators inputs. These are described in Fig. 12, with the longitudinal actuators elevator δ_e, total thrust T_X and vectoring angle δ_v in Fig. 12(a) and the lateral ones, aileron δ_a, rudder δ_r and differential thrust T_D, in Fig. 12(b). The noise levels in Fig. 12 are justified by the high value of turbulence considered in the simulation.

The elevator δ_e shows a higher demand at the beginning of the mission, corresponding to the ascent where the vertical rate is higher, while the rudder δ_r has a higher command during the curves and with tail wind, due to a lower airspeed. The vectoring angle δ_v is responsible for the airship lift when the airspeed V_t is too low to provide the necessary aerodynamic lift: if we compare the δ_v and the V_t graphics, this correlation is obvious.

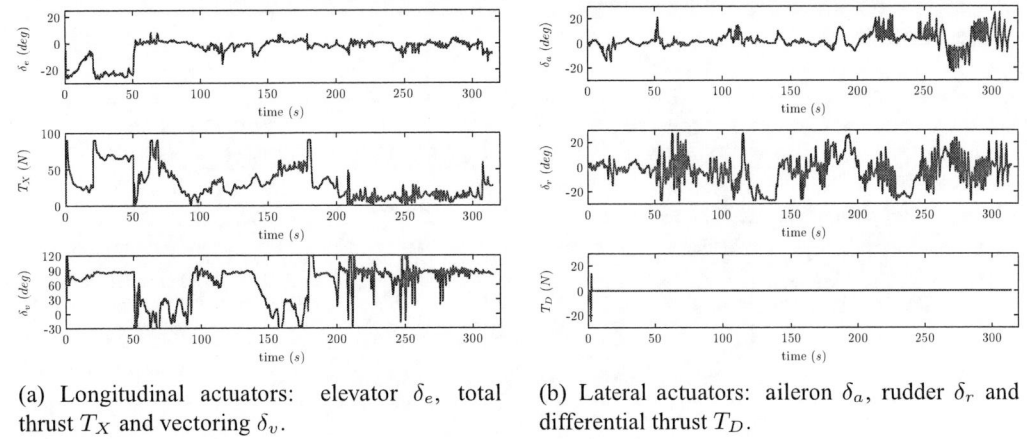

(a) Longitudinal actuators: elevator δ_e, total thrust T_X and vectoring δ_v.

(b) Lateral actuators: aileron δ_a, rudder δ_r and differential thrust T_D.

Figure 12. Airship actuators input.

This mission was defined to be representative and illustrative of a realistic behavior. It clearly represents a challenge for the automatic control system, as (i) the dynamics varies from the hovering to the aerodynamic flight during the path-tracking, (ii) we have a wind input with different incidence angles (as the trajectory is circular) and also stochastic gust, and (iii) the mission includes vertical maneuvers. These simulation results show that the approach is a strong and robust tool for the design of a single global control scheme for such an underactuated airship and surely for other Unmanned Vehicles.

4. Multi-Sensing for Mapping and Surveillance

The airship is equipped with several sensing devices where vision plays an important role. The information provided by the vision sensor is complemented with other sensing devices towards a complete and robust mapping and surveillance system. In this chapter, vision systems are rigidly coupled with Inertial Measurement Units (IMUs), which complement it with sensors providing direct measures of orientation relative to the world NED frame, such as magnetometers and inclinometers (that measure gravity components).

A novel calibration technique [39] finds the rigid body rotation between the camera and IMU frames. The camera orientation in the world is then obtained rotating the IMU orientation measurement. The approximation of the rotational degrees of freedom should allow faster processing or the use of simpler movement models in computer vision tasks. For example, it can be explored to improve robustness on image segmentation and 3D structure recovery [40, 41].

Sections 4.1. and 4.2. aim on exploiting the calibrated camera-inertial system in two other domains. With direct measurement of camera orientation, rotational and translational components of camera motion can be separated. Therefore, simpler movement models can be assumed, resulting in increased performance or accuracy. Images obtained from a camera on-board the DIVA airship (see Fig. 13) are used in all experiments. Note that the blimp envelope is transparent to GPS signals, thus the GPS receiver can be safely mounted over the gondola.

Figure 13. The DIVA airship, with details showing the vision-inertial system and GPS receiver mounted on the gondola.

4.1. Vision-Based Trajectory Recovery and Mapping

In this section we discuss trajectory recovery from images, presenting the results in section 4.1.2.. Previous results [42] have shown that the pure translation model yields more accurate height estimation that the usual homography model into a controlled laboratory environment with hand-measured ground truth. The vertical component of the airship trajectory can also be recovered this way. The other two horizontal components of the trajectory are recovered by estimating the Focus Of Expansion (FOE), and using the known vertical component to resolve scale.

Furthermore, only one solution is recovered by the process shown in this section, as opposed to the four solutions for the rotational and translational parameters recovered by the homography model. Often two of these are potentially viable, and geometric constraints are used to recover the right motion. This last step is not necessary here once the orientation and FOE are directly measured.

Experimental Platforms and Calibration

The hardware used on-board the airship is shown in Fig. 13 and described in section 2.2.. The camera is a Point Gray Flea [43], which is rigidly mounted with the inertial and magnetic sensor package Xsens MT9-B [44]. During an experimental flight, images with resolution of 1024×768 pixels were captured at $2 fps$. The camera is calibrated, its intrinsic parameter matrix K is known, and f is its focal length. Its optical center is indicated by C.

The rotation between the camera and the inertial frames is calculated by a novel calibration process [39, 45], which is available as a Matlab® toolbox for download [46]. Two examples of calibration images are shown in Fig. 14, where a chessboard is placed in the vertical position, so that its vertical lines provide an image-based measurement of the gravity direction to be registered with the gravity measurements provided by the accelerometers. The same images are also used to calibrate the camera with the Camera Calibration Toolbox [47].

Figure 14. Two examples of calibration images used to calibrate the camera-inertial system.

A Virtual Leveled Plane

The camera inertial calibration outputs the constant rotation between the camera and the IMU frames, supposing that both are rigidly mounted together. The absolute camera orientation is thus provided directly by the IMU absolute orientation measurements. Therefore, the images can be projected on entities defined on the absolute NED frame, such as a virtual horizontal plane (with normal parallel to the gravity), at a distance f (the focal length) below the camera center, named *virtual leveled plane* (see Fig. 15). Projection rays from 3D points to the camera center intersect this plane, projecting the 3D point into the plane. This projection corresponds to the image of a virtual camera with the same optical center as the original camera, but with optical axis coincident with the gravity vector, named here *virtual downwards camera*. Therefore the principal point of the virtual downward camera is the direction of gravity, named as the *nadiral point* $\mathbf{N}$. See [42, 48] for details.

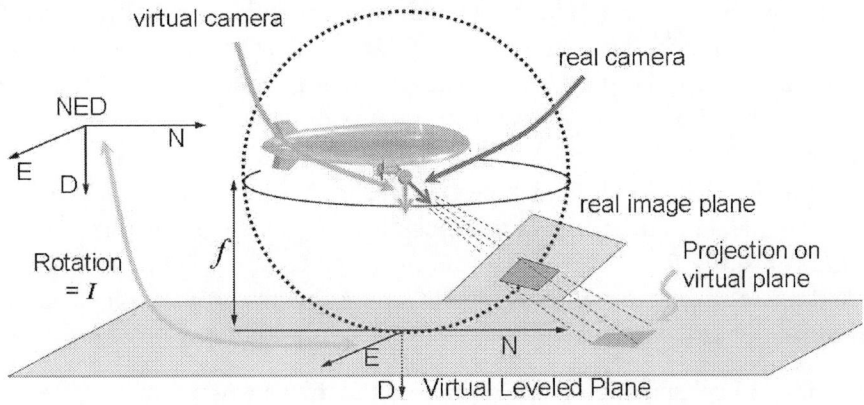

Figure 15. The virtual leveled plane concept.

4.1.1. Planar Homographies and Homologies

In computer vision, given an image pair, pixel correspondences are used to recover the relative camera poses corresponding to these two images, i.e., the rotation between both camera frames and the translation between the camera centers, albeit the translation is retrieved only up to scale. If the images are taken from a moving observer, it is possible to reconstruct the observer trajectory from a sequence of such relative poses.

This section defines and explores a pure translation model for planar scenes to achieve more accurate measurements of the camera height and to recover the camera trajectory from an image sequence. First the vertical component of the trajectory will be recovered, followed by the other two horizontal components.

The simplest case is when the 3D plane is parallel to the image plane. To simulate cameras under pure translation, and with image plane parallel to the horizontal plane, the image sequence is projected on the ground plane as exposed above using the IMU orientation measurements. The camera is allowed to move freely but the ground plane is required to be horizontal. Interesting point algorithms find pixel correspondences on the rotation-compensated images - the SURF algorithm [49] is used through this chapter.

Definition of Homography and Planar Homology

Consider a 3D plane imaged in two views. Consider also a set of pixel correspondences belonging to that plane in the form of pairs of pixel coordinates $(\mathbf{x}, \mathbf{x}')$, corresponding to the projection of the same 3D point into each view. An homography represented by a 3×3 matrix H relates these two sets of homogeneous pixel coordinates such that $\mathbf{x}' = H\mathbf{x}$. The homography can be recovered from pixel correspondences, and it is related to the 3D plane normal $\mathbf{n}$, the distance from the camera center to the plane d, and to the relative camera poses represented by the two camera projection matrices $P = [I|\mathbf{0}]$ and $P' = [R|\mathbf{t}]$, by:

$$\lambda H = \lambda \left(R - \mathbf{t}\mathbf{n}^T/d \right) \tag{75}$$

where R is rotation matrix and $\mathbf{t}$ a translation vector [50]. The scale module λ of the recovered homography $H_\lambda = \lambda H$ is the second largest singular value of H_λ, and the correct signal can be recovered by imposing a positive depth constraint. Then, the matrix H can be decomposed into R, $\mathbf{n}$, and $\mathbf{t}/d$ [51]. The translation magnitude is not recovered, only the ratio $\mathbf{t}/d$. In the translation-only case, $R = I$ and the homography becomes a *planar homology*.

A planar homology G is a planar perspective transformation that has a line of fixed points (the *axis*), and another fixed point outside of the axis (the *vertex*)[3] The axis is the image of the vanishing line of the plane (the intersection of the 3D plane and the plane at infinity), and the vertex is the epipole, or the FOE. Among the properties of homologies [50, 52], we recall:

- Lines joining corresponding points intersect at the vertex;

- The cross-ratios defined by the vertex, a pair of corresponding points, and the intersection of the line joining this pair with the axis, have the same value μ for all points;

- The homology matrix G may be defined from its axis, vertex, and the cross-ratio μ by:

$$G = I + (\mu - 1)\frac{\mathbf{v}\mathbf{a}^T}{\mathbf{v}^T\mathbf{a}} \tag{76}$$

where $\mathbf{v}$ is the vertex, and $\mathbf{a}$ the axis line.

---

[3]A fixed point is a point $\mathbf{x}$ such that $G\mathbf{x} = \mathbf{x}$, i.e., a point that is not changed by the transformation

Homology for 3D Planes Parallel to Image Plane

If the 3D plane is parallel to the image planes and the axis is the infinite line $\mathbf{a} = (0, 0, 1)^T$, then (76) becomes:

$$G = \begin{bmatrix} 1 & 0 & (\mu - 1) \cdot v_x \\ 0 & 1 & (\mu - 1) \cdot v_y \\ 0 & 0 & \mu \end{bmatrix} \qquad (77)$$

where v_x, v_y are the *inhomogeneous* image coordinates of the vertex $\mathbf{v} = (v_x, v_y, 1)$. The cross-ratio μ depends only of the relative depths of the 3D plane on the two views. To analyze this relation, we recall from Arnspang's paper [53] that the relative scene depth of two points equals the reciprocal ratio of the image plane distances to the vanishing point of their connecting line.

Taking two images of the same 3D point $\mathbf{X}$ under pure translation, and defining Z and Z' as the depth of $\mathbf{X}$ in the first and second views, and $\mathbf{x}$ and $\mathbf{x}'$ as its respective image coordinates, as in Fig. 16, we have:

$$\frac{Z'}{Z} = \frac{dist(\mathbf{x}, \mathbf{v})}{dist(\mathbf{x}', \mathbf{v})} \qquad (78)$$

where $dist$ means Euclidean distance on the image. Therefore, the relative depth of the same point in two views is calculated from image measures.

The relation between scene depths and image distances is valid for every single point, and it only requires an image of the same point in two views, and the FOE. However, if a 3D plane is parallel to the image planes, all points in the plane have the same depth, and are transferred between the two views by the same homology.

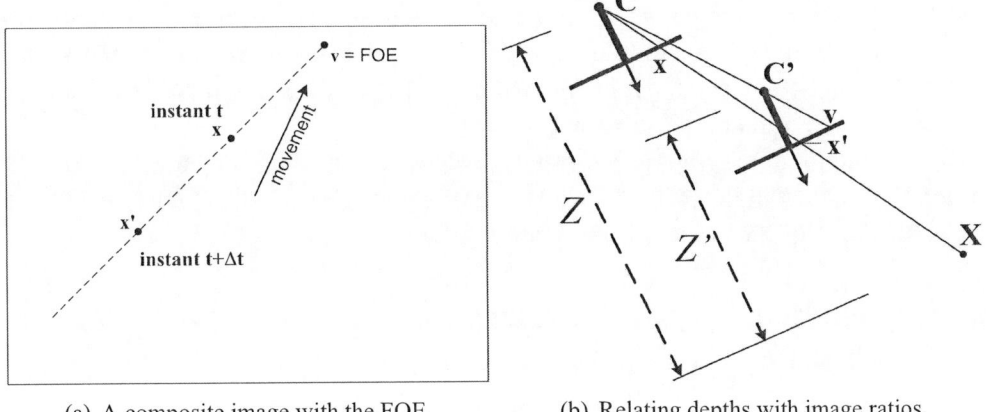

(a) A composite image with the FOE.   (b) Relating depths with image ratios.

Figure 16. A pair of cameras under pure translation imaging the same 3D point.

Therefore, the homology calculation involves many corresponding pixel pairs, and thus is potentially more stable than an image measure involving just one pair. Going back to the relation between the homology parameter μ and the depth of the 3D plane, applying equation (77) allows us to find:

$$\mathbf{x}' = G\mathbf{x} = \begin{bmatrix} \frac{x_x}{\mu} + v_x - \frac{v_x}{\mu} \\ \frac{x_y}{\mu} + v_y - \frac{v_y}{\mu} \\ 1 \end{bmatrix} \quad (79)$$

where $\mathbf{x} = (x_x, x_y, 1)^T$. Now, by calculating $|\mathbf{x} - \mathbf{x}'| = |\mathbf{x} - G\mathbf{x}|$, we relate this difference with $|\mathbf{x} - \mathbf{v}|$, in image coordinates:

$$|\mathbf{x} - \mathbf{x}'| = \left[\left(x_x - \left(\frac{x_x}{\mu} + v_x - \frac{v_x}{\mu} \right) \right)^2 + \left(x_y - \left(\frac{x_y}{\mu} + v_y - \frac{v_y}{\mu} \right) \right)^2 \right]^{1/2}$$

$$= \left[\left((x_x - v_x) \left(1 - \frac{1}{\mu} \right) \right)^2 + \left((x_y - v_y) \left(1 - \frac{1}{\mu} \right) \right)^2 \right]^{1/2}$$

$$= \left[\left(((x_x - v_x))^2 + ((x_y - v_y))^2 \right) \left(1 - \frac{1}{\mu} \right)^2 \right]^{1/2} \quad (80)$$

and then:

$$|\mathbf{x} - \mathbf{x}'| = |\mathbf{x} - \mathbf{v}| \left(1 - \frac{1}{\mu} \right) \quad (81)$$

As $\mathbf{x}, \mathbf{x}', \mathbf{v}$ are collinear, $|\mathbf{x} - \mathbf{x}'| + |\mathbf{v} - \mathbf{x}'| = |\mathbf{x} - \mathbf{v}|$. So, from (81) we find the image distances of equation (78) and we can update that equation as:

$$\frac{Z'}{Z} = \frac{dist(\mathbf{x}, \mathbf{v})}{dist(\mathbf{x}', \mathbf{v})} = \mu \quad (82)$$

Therefore, the relative depth of the plane is equal to μ, a parameter of the homology matrix. This relation agrees with the known fact that given the homography matrix induced by a 3D plane in two views, the relative distance between the camera centers and the plane is equal to the determinant of the homography [54, 55].

This is valid for homographies (correctly scaled), thus also for planar homologies. From equation (77), note that $\det(G) = \mu$, and as the distance between the camera center and the plane is the depth of the plane, (82) is again verified.

Calculating Relative Depth for Horizontal Planes

This section describes the process to calculate the depth ratio, in two views related by a pure translation, of a 3D plane parallel to the two image planes. This process exploits (82) in a practical implementation.

First, the images are projected on the virtual leveled plane, and pixel correspondences are established. Then an initial FOE estimate $\mathbf{v}_0$ is obtained from the pixel correspondences using a robust linear estimation with RANSAC (Random Sample Consensus) [56] followed by an optimization step [57]. The FOE estimation also excludes outliers from the set of corresponding pixel pairs, as not to hamper the final optimization.

From the pixel correspondences and the FOE estimate, an initial estimate μ_0 of the cross-ratio parameter μ is obtained by measuring and averaging for all pairs of corresponding pixels the ratios of image distances to the FOE as in (82).

Given the initial estimates $\mathbf{v}_0$ and μ_0, an optimization routine minimizes the projection error of the pixel correspondences when projected by the homology $G(\mathbf{v}, \mu, \mathbf{a} = [0, 0, 1]^T)$, finding improved estimates for $\mathbf{v}$ and μ. The error metric is Sampson distance, also used to estimate full homographies [50]. The optimization is performed by the Levenberg-Marquardt algorithm, using the same implementation used to estimate full homographies [58], but parameterized by the homology parameters $\mathbf{v}$ and G. As it is advisable to over-parameterize the optimization [50], $\mathbf{v}$ is considered as a 3-element vector, which is normalized once its final value is known. The relative depth is the determinant of G, i.e., μ.

Figure 17 summarizes this process. Notice that there is no need to project all the image on the virtual plane, but only the coordinates of the pixel correspondences. Sensor data could provide directly an initial FOE estimate. The initial μ estimate is trivial, and the final optimization roughly takes as much time as the optimization necessary to calculate an homography. Therefore, potentially this process can be fast enough for robotic applications.

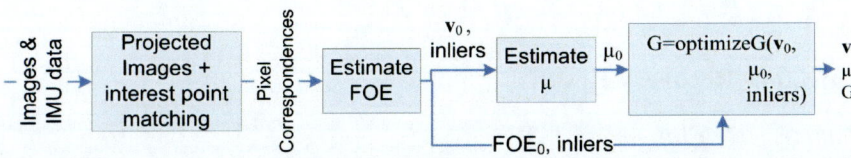

Figure 17. Finding the homology transformation between two warped images.

Reconstructing the Complete Relative Pose

The relative depth between the two virtual cameras has been recovered in section 4.1.1.. As the rotation is compensated, the virtual cameras may be represented by $P = [I|\mathbf{0}]$ and $P' = [I|\mathbf{t}]$. The relative height corresponds to the z component of the vector $\mathbf{t}$, although its scale depends on the height of the first camera.

The FOE, that is already calculated by the above process, is the direction of the other two components of $\mathbf{t}$, although it does not indicate the scale. The correct scale of translation must come from another independent measurement. Nevertheless, the scale of the x and y components given by the FOE may be calculated in function of the altitude component.

Given the FOE $\mathbf{v} = (v_x, v_y, 1)^T$, the nadiral point $\mathbf{N} = (N_x, N_y, 1)^T$, and the camera intrinsic parameters focal length f and pixel size dpx and dpy, in the x and y directions, the vector $\mathbf{t}$ is calculated as a function of its vertical component t_z, as:

$$\mathbf{t} = \begin{bmatrix} \frac{(v_x - N_x) \cdot dpx \cdot t_z}{f} \\ \frac{(v_y - N_y) \cdot dpy \cdot t_z}{f} \\ t_z \end{bmatrix} \qquad (83)$$

This relation is derived from the similar triangles shown in Fig. 18 for the x component, and a similar relation exists for the y component. The figure omits the change of coordinates

(N_x) and units (dpx) that must be applied to v_x. Therefore, the trajectory of a mobile observer may be reconstructed up to scale, as in the usual homography model, by adding the relative pose vectors over an image sequence.

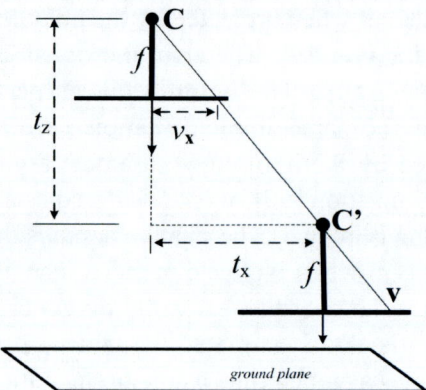

Figure 18. Finding the scale of translation from the difference on height.

4.1.2. Trajectory Recovery Results

Visual Odometry: Heights for an UAV

This experiment uses images taken by the remotely controlled airship of Fig. 13 carrying the IMU-camera system and a Global Positioning System (GPS) receiver, flying over a planar area. The GPS measured height is shown in Fig. 19 compared against visual odometry based on the μ value of homologies calculated for the sequence of images using the process described in section 4.1.1..

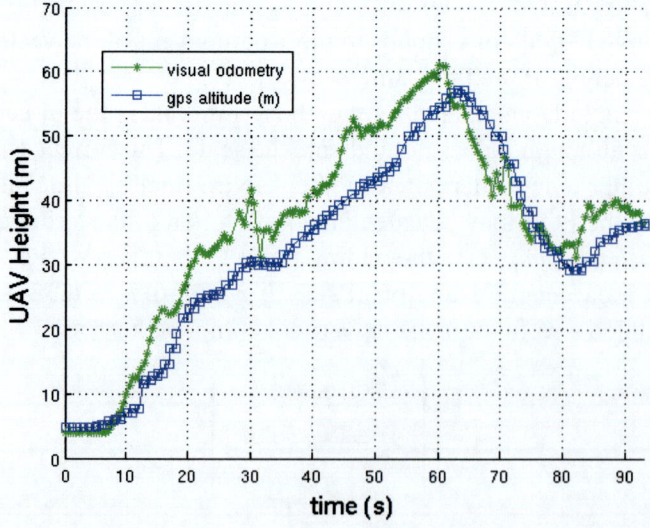

Figure 19. Visual odometry based on homology compared with GPS altitude measurements.

For the first image the height is manually set as $h_1 = 4m$, and for the ith image the height is $h_i = \left(\prod_{j=1}^{i-1} {}^{j+1}\mu_j\right) \cdot h_1$, where ${}^{j+1}\mu_j$ represents the cross-ratio for the homology that transforms the jth image into the image $j + 1$. For a few image pairs, the homology could not be calculated due to corrupted or missing images, or because too few interesting points were matched. On this case the last valid μ value is assumed to be the current one. Except by that, no other attempt is made to filter the data to avoid the drift from successive multiplication of relative heights. The scale of the visual odometry depends on the manually set height of the first image. As GPS altitude measurements are not very accurate, the comparison mainly demonstrates the existence of a correlation.

UAV Trajectory from Relative Heights and FOE

As exposed on section 4.1.1., for an image pair projected on the leveled plane, the camera projection matrices may be written as $P = [I|0]$ and $P' = [I|\mathbf{t}]$, as the rotation has been compensated. Therefore, by recovering $\mathbf{t}$ for each image pair on the sequence, the UAV trajectory can be reconstructed by adding the sequence of translation vectors.

Given an image pair I_i, I_{i+1}, the vertical component of the vector $\mathbf{t}$ is calculated from the heights computed above as $t_z = h_{i+1} - h_i$. The other two components are given by the direction of the FOE. As the FOE is only a direction without scale, the scale of $\mathbf{t}$ must be given by the known altitude component (83).

During flight, the airship height changes very little between consecutive frames. Thus, the lines connecting corresponding pixels are almost parallel. Therefore, when calculating the FOE from the correspondences, the direction recovered may be the direction opposite to the real FOE. To solve this ambiguity, the measured FOE is compared with the vehicle heading indicated by the IMU. If the FOE points are behind the vehicle, the measurement is inverted, i.e., $\mathbf{v} = -\mathbf{v}_{measured}$. The airship is never moving backwards in our data sets.

The trajectory is thus reconstructed for the same UAV data set used above. The height data are the same as in Fig. 19, but the other two dimensions are interpolated for the images where the homology calculation failed (indicated with red squares in Fig. 20).

The blue squares show the trajectory reconstructed by adding up all translation vectors, with no filtering applied. The pink diamonds show the smoother trajectory reconstructed after applying a Kalman filter on the translation measurements. Both 2D and 3D plots of the same data are provided.

4.2. Digital Elevation Map from Image Correspondences

In this section we further explore the relation between scene depths and image ratios with the FOE for pixel correspondences [53]. For rotation compensated images projected on the horizontal ground plane, scene depth indicates height, and is used to build a coarse Digital Elevation Map (DEM) grid, performing 3D mapping from monocular aerial images.

4.2.1. Calculating Height For Each Pixel Correspondence

Recall that (78) is valid for each individual corresponding pixel pair. Therefore, if an image contains regions above the ground plane, the relative height can be directly recovered for

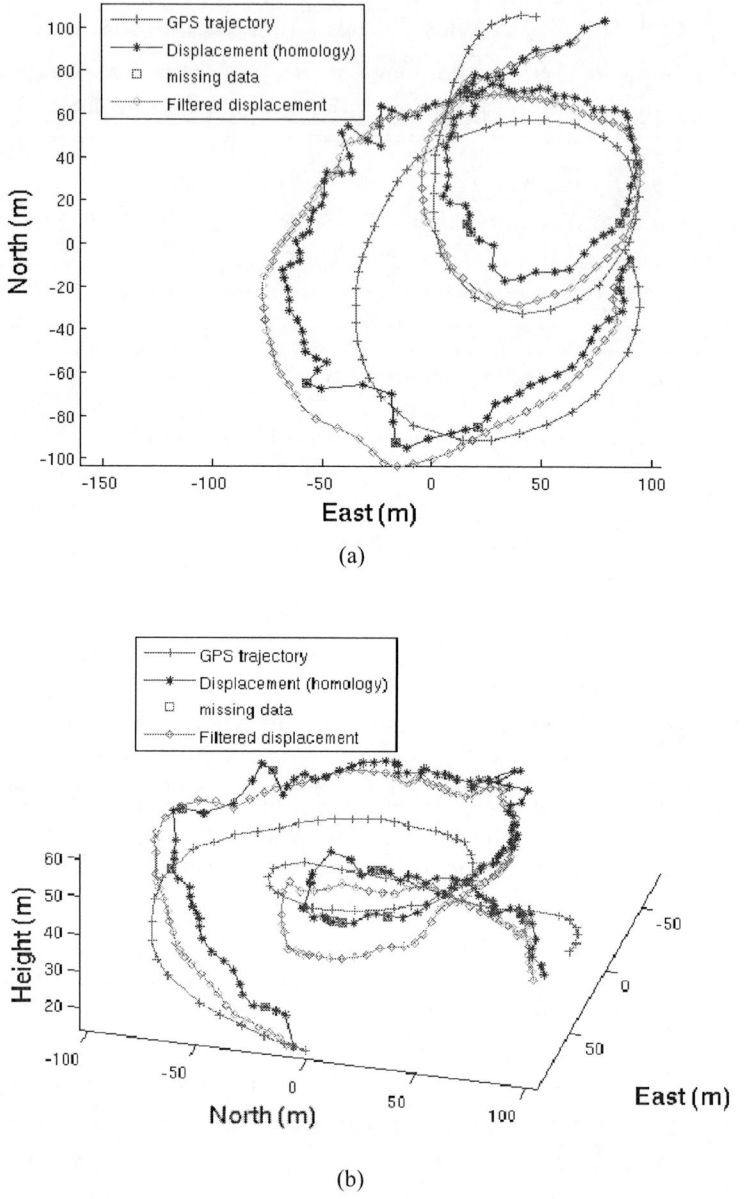

Figure 20. 2D and 3D plots comparing visual odometry based on homology compared with GPS trajectory measurements.

each corresponding pixel. This suffices to order these pixels by their height, but it is not an absolute measurement. Again, additional information is needed to recover scale from imagery. The absolute height of these points may be recovered if the absolute height on both views is known – case (a), or, equivalently, if the height of one view and relative height corresponding to the ground plane is known – case (b).

In case (a), defining $\mu_i = \frac{dist(\mathbf{x}_i, \mathbf{v})}{dist(\mathbf{x}'_i, \mathbf{v})} = \frac{Z'_i}{Z_i} = \frac{h'-hp_i}{h-hp_i}$ as the relative height for the corresponding pixel pair $(\mathbf{x}_i, \mathbf{x}'_i)$, the height of the 3D point $\mathbf{X}_i$ imaged by $\mathbf{x}_i$ as hp_i, and h and h' as the known camera heights, as shown in Fig. 21, and then solving for hp_i, we have:

$$hp_i = \frac{\mu_i h - h'}{\mu_i - 1} \qquad (84)$$

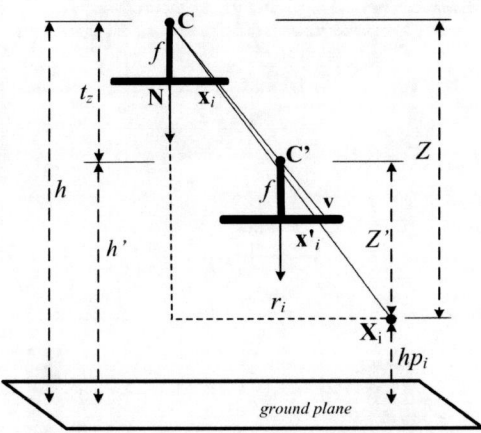

Figure 21. Calculating the height of a 3D point from an image pair under pure translation.

In case (b), we substitute $h' = \mu h$, where μ is the relative camera height over the ground plane, into (84) to obtain:

$$hp_i = \frac{(\mu_i - \mu)h}{\mu_i - 1} \qquad (85)$$

Supposing that the ground plane is visible in the majority of the image, the term $(\mu_i - \mu)$ will be used to compensate for errors in the IMU orientation measurements. An image wrongly projected on the ground plane due to measurement error results on a deviation on the image ratios calculated for the pixel correspondences, related to the difference between the real ground plane orientation and the one given by the orientation measurement. Therefore, given all image ratios for all pixel correspondences, a 3D plane is fitted over the (x_i, y_i, μ_i) triplets, where $\mathbf{x}_i = (x_i, y_i, 1)$ in inhomogeneous image coordinates, using RANSAC to look for the dominant plane on the scene, that should be the ground plane.

This fitted 3D plane is used to compensate for a linear deviation induced by orientation errors, by calculating, for each $\mathbf{x}_i$, the value $(\mu_i - \mu_0(x_i, y_i))$, where $\mu_0(x_i, y_i)$ is the μ value corresponding to the (x_i, y_i) coordinates on the fitted plane, and taking this value as $(\mu_i - \mu)$ on (85).

Figure 22 shows the image ratios calculated before and after correction with a fitting 3D plane. These points are very fast to obtain, as no homography or homology is recovered, and just image ratios are needed for individual points. The plane fitting is a simple model for RANSAC, and potentially it can be skipped if better measurements are available.

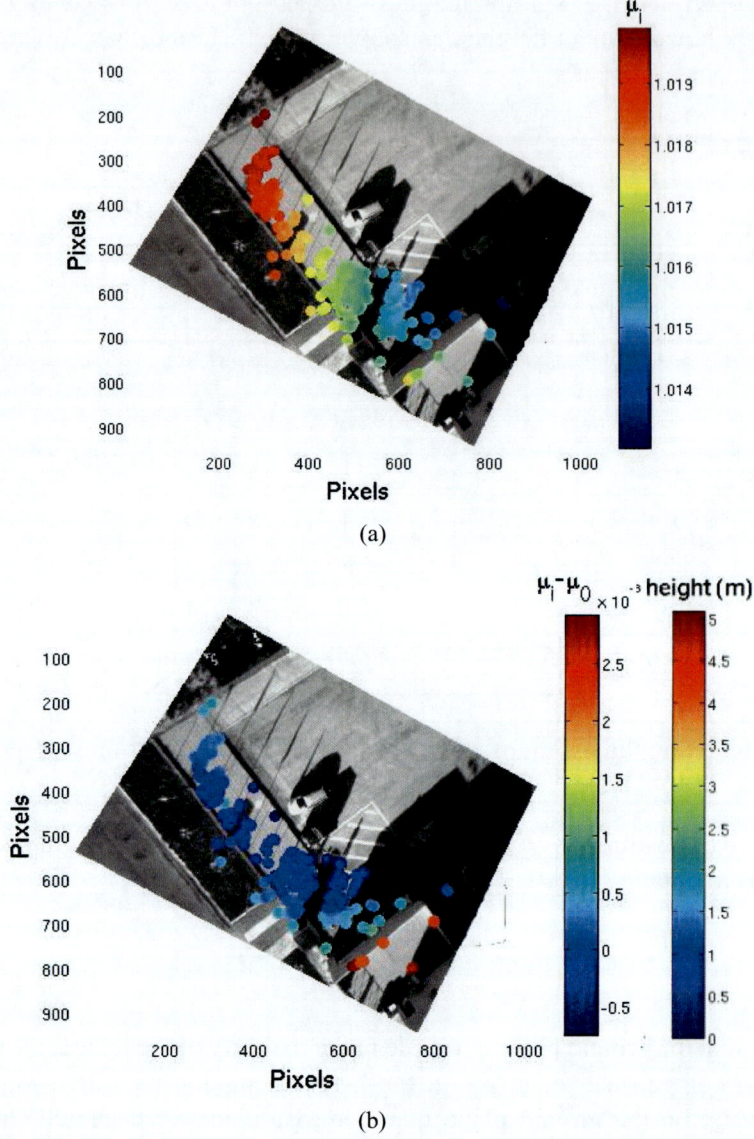

Figure 22. Image ratios before (a) and after (b) compensation by fitting a plane. The points on the building are the only ones above the ground plane.

4.2.2. Constructing a DEM

A DEM is a 2D grid dividing the ground plane into equal square regions called cells, and storing the height of each cell represented as a probability density. A DEM can be constructed from punctual height measurements if the 3D coordinates of each measured point are known. The camera geometry allows us to find the remaining two horizontal coordinates after the height from the ground plane is calculated as exposed in section 4.2.1..

The distance r_i between the principal axis of the virtual downward camera and the 3D point $\mathbf{X}_i = (X_i, Y_i, hp_i)$ may be calculated by similarity of triangles (see Fig. 21) as:

$$r_i = \frac{dist(\mathbf{x}_i, \mathbf{N})}{f}(h - hp_i) \qquad (86)$$

where $\mathbf{x}_i$ is the image projection of $\mathbf{X}_i$ and $\mathbf{N}$ the principal point of the virtual downward camera. The angle θ_i between the line $\overline{\mathbf{x}_i \mathbf{N}}$ and the east axis is directly calculated from their coordinates. Then, transforming (r_i, θ_i) from polar to rectangular coordinates yields (X_i, Y_i), the remaining two components of $\mathbf{X}_i = (X_i, Y_i, hp_i)$, in the local NED frame. If the camera pose in the world frame is known, these coordinates may be registered onto the world frame, incorporating the height measurements in a global DEM.

Each point $\mathbf{X}_i$ represents a height measurement for an individual DEM cell. The height of each cell and the heights measurements are represented by random variables with gaussian distribution, and the cell update follows the Kalman Filter update rule. There is not an initialization and the cell takes its first value when incorporating its first measurement.

The position of each point $\mathbf{X}_i$ has also an uncertainty on the xy axes, i.e., it may be uncertain which cell the measurement belongs to. Currently the influence of measurements on neighboring cells is approximated by considering all measurements exact on the North-East axes, and then smoothing each local DEM with a gaussian kernel with variance similar to the measurements.

Figure 23 shows a DEM constructed over a 10 frame sequence (a $5s$ portion of the flight), with the GPS-measured vehicle localization shown as red stars. The cell size is $3m$, and the gaussian convolution kernel has standard deviation of $2m$. The highest cells correspond to the building, and the smaller blue peaks correspond to the airplanes and vehicles. Points more than $1m$ below the ground plane are discarded as obvious outliers. The mosaicing on the left - included just to allow the reader to compare visually the covered area and the DEM grid - is built from homographies calculated between each successive image pair (see [42, 48] for details).

4.2.3. Applications and Future Improvements

In section 4.1.2. the complete UAV trajectory was reconstructed from rotation compensated imagery. In this section, a 3D map of the environment, in the form of a sparse DEM, is built from punctual height measurements generated from a sequence of images and the vehicle pose. These two techniques assume mutually exclusive conditions: the trajectory recovery requires imaging a planar area, and the mapping procedure is meaningful only if there are obstacles above the ground plane. Therefore, they could not make a complete SLAM (Simultaneous Localization and Mapping) scheme on their own, but they may be complimentary methods to other SLAM approaches. A SLAM scheme also needs to obtain, from the maps built at each step, another displacement measurement to constrain the vehicle pose estimation, by matching local maps with the previous local map or with the global map built so far. A comparison of different approaches for matching 2D occupancy grids built from sonar data indicates that both options are feasible [59], and the DEM grids could be correlated this way.

Figure 22 shows that all pixel correspondences are concentrated into one small portion of the image. Outside this region, there are other correspondences found by the interesting

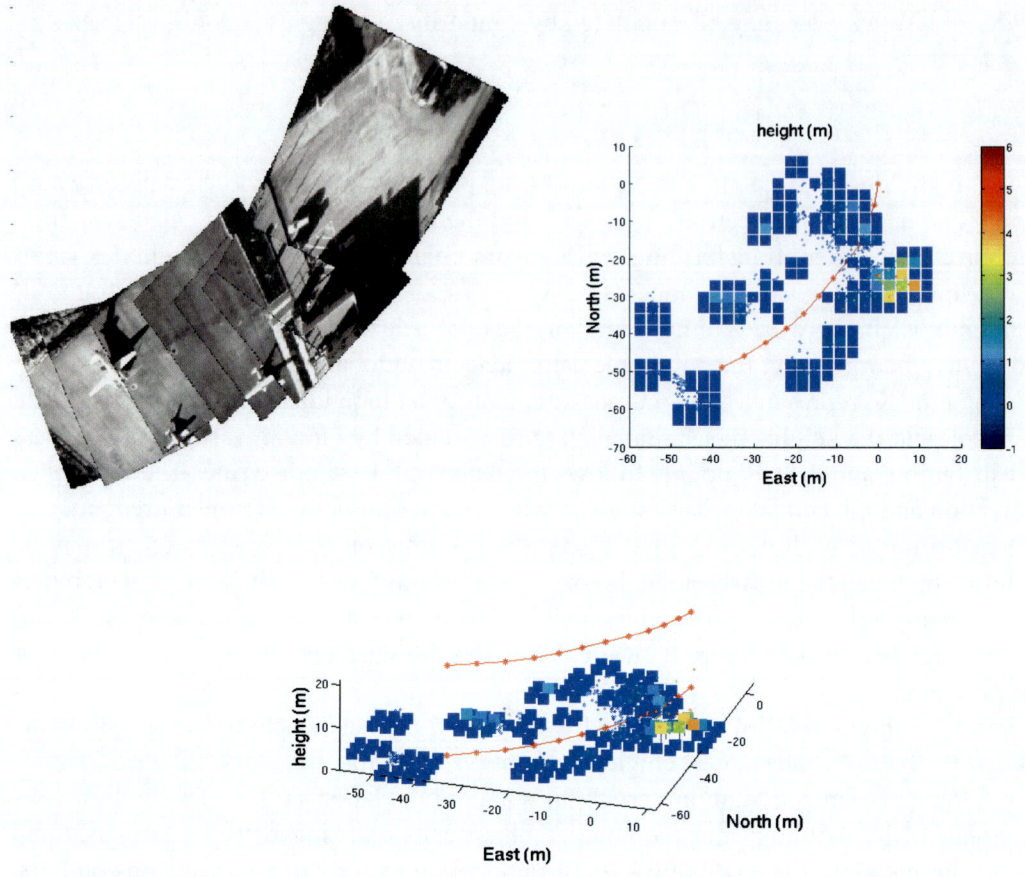

Figure 23. 2D and 3D plots of the DEM generated from the 10 frames mosaiced on the left. The red stars indicate the vehicle trajectory.

point matching algorithm, but they are considered outliers during the FOE estimation and rejected, probably due to errors on the orientation measurement. For the estimation of the homology, and specially of its cross ratio used on the height component of the trajectory estimation, usually there are enough correspondences for a reliable estimation. But for the mapping problem, the area covered by height measurements is quite smaller than the total area imaged (see Fig. 23), and thus the map becomes too sparse for a reliable correlation between the DEM grid maps. This problem could be addressed by improving the accuracy of the orientation measurements, by increasing the image frame rate, or by developing models that take into account small uncompensated rotations. A Technical Report is available with more details on the theory and usage of the homology model, and on the mosaicing procedure [48].

5. Conclusion

This chapter introduces the Portuguese Project DIVA, which focuses on the establishment of the technologies required to substantiate autonomous operation of unmanned robotic airships for environmental monitoring and aerial inspection missions. The research is at the moment divided in two pathways - guidance and control, and vision perception - for which significant results are presented.

we have compiled a list of requirements for a system to be able to collect the data sets used on this chapter and also to perform automatic flight tests, even if we have not satisfied all these requirements ourselves, neither performed automatic flight. This list has evolved through various years and two different projects and should provide an overall road map to similar endeavors. A short overview of the system developed so far is also provided.

A novel approach for the airship automatic control problem is described. The main outcome in this area is the resulting asymptotically stable backstepping control law which takes into account input saturations and bounded wind estimation errors. Important implementation issues are considered in the design process, namely control allocation and reference shaping to deal with airship underactuation. The simulation results obtained for a representative mission covering all the usual tasks like take-off and landing, stabilization and path-tracking, illustrate the controller performance even in the presence of realistic wind disturbances. The presented single global control law proves to be a strong and robust solution for the automatic guidance and control of an airship, valid over a wide flight envelope ranging from hover to aerodynamic flight.

The complete UAV trajectory was reconstructed from rotation compensated imagery. The reconstructed trajectory has visible errors and drift in the long term, but it may be useful in the case of temporary GPS dropout. The process is relatively fast, and can be made more accurate by having more accurate orientation measurements or incorporating other sensors to measure the speed of the vehicle, i.e., the direction of the FOE. A 3D map of the environment, in the form of a sparse DEM, is built from punctual height measurements generated from a sequence of images and the vehicle pose. After interesting point matching is performed, the mapping process is very fast, measuring the height for each corresponding pixel pair with just a simple calculation. These two techniques may be useful into a SLAM context, although they can not build a complete SLAM scheme on their own.

Autonomous flight and visual perception are two key issues to be addressed in order to define an aerial surveillance system. The approaches currently under development in project DIVA have been presented in this chapter and are gradually being validated and integrated in the airship prototype.

References

[1] Cox, T. H., Nagy, C. J., Skoog, M. A., and Somers, I. A. (2004) Civil UAV capability assessment - draft version. *NASA - Dryden Flight Research Center*, http://www.nasa.gov/centers/dryden/pdf/111760main_UAV_Assessment_Report_Overview.pdf.

[2] Cambone, S. A., Krieg, K. J., Pace, P., and Wells II, L. (2005) Unmanned aircraft systems roadmap 2005 2030. *Department of Defense, USA*, http://www.acq.osd.mil/usd/Roadmap%20Final2.pdf.

[3] Bueno, S. S. et al. (2002) Project AURORA: Towards an autonomous robotic airship. *Workshop on Aerial Robotics, IEEE International Conference on Intelligent Robots and Systems*, Lausanne, Switzerland, October, pp. 43–54.

[4] Elfes, A., Bueno, S. S., Bergerman, M., and Ramos, J. J. G. (1998) A semi-autonomous robotic airship for environmental monitoring missions. *IEEE International Conference on Robotics and Automation - ICRA'98*, Leuven, Belgium, May, pp. 3449–3455.

[5] Wimmer, D.-A., Bildstein, M., Well, K. H., Schlenker, M., Kungl, P., and Kroplin, B.-H. (2002) Research airship "Lotte": Development and operation controllers for autonomous flight phases. *Workshop on Aerial Robotics, IEEE International Conference on Intelligent Robots and Systems*, Lausanne, Switzerland, October, pp. 55–68.

[6] Hygounenc, E. and Soueres, P. (2002) Automatic airship control involving backstepping techniques. *Proceedings of the IEEE International Conference on Systems, Man and Cybernetics*, Hammamet, Tunisia, October.

[7] Hygounenc, E., Jung, I.-K., Soures, P., and Lacroix, S. (2004) The autonomous blimp project of LAAS-CNRS: Achievements in flight control and terrain mapping. *The International Journal of Robotics Research*, **23**, 473–511.

[8] Beji, L. and Abichou, A. (2005) Tracking control of trim trajectories of a blimp for ascent and descent flight manoeuvres. *Int. J. Control*, **78**, 706–719.

[9] Cox, T. H., Nagy, C. J., Skoog, M. A., and Somers, I. A. (2002) STWing-SEAS of University of Pensilvania in partnership with the EnviroBLIMP project of CMU. *Carnegie Mellon University*, http://www.stwing.org/blimp/ and http://www.frc.ri.cmu.edu/projects/enviroblimp/.

[10] de Paiva, E. C., Azinheira, J. R., Josué G. Ramos, J., Moutinho, A., and Bueno, S. S. (2006) Project AURORA: Infrastructure and flight control experiments for a robotic airship. *Journal of Field Robotics*, **23**, 201–222.

[11] Tan, S. B. and Nagabhushan, B. L. (1997) Robust heading-hold autopilot for an advanced airship. *Proceedings of the 12th AIAA Lighter-Than-Air Technology Conference*, July.

[12] de Paiva, E. C., Bueno, S. S., Gomes, S. B. V., Ramos, J. J. G., and Bergerman, M. (1999) A control system development environment for AURORA's semi-autonomous robotic airship. *Proceedings of the IEEE International Conference on Robotics and Automation*, Detroit, USA, May, vol. 3, pp. 2328–2335.

[13] Ramos, J. J. G., de Paiva, E. C., Azinheira, J. R., Bueno, S. S., Maeta, S. S., Mirisola, L. G. B., and Bergerman, M. (2001) Autonomous flight experiment with a robotic unmanned airship. *IEEE International Conference on Robotics and Automation - ICRA'2001*, Seoul, Korea, May, pp. 4152–4157.

[14] Azinheira, J. R., Moutinho, A., and de Paiva, E. C. (2006) Airship hover stabilization using a backstepping control approach. *Journal of Guidance, Control, and Dynamics*, **29**, 903–914.

[15] Slotine, J.-J. E. and Li, W. (1991) *Applied Nonlinear Control*. Prentice-Hall.

[16] Healey, A. J. and Lienard, D. (1993) Multivariable sliding mode control for autonomous diving and steering of unmanned underwater vehicles. *IEEE Journal of Oceanic Engineering*, **18**.

[17] Enns, D., Bugajski, D., Hendrick, R., and Stein, G. (1994) Dynamic inversion: an evolving methodology for flight control design. *International Journal of Control*, **59**, 71–91.

[18] Moutinho, A. and Azinheira, J. R. (2005) Stability and robustness analysis of the AURORA airship control system using dynamic inversion. *Proceedings of the IEEE International Conference on Robotics and Automation*, Barcelona, Spain, April.

[19] Khalil, H. K. (2002) *Nonlinear Systems*, chap. 14, pp. 589–603. Prentice-Hall, third edn.

[20] Teel, A. R. (1992) Global stabilization and restricted tracking for multiple integrators with bounded controls. *Systems & Control Letters*, **18**, 165–171.

[21] Kim, K.-S. and Kim, Y. (2003) Robust backstepping control for slew maneuver using nonlinear tracking function. *IEEE Transactions on Control Systems Technology*, **11**, 822–829.

[22] Frazzoli, E., Dahleh, M. A., and Feron, E. (2000) Trajectory tracking control design for autonomous helicopters using a backstepping algorithm. *Proceedings of the American Control Conference*, Chicago, Illinois, USA, June, pp. 4102–4107.

[23] Beji, L., Abichou, A., and Bestaoui, Y. (2002) Stabilization of a nonlinear underactuated autonomous airship - a combined averaging and backstepping approach. *Proceedings of the Third International Workshop on Robot Motion and Control*, Bukowy Dworek, Poland, November, pp. 223–229.

[24] Freeman, R. and Praly, L. (1998) Integrator backstepping for bounded controls and control rates. *IEEE Transactions on Automatic Control*, **43**, 258–262.

[25] Metni, N., Hamel, T., and Derkx, F. (2004) A UAV for bridge's inspection: Visual servoing control law with orientation limits. *Proceedings of the 5th IFAC/EURON Symposium on Intelligent Autonomous Vehicles*, Lisbon, Portugal, July.

[26] Caballero, F., Merino, L., Ferruz, J., and Ollero, A. (2006) Improving vision-based planar motion estimation for unmanned aerial vehicles through online mosaicing. *IEEE Int. Conf. on Robotics and Automation (ICRA06)*, Orlando, FL, USA, May, pp. 2860–2865.

[27] Saripalli, S., Montgomery, J., and Sukhatme, G. (2003) Visually-guided landing of an unmanned aerial vehicle. *IEEE Transactions on Robotics and Automation*, **19**, 371–381.

[28] Corke, P., Sikka, P., and Roberts, J. M. (2001) Height estimation for an autonomous helicopter. *ISER '00: Experimental Robotics VII*, London, UK, pp. 101–110, Springer-Verlag.

[29] Hygounenc, E., Jung, I.-K., Soueres, P., and Lacroix, S. (2004) The Autonomous Blimp Project at LAAS/CNRS: Achievements in Flight Control and Terrain Mapping. *Int. J. of Robotics Research*, **23**, 473–512.

[30] Huguet, A. B., Carceroni, R. L., and de A. Arajo, A. (2003) Towards automatic 3d reconstruction of urban scenes from low altitude aerial images. *IEEE 12th Int. Conf. on Image Analysis and Processing (ICIAP '03)*, Mantova, Italy, September.

[31] Cunningham, D., Grebby, S., Tansey, K., and Gosar, A. (2006) Application of airborne LiDAR to mapping seismogenic faults in forested mountainous terrain, southeastern Alps, Slovenia. *Geophysical Research Letters*, **33**.

[32] Tupin, F. and Roux, M. (2005) Markov random field on region adjacency graph for the fusion of sar and optical data in radargrammetric applications. *IEEE Trans. on Geoscience and Remote Sensing*, **43**, 1920–1928.

[33] Schmidt, D. C. (2007), TAO - The ACE ORB. http://www.cs.wustl.edu/~schmidt/TAO.html.

[34] Gomes, S. B. V. and Ramos, J. G. (1998) Airship dynamic modeling for autonomous operation. *Proceedings of the IEEE International Conference on Robotics and Automation*, Leuven, Belgium, May, pp. 3462–3467.

[35] Azinheira, J. R., de Paiva, E. C., and Bueno, S. S. (2002) Influence of wind speed on airship dynamics. *Journal of Guidance, Control, and Dynamics*, **25**, 1116–1124.

[36] Thomasson, P. G. (2000) Equations of motion of a vehicle in a moving fluid. *Journal of Aircraft*, **37**, 630–639.

[37] Stevens, B. L. and Lewis, F. L. (1992) *Aircraft Control and Simulation*. John Wiley and Sons, Inc.

[38] de Paiva, E. C. and Azinheira, J. R. (2001) Airship AS800 dynamic model and simulation. Internal report, CenPRA, Brazil and IDMEC-IST / TULisbon, Portugal.

[39] Lobo, J. and Dias, J. (2007) Relative pose calibration between visual and inertial sensors. *International Journal of Robotics Research*, (in press).

[40] Lobo, J., Ferreira, J. F., and Dias, J. (2006) Bioinspired visuo-vestibular artificial perception system for independent motion segmentation. *ICVW06 (2nd Int. Cognitive Vision Workshop)*, Graz, Austria, May.

[41] Mirisola, L. G. B., Lobo, J., and Dias, J. (2006) Stereo vision 3D map registration for airships using vision-inertial sensing. *The 12th IASTED Int. Conf. on Robotics and Applications (RA 2006)*, Honolulu, HI, USA, August.

[42] Mirisola, L. G. B. and Dias, J. M. M. (2007) Exploiting inertial sensing in mosaicing and visual navigation. *6th IFAC Symposium on Intelligent Autonomous Vehicles (IAV07)*, Toulouse, France, September.

[43] Point Gray Inc. (2006). http://www.ptgrey.com.

[44] XSens Tech. (2006). http://www.xsens.com.

[45] Lobo, J. and Dias, J. (2005) Relative pose calibration between visual and inertial sensors. *ICRA Workshop on Integration of Vision and Inertial Sensors - 2nd InerVis*, Barcelona, Spain, April 18.

[46] Lobo, J. (2006), InerVis Toolbox for Matlab. http://www.deec.uc.pt/~jlobo/InerVis_WebIndex/.

[47] Bouguet, J. (2006), Camera Calibration Toolbox for Matlab. http://www.vision.caltech.edu/bouguetj/calib_doc/index.html.

[48] Mirisola, L. G. B. and Dias, J. M. M. (2007) Exploiting inertial sensing in vision-based mapping and navigation. Tech. rep., Institute of Systems and Robotics, Univ. of Coimbra, Portugal, http://paloma.isr.uc.pt/~lgm/TR-ISR0703.pdf.

[49] Bay, H., Tuytelaars, T., and van Gool, L. (2006) SURF: Speeded Up Robust Features. *the Ninth European Conference on Computer Vision*, Graz, Austria, May.

[50] Hartley, R. and Zisserman, A. (2000) *Multiple View Geometry in Computer Vision*. Cambridge University Press.

[51] Ma, Y., Soatto, S., Kosecka, J., and Sastry, S. (2004) *An Invitation to 3D Vision*. Springer.

[52] van Gool, L., Proesmans, M., and Zisserman, A. (1998) Planar homologies for grouping and recognition. *Image and Vision Computing*, **16**.

[53] Arnspang, J., Henriksen, K., and Bergholm, F. (1999) Relating scene depth to image ratios. Solina, F. and Leonardis, A. (eds.), *Proc. of the 8th Int. Conf. on Computer Analysis of Images and Patterns (CAIP'99)*, Ljubljana, Slovenia, Sep, vol. 1689 of *Lecture Notes in Comp. Science*, pp. 516–525, Springer.

[54] Malis, E., Chaumette, F., and Boudet, S. (1999) 2-1/2-D Visual Servoing. *IEEE Trans. on Robotics and Automation*, **15**, 238–250.

[55] Malis, E. (1998) *Contributions ' la modilisation et ' la commande en asservissement visuel*. Ph.D. thesis, L'Universiti de Rennes, Icole de Informatique, Traitement du Signal et Telecommunications, Rennes, France.

[56] Fischler, M. A. and Bolles, R. C. (1981) Random sample consensus: A paradigm for model fitting with applications to image analysis and automated cartography. *Comm. of the ACM*, **24**, 381–395.

[57] Chen, Z., Pears, N., McDermid, J., and Heseltine, T. (2003) Epipole estimation under pure camera translation. Sun, C., Talbot, H., Ourselin, S., and Adriaansen, T. (eds.), *DICTA*, pp. 849–858, CSIRO Publishing.

[58] Capel, D., Fitzgibbon, A., Kovesi, P., Werner, T., Wexler, Y., and Zisserman, A. (2006), Matlab functions for multiple view geometry. `http://www.robots.ox.ac.uk/~gg/hzbook/code/`.

[59] Schiele, B. and Crowley, J. L. (1994) A comparison of position estimation techniques using occupancy grids. *Int. Conf. on Robotics and Automation (ICRA94)*, San Diego,CA, USA, May, pp. 1628– 1634.

Chapter 3

AN ARCHITECTURE FOR ADAPTIVE SWARMS

Suranga Hettiarachchi, Paul Maxim and William M. Spears
University of Wyoming, Laramie, WY 82071, USA

Abstract

The focus of our research is to design and build rapidly deployable, adaptive, cost-effective, and autonomous distributed robot swarms. Our objective is to provide a scientific, yet practical, approach to the design and analysis of swarm behaviors. This chapter provides an overview of our work in this area. First, we summarize the basis for our robot control algorithms, which we call *artificial physics* or *physicomimetics*. Unlike biomimetic approaches, we focus on robotic behaviors that are similar to those shown by solids, liquids, and gases. Solid formations are useful for distributed sensing tasks, while liquids are for obstacle avoidance tasks. Gases are practical for coverage tasks, such as surveillance and sweeping. Physicomimetics is scalable, robust, and fault-tolerant.

Despite the fact that physicomimetics is amenable to theoretical analyses that guide its use, the fact remains that a real-world environment will often have unanticipated qualities that hurt the performance of the robot swarm. Hence, we also describe our novel technique for adaptive swarms. Unlike prior off-line approaches that attempt to re-train the behavior of the swarms in a simulation environment, our on-line approach adapts the behavior of the swarm in real time, while the swarm is performing the task.

In order to function properly the robots in the swarm must be able to accurately localize their local neighbors and to share information. Hence we also outline our enabling hardware technology for swarms of robots. Our plug-in hardware module provides the capability to accurately localize neighboring robots, without using global information and/or the use of vision systems. It also couples localization with data exchange, allowing physicomimetics and adaptation to be fully integrated onto physical robots.

1. Introduction

The focus of our research is to design and build rapidly deployable, adaptive, cost-effective, and autonomous distributed robot swarms. Our objective is to provide a scientific, yet practical, approach to the design and analysis of swarm behaviors.

The team of robots could vary widely in type, as well as size, e.g., from nanobots to micro-air vehicles (MAVs) and micro-satellites. A robot's sensors perceive the world, including other robots, and a robot's effectors make changes to that robot and/or the world, including other robots. It is assumed that robots can only sense and affect nearby robots; thus, a key challenge has been to design "local" control rules. Not only do we want the desired global behavior to emerge from the local interaction between robots (self-organization), but we also require fault-tolerance, that is, the global behavior degrades very gradually if individual robots are damaged. Self-repair is also desirable, in the event of damage. Self-organization, fault-tolerance, and self-repair are precisely those principles exhibited by natural physical systems. Thus, many answers to the problems of distributed control can be found in the natural laws of physics.

This chapter provides an overview of our framework for distributed control, called "physicomimetics" or "artificial physics" (AP). We use the term "artificial" (or virtual) because although we are motivated by natural physical forces, we are not restricted to them [1]. Although the forces are virtual, robots act as if they were real. Thus the robot's sensors must see enough to allow it to compute the force to which it is reacting. The robot's effectors must allow it to respond to this perceived force.

There are two potential advantages to this approach. First, in the real physical world, collections of small entities yield surprisingly complex behavior from very simple interactions between the entities. Thus there is a precedent for believing that complex control is achievable through simple local interactions. This is required for very small robots, since their sensors and effectors will necessarily be primitive. Second, since the approach is largely independent of the size and number of robots, the results scale suitably to larger robots and larger sets of robots.

In this chapter we focus on the application of physicomimetics to swarms of robots moving through obstacle fields [2]. Our objective was two-fold. Prior research in this area has generally focused either on a small number of robots moving through a large number of obstacles, or a large number of robots moving through a small number of obstacles [4, 3]. However, the more difficult task of moving a large number of robots in formation through a large number of obstacles is generally not addressed. Also, proposed metrics of performance are not complete, ignoring criteria such as the number of collisions between robots and obstacles, the distribution in time of the number of robots that reach the goal, and the connectivity of the formation as it moves. Hence, one objective was to provide a more complete set of metrics from which meaningful comparisons could be made. Second, we used these metrics, coupled with a more complete experimental methodology, to examine (a) different strategies for performing the task, and (b) trade-offs between different criteria.

Two other issues must be addressed before swarms of robots can be successfully deployed. First, on-line learning must occur, when unanticipated events are encountered in the real world. Second, a rapid and accurate robot localization technology is required, in order for the robots to see their local neighbors and to share information with them. This chapter also summarizes our work in these two areas.

2. The Artificial Physics Framework

In our artificial physics framework, virtual physics forces drive a swarm robotics system to a desired configuration or state. The desired configuration is one that minimizes overall system potential energy, and the system acts as a molecular dynamics ($\vec{F} = m\vec{a}$) simulation.

Each robot has position $\vec{p}$ and velocity $\vec{v}$. We use a discrete-time approximation to the continuous behavior of the robots, with time-step Δt. At each time step, the position of each robot undergoes a perturbation $\Delta \vec{p}$. The perturbation depends on the current velocity, i.e., $\Delta \vec{p} = \vec{v}\Delta t$. The velocity of each robot at each time step also changes by $\Delta \vec{v}$. The change in velocity is controlled by the force on the robot, i.e., $\Delta \vec{v} = \vec{F}\Delta t/m$, where m is the mass of that robot and $\vec{F}$ is the force on that robot. F and v denote the magnitude of vectors $\vec{F}$ and $\vec{v}$. A frictional force is included, for self-stabilization.

From the start, we intended to have our framework map easily to physical hardware, and our model reflects this design philosophy. Having a mass m associated with each robot allows our simulated robots to have momentum. Robots need not have the same mass. The frictional force allows us to model actual friction, whether it is unavoidable or deliberate, in the real robotic system. With full friction, the robots come to a complete stop between sensor readings and with no friction the robots continue to move as they sense. The time step Δt reflects the amount of time the robots need to perform their sensor readings. If Δt is small, the robots get readings very often whereas if the time step is large, readings are obtained infrequently. We have also included a parameter F_{max}, which provides a necessary restriction on the acceleration a robot can achieve. Also, a parameter V_{max} restricts the maximum velocity of the robots (and can always be scaled appropriately with Δt to ensure smooth path trajectories).

3. Hexagonal Lattice Sensing Grids

In prior work we have shown how AP can be applied to self-organize swarms of robots into hexagonal lattices [1], while they move toward a goal [5, 6]. In order to accomplish this, robots must be able to sense the range and bearing to nearby robots, as well as the goal location (Figure 1). All movement is controlled via the $\vec{F} = m\vec{a}$ control law.

In this chapter we compare two different force laws in the context of moving formations of robots through obstacle fields. The first has been used in our prior work and is a generalization of the "Newtonian" gravitational force law to include both attraction and repulsion. The force law is:

$$F_{i,j} = \frac{m_i m_j G}{r^p} \tag{1}$$

$F \leq F_{max}$ is the magnitude of the force between two robots i and j, and r is the distance between the two robots. The masses of the robots are denoted as m_i and m_j, and are assumed to be set to 1.0 in this chapter. The variable G affects the strength of the force. The variable p is a user-defined power that controls the reduction in strength with distance. The force is repulsive if $r < R$, attractive if $r > R$, and is zero beyond a certain range (e.g., 1.5R), to enforce the local nature of the force law. R is the desired separation between that robot and neighboring robots. In order to achieve optimal behavior, the values of G, p, and F_{max}

Figure 1. Seven robots form a hexagon and move towards a light source.

must be determined as well as the amount of friction. The Newtonian force law generally creates rigid formations that act as solids, even in the presence of sensor and locomotion uncertainty (Figure 1).

In this chapter we are also investigating the utility of a second force law, which is a generalization of the Lennard-Jones (LJ) force law (which models forces between molecules and atoms):

$$F_{i,j} = 24\varepsilon \left[\frac{2dR^{12}}{r^{13}} - \frac{cR^6}{r^7} \right] \quad (2)$$

Again, $F \leq F_{max}$ is the magnitude of the force between two robots, and r is the distance between the two robots. The variable ε affects the strength of the force, while c and d control the relative balance between the attractive and repulsive components. In order to achieve optimal behavior, the values of ε, c, d, and F_{max} must be determined as well as the amount of friction. Our motivation for trying the LJ force law is that (depending on the parameter settings) it can easily model crystalline solid formations, liquids, and gases.

4. The Simulation

Our 2D simulation world is 900×700, and contains a goal, obstacles and robots. Up to a maximum of 100 robots and 100 static obstacles with one static goal are placed in the environment. The goal is always placed at a random position in the right side of the world, while the robots are initialized in the bottom left area. The obstacles are randomly distributed throughout the environment, but are kept 50 units away from the initial location of the robots, to give the robots the opportunity to first get into formation. Each circular obstacle has a radius R_o of 10, and the square shaped goal is 20×20. When 100 obstacles are placed in the environment, roughly 5% of the environment is covered by the obstacles (similar to [3]). The desired separation between robots R is 50, and the maximum velocity V_{max} is 20. Figure 2 shows 40 robots navigating through randomly positioned obstacles. The larger circles are obstacles and the square to the right is the goal. Robots can sense other robots within a distance of $1.5R$, and can sense obstacles within a distance of $R_o + 1$ (the minimum sensing distance). The goal can be sensed at any distance.

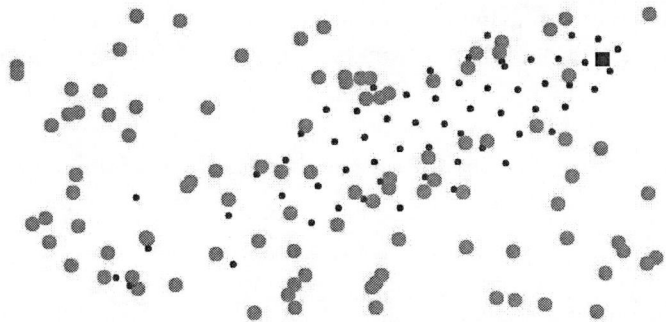

Figure 2. 40 robots moving to the goal. The larger circles represent obstacles, while the square in the upper right represents the goal.

The simulation tool consists of training and performance modules. The training module is used to evolve parameter sets for either the Newtonian or the LJ force laws. The performance module evaluates the optimized force laws with respect to the following metrics: collisions, connectivity, reachability, and time to goal (to be explained later).

4.1. Optimization using Evolutionary Algorithms

Given generalized force laws such as the Newtonian force law or Lennard-Jones force law, it is necessary to optimize the parameters to achieve the best performance. We achieve this task using an Evolutionary Algorithm (EA). EAs are optimization algorithms inspired by natural evolution. We mutate and recombine a population of candidate solutions (individuals) based on their performance in our environment. One of the major reasons for using this population-based stochastic algorithm is that it quickly generates individuals that have robust performance. Every individual in the population is a vector of L real-valued parameters, representing an instantiation of either the Newtonian or LJ force law (depending on the force law being optimized).

In addition to friction, the evolving parameters of the Newtonian force law are:

- G - gravitational constant of robot-robot interactions,

- p - power of the force law for robot-robot interactions,

- F_{max} - maximum force of robot-robot interactions,

and similar 3-tuples for obstacle/goal-robot interactions. The evolving parameters of the LJ force law are:

- ε - strength of the robot-robot interactions,
- c - non-negative attractive robot-robot parameter,
- d - non-negative repulsive robot-robot parameter,
- F_{max} - maximum force of robot-robot interactions,

and similar 4-tuples for obstacle/goal-robot interactions.

Offspring are generated using one-point crossover with a crossover rate of 60%. Mutation adds/subtracts an amount drawn from a $N(0, \delta)$ Gaussian distribution. Each parameter has a $1/L$ probability of being mutated. Mutation ensures that parameter values stay within accepted ranges.

Since we are using an EA that minimizes, the performance of an individual is measured as a weighted sum of penalties:

$$Perf = w_1 \times P_{Collision} + w_2 \times P_{NoCohesion} + w_3 \times P_{NotReachGoal}$$

The weighted fitness function consists of three components: a penalty for collisions, a penalty for lack of cohesion, and a penalty for robots not reaching the goal. Since there is no safety zone around the obstacles [3], a penalty is added to the score if the robots collide with obstacles. The cohesion penalty is derived from the fact that in a good hexagonal lattice, interior robots should have six local neighbors. A penalty occurs if a robot has more or less neighbors. If no robot reaches the goal within the time limit, a penalty occurs.

4.2. Performance Metrics

After optimization, the best force laws are evaluated with our performance module. The performance module consists of four metrics:

- Collisions: the number of robots colliding with obstacles. We consider such robots to be damaged, but they can still move with the formation.

- Swarm connectivity: the maximum number of robots in the swarm that are connected via a communication path. Two robots are connected if their separation is $\leq 1.5R$.

- Reachability: the percentage of robots that reach the goal. A robot has reached the goal if it is within $4R$ distance of the goal.

- Time to goal: the amount of time taken by the last robot to reach the goal.

The importance of the collision, connectivity, and time to goal metrics is obvious. We also consider connectivity, since this is an important metric for the quality of a swarm of robots acting as a sensor grid. The connectivity result we will provide is the minimum size of the largest connected swarm as the swarm moves to the goal. Although each metric provides useful information, a more complete picture arises by considering all.

Table 1. Summary of results for 100 obstacles with 40 to 100 robots.

	Newtonian Force Law				LJ Force Law			
Number of Robots	40	60	80	100	40	60	80	100
Number of Collisions	0	1	2	4	0	2	4	5
Connectivity	27	60	80	100	23	37	52	64
Reachability%	28	0	0	0	98	97	97	97
Time to Goal	1500	-	-	-	940	970	990	1180

4.3. Simulation Results: Solid and Fluid Behaviors

Both Newtonian and LJ force laws were evolved using our training module. The population size was 100 and the EA was run for 100 generations. We first trained over scenarios with 15 robots and 50 obstacles. Each individual (an instance of the force law) was evaluated for 1500 time steps, averaged over 50 random instantiations of the environment. However, the resulting optimized force laws did not scale well to higher numbers of robots and/or obstacles. Training with 100 robots and 100 obstacles was time prohibitive, since the simulation runs in time $O(N^2)$, where N is the total number of robots and obstacles. As a consequence, we settled on a compromise of 40 robots and 90 obstacles.

To measure the performance of the optimized force laws, experiments were carried out with 20 to 100 robots (in increments of 20), and 20 to 100 obstacles (in increments of 20). Each experiment was averaged over 100 runs of different robot, goal and obstacle placements. A '–' entry indicates that the robots did not make it to the goal within the allotted time period.

Table 1 shows a summary of results. It is clear that collisions in both force laws are not a primary concern. Interestingly, the number of obstacles does not appear to be the important factor here, although the number of robots is.

With the Newtonian force law, when there are 40 robots, 28% of the robots reach the goal. However, it is clear that this is achieved by fragmenting the formation into small parts. When there are more than 40 robots, none reach the goal (within the time period). Instead, the structure remains connected, but the strict rigidity of the structure prevents it from making good progress through the obstacle field. It is clear from these results that training with 40 robots does not yield a Newtonian force law that scales to a larger number of robots.

With the LJ force law, 97% or more of the robots make it to the goal in all circumstances. The time to reach the goal increases slowly as the number of obstacles and robots increases (with the number of robots having a larger effect). Finally, swarm connectivity remains reasonably high, ranging from 58% to 65%. Interestingly, swarm connectivity increases as the number of robots increases, and is almost totally unaffected by the number of obstacles. In contrast with the Newtonian force law, the LJ force law (which is also trained with 40 robots) scales well with larger numbers of robots. This provides evidence that the LJ force law is a good model for the swarm behavior that we desire.

Observation of the system behavior shows that the LJ formation acts like a viscous fluid, rather than a solid. Although the formation is not rigid, it does tend to retain much of the hexagonal structure. Deformations and rotations of portions of the fluid are temporary manifestations imposed by the obstacles. Hence, the added flexibility of this formation (over that achieved by the Newtonian force law) has a significant impact on behavior. The optimized LJ force law provides low collision rates, very high goal reachability rates within a reasonable period of time, and high swarm connectivity.

4.4. Discussion and Elaboration

To further analyze our system, we also collected data concerning the change in the connectivity and the percentage of robots reaching the goal over time. The resulting graphs are far too numerous to present here, but we present representative examples. All graphs are averaged over 50 independent runs.

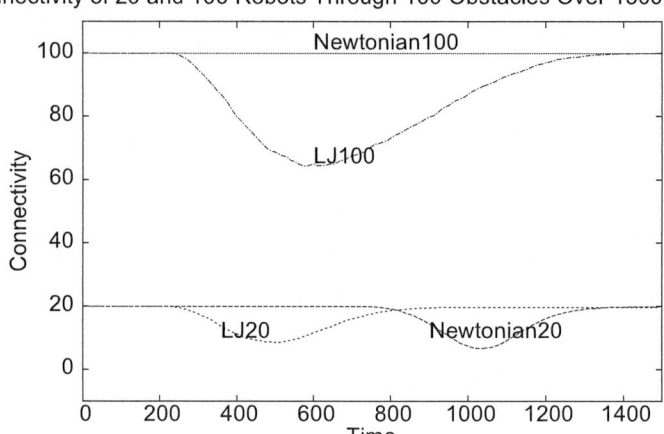

Figure 3. Change in connectivity over 1500 time steps for 20 and 100 robots through 100 obstacles using Newtonian and LJ force laws

Figure 3 illustrates the change in connectivity of the swarm over time. Two sets of results are presented in this graph. The curves at the top are for 100 robots moving through 100 obstacles. The robots controlled by the Newtonian force law remain fully connected (although, as we know from the prior results, this is because the formation has not succeeded in reaching the goal). However, the swarm connectivity for the LJ-controlled robots drops after 200 time steps, as the formation begins to move through the obstacle field. After 600 time steps, the formation connectivity increases as the robots reach the goal.

The curves at the bottom are for 20 robots moving through 100 obstacles. In this situation the Newtonian-controlled robots arrive at the goal, and the swarm connectivity drops after 800 time steps and then increases after roughly 1100 steps. Because the LJ-controlled formation moves much more quickly, the formation connectivity drops after 200 time steps and then increases after roughly 500 steps. It is interesting to note that the LJ-controlled swarm does not break apart quite as much as the Newtonian-controlled swarm.

Figure 4 shows how the number of robots reaching the goal changes with time. Again, two sets of results are presented, for 20 and 100 robots moving through 100 obstacles. The

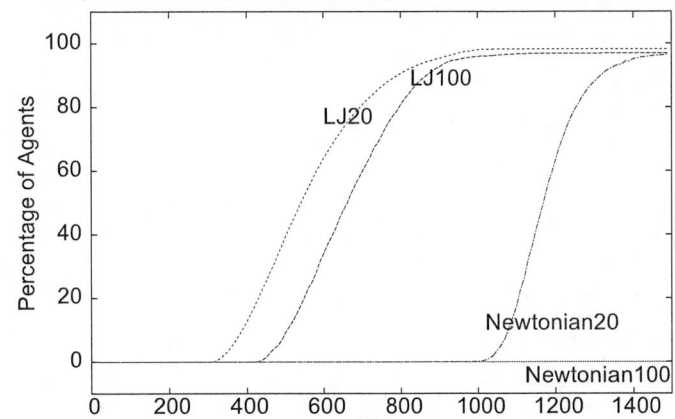

Figure 4. The percentage of 20 and 100 robots reaching the goal through 100 obstacles over 1500 time steps using Newtonian and LJ force laws

two left-most curves are for the LJ-controlled robots. Note that, regardless of the number of obstacles, robots start to arrive at the goal at roughly 400 time steps. With 20 robots, they have all arrived at the goal by about 900 time steps. This indicates that all robots arrived at the goal within a 500 time step interval – a relatively narrow band in time. Increasing the number of robots to 100 increases the time interval to only 800 steps. The other two curves are for the Newtonian-controlled robots. With 20 robots, they start to reach the goal at 1000 time steps, and the interval is approximately 400 time steps. When there are 100 robots, none reach the goal within the allotted time period.

Figure 5 shows the evolved LJ robot-robot force law. It is strongly repulsive when the distance between robots is less than 50, and weakly attractive when the distance is greater than 50. This weak attractive force provides the "stickiness" that manifests itself as a viscous fluid in the aggregate.

4.5. Summary

We presented a novel extension to our artificial physics framework with the use of a generalized Lennard-Jones force law. We then summarized how we used evolutionary algorithms to optimize the parameters of the force laws. These force laws were tested within the context of moving robotic swarm formations through obstacle fields to a goal.

In addition, we presented novel metrics of performance, namely, the number of robots that collide with obstacles, their connectivity, the number of robots that reach the goal, and the time to the goal. Although each metric provides useful information, a much better picture arises by considering all metrics. Our empirical analysis is methodical ranging from 20 to 100 robots and 20 to 100 obstacles.

Our results indicate that LJ-controlled robots have far superior performance to our more "classic" Newtonian-controlled robots. This is because the emergent behavior of the LJ-controlled swarm is to act as a viscous fluid, generally retaining good connectivity while allowing for the deformations necessary to smoothly flow through the obstacle field. De-

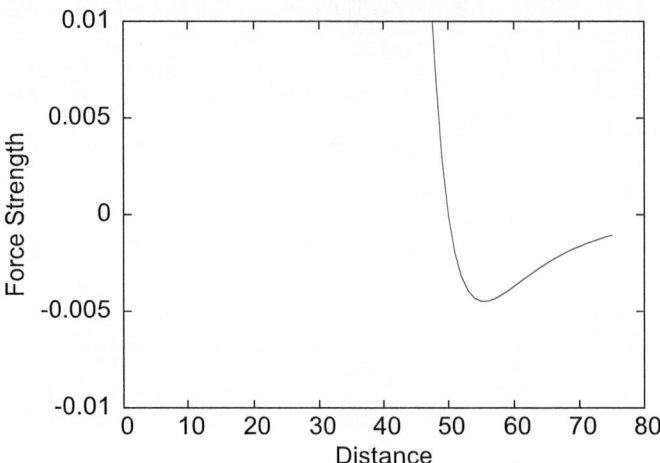

Figure 5. The evolved LJ force law, which is repulsive when distance is less than 50, and weakly attractive when distance is greater than 50

spite being trained with only 40 robots, the emergent behavior scales well to larger numbers of robots. In contrast, the Newtonian-controlled swarm produces more rigid structures that have much more difficulty maneuvering through the obstacles. Furthermore, performance drops dramatically when there are more than 40 robots.

5. Distributed Agent Evolution with Dynamic Adaptation to Local Unexpected Scenarios

We have shown how our artificial physics framework can be used to control formations of mobile robots that move towards a goal while avoiding obstacles (see Figure 2). An *offline* EA evolved an agent-level force law, such that robots maintained network cohesion, avoided the obstacles, and reached the goal. The emergent behavior was that the collective moved as a viscous fluid [2].

Engineering multi-agent systems is difficult due to numerous constraints, such as noise, limited range of interaction with other agents, delayed feedback, and the distributed autonomy of the agents. One potential solution is to automate the design of multi-agent systems in simulation using evolutionary algorithms (EAs) [7, 8]. In this paradigm, the EA evolves the behaviors of the agents (and their local interactions) such that the global task behavior emerges. A global observer monitors the collective and provides a measure of performance to the individual agents. Agent behaviors that lead to desirable global behavior are hence rewarded, and the collective system is gradually evolved to provide optimal global performance.

There are several difficulties with this approach. First, a global observer may not exist. Second, some (but not all) agents may experience some form of reward for achieving task behavior, while others do not. Third, this reward may be delayed, or may be noisy. Fourth, the above paradigm works well in simulation (offline), but is not feasible for real-world online applications where unexpected events occur. Finally, the above paradigm may have difficulty evolving different individual behaviors for different agents (heterogeneity versus homogeneity).

We propose a novel framework called "Distributed Agent Evolution with Dynamic Adaptation to Local Unexpected Scenarios" (DAEDALUS) [9] for engineering multi-agent systems that can be used either offline or online. We will explore how DAEDALUS can be used to achieve global aggregate behavior by examining a case study.

With the DAEDALUS paradigm, we assume that agents (whether software or hardware) move throughout some environment. As they move, they interact with other agents. These agents may be of the same species or of some other species [10]. Agents of different species have different roles in the environment. The goal is to evolve agent behaviors and interactions between agents, in a distributed fashion, such that the desired global behavior occurs.[1]

Let us further assume that each agent has some procedure to control its own actions in response to environmental conditions and interactions with other agents. The precise implementation of these procedures is not relevant, thus they may be programs, rule sets, finite state machines, real-valued vectors, force laws, or any other procedural representation. Agents have a sense of self-worth, or "fitness". Agents that experience direct performance rewards have higher fitness. Other agents may not experience any direct reward, but may in fact have contributed to the agents that did receive direct reward. This "credit assignment" problem can be addressed in numerous ways, including the "bucket brigade" algorithm or the "profit sharing" algorithm [12]. Assuming that a set A of agents has received some direct reward, both algorithms provide reward to the set B of agents that have interacted (and helped) those in A. The rewards that trickle-back further are also given to those agents in set C that helped those in B and so on. Agents that receive no rewards lose fitness. If fitness is low enough, agents stop moving or die.

Evolution occurs when individuals of the same species interact. Those agents with high fitness give their procedures to agents with lower fitness. Evolutionary recombination and mutation provide necessary perturbations to these procedures, providing increasing performance and the ability to respond to environmental changes. Different species may evolve different procedures, reflecting the different niches they fill in the environment.

5.1. Online Approach with DAEDALUS

Each robot of the swarm is an individual in a population that interacts with its neighbors. Each robot contains a *slightly mutated* copy of the optimized LJ force law rule set found with offline learning. This ensures that our robots are not completely homogeneous. We allowed this slight heterogeneity because when the environment changes, some mutations perform better than others. The robots that perform well in the environment will have higher fitness than the robots that perform poorly. When low fitness robots encounter high fitness robots, the low fitness robots ask for the high fitness robot's rules. Hence, better performing robots share their knowledge with their poorer performing neighbors.

When we apply DAEDALUS to obstacle avoidance, we focus on two aspects of our swarm: reducing obstacle-robot collisions and maintaining the cohesion of the swarm.

---

[1] The work by [11] is conceptually similar and was developed independently.

Robots are penalized if they collide with obstacles and/or if they leave their neighbors behind. The second scenario arises when the robots are left behind in cul-de-sacs. This causes the cohesion of the formation to be reduced.

5.2. Experimental Methodology of Online Adaptation

Each robot of the swarm contains a slightly mutated copy of the optimized LJ force law rule set found with offline learning and all robots have the same fitness at the start. There are five goals to achieve in a long corridor, and between each randomly positioned goal is a different obstacle course with 90 randomly positioned obstacles. The online 2D world is 1650 × 950, which is larger than the offline world. In our changed environment, each obstacle has a radius of 30 compared to the offline obstacle radius of 10. So more than 16% of the online environment is covered with the obstacles. Compared to the offline environment, the online environment triples the obstacle coverage. We also increase the maximum velocity of the robots to 30 units/sec, making the robots moves 1.5 times faster than in the offline environment. The LJ force law learned in offline mode is not sufficient for this more difficult environment, producing collisions with obstacles (due to the higher velocity), and robots that never reach the goal (due to the high percentage of obstacles). Figure 6 shows an example of the more difficult environment.

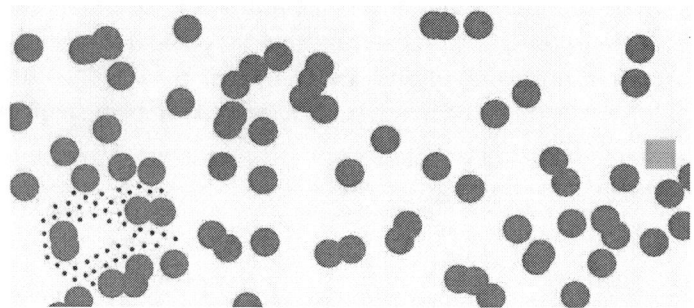

Figure 6. 60 robots moving to the goal. The larger circles represent obstacles, while the square in the upper right represents the goal. The larger obstacles make this environment far more difficult for the robots to traverse.

Robots that are left behind (due to obstacle cul-de-sacs) do not proceed to the next goal, but the robots that had collisions and made it to the goal are allowed to proceed to the next goal. We assume that damaged robots can be repaired once they reach a goal.

5.3. DAEDALUS Results

To measure the performance of the DAEDALUS approach, an experiment was carried out with 60 robots, 5 goals in the long corridor, and 90 obstacles in between each goal. The experiment was averaged over 50 runs of different robot, goal, and obstacle placements. Each robot is given equal initial fitness and "seeded" with a mutated copy of the optimized LJ force law learned in offline mode. If a robot collides with an obstacle, its fitness is reduced. Whenever a robot encounters another robot with higher fitness, it takes the relevant parameters pertaining to the obstacle-robot interaction of the better performing robot.

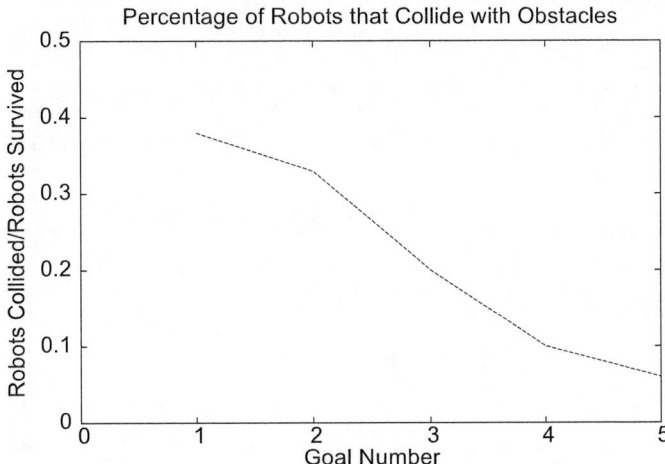

Figure 7. The ratio of colliding robots versus the number of surviving robots for 60 robots moving through 5 goals with 90 obstacles in between each goal.

Figure 7 shows the ratio of the number of robots that collided with obstacles versus the number of robots that survived to reach the goals. The graph indicates that after only 3 goals, the percentage of robots that collide with obstacles has dropped from about 38% to well under 10%. Inspection of the obstacle-robot parameters indicates that the repulsive component increased through the online process of mutation and the copying of superior force laws (this was confirmed via inspection of the mutated force laws).

This first experiment did not attempt to alleviate the situation where robots are left behind; in fact, only roughly 48% of the original 60 robots reach the final goal (see Figure 8, lower line). This is caused by the large number of cul-de-sacs produced by the large obstacle density. Our second experiment attempts to alleviate this problem by focusing on the robot-robot interactions. Our assumption was that the LJ force law needs to provide stronger cohesion, so robots aren't left behind.

If robots are stuck behind in cul-de-sacs (i.e. they make no progress towards the goal) and they sense neighbors, they slightly mutate the robot-robot interaction parameters of their force laws. In a situation in which they do not sense the presence of neighbors and do not progress towards the goal, they rapidly mutate their robot-goal interaction causing a "panic behavior". These relatively large perturbations of the force law allow the robots to escape their motionless state.

Figure 8 shows the results of this second experiment. In comparison with the first experiment (with survival rates of 48%), the survival rates have increased to 63%. As a control experiment, we ran our offline approach on this more difficult task. After five goals, the survival rate is about 78%. Recall that the offline results are obtained by running an EA with a population size of 100 for 100 generations with each individual averaged over 50 random instantiations of the environment. As can be seen, the DAEDALUS approach provides results only somewhat inferior to the offline approach, in real time, while the robots are in the environment.

Although not shown in the graph, it is important to point out that the collision rates

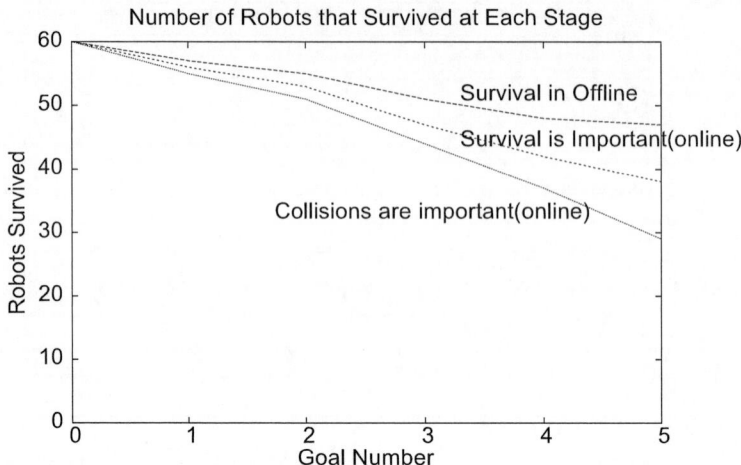

Figure 8. A comparison of (a) the number of robots that survive when rules are learned using offline learning, (b) the number of robots that survive when using online learning (where the focus is on reducing collisions), and (c) the number of robots that survive when using online learning (and the focus in on survivability).

were not affected in the second experiment. Hence, we believe that it is quite feasible to combine both aspects in the future. Collision avoidance can be improved via mutation of the obstacle-robot interaction, while survival can be improved via mutation of the robot-robot interaction and robot-goal interaction.

5.4. Obstructed Perception

When a robot can not see another robot, due to the presence of obstacles, we call this "obstructed perception." When the robot's line of sight lies along an edge of an obstacle, the robots are capable of sensing each other. Surprisingly, this is not generally modeled in prior work in this area [3]. In our prior work [2, 9] obstacles did not obstruct perception. The addition of obstructed perception makes the task far more difficult, especially as obstacle size increases [14]. Figure 9 shows an example scenario of obstructed perception. The larger circle represents an obstacle, and A and B are robots. We define *minD* to be the minimum distance from the center of the obstacle to the line of sight of robot A and robot B, and r is the radius of an obstacle. If $r > minD$, then robot A and robot B have their perception obstructed.

We utilize a parameterized description of a line segment [13] to find the *minD*.

$$minD = \sqrt{(((1-q) \times X_a + q \times X_b) - X_c)^2 + (((1-q) \times Y_a + q \times Y_b) - Y_c)^2} \quad (3)$$

where (X_a, Y_a) and (X_b, Y_b) are the x,y positions of robots A and B, (X_c, Y_c) is the position of the center of an obstacle, and q is the minimum function that is defined by:

$$\frac{((X_c - X_a) \times (X_b - X_a) + (Y_c - Y_a) \times (Y_b - Y_a))}{\left((X_b - X_a)^2 + (Y_b - Y_a)^2\right)} \quad (4)$$

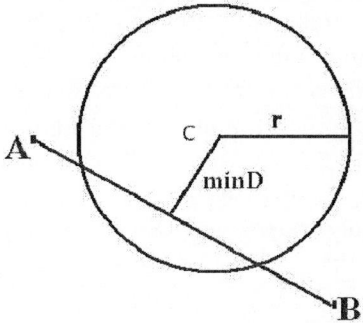

Figure 9. Sensing capability of two robots (A, B) is obstructed by a large obstacle (C).

5.5. Results with Obstructed Perception

We compared DAEDALUS to three control studies. In the first control study, we train the robots with an offline EA on small obstacles, and test them again on small obstacles to verify their performance. In the second control study, we train the robots with an offline EA on large obstacles and test them on large obstacles. The purpose of this control study is to clarify the difficulty of the task. Finally, in the third control study, we train the robots with an offline EA on small obstacles and test them on large obstacles. The purpose of this study was to see how well the knowledge learned while avoiding small obstacles transferred to large obstacles.

Figure 10 shows the results. The y-axis gives the number of robots that survived to reach the goal at each stage for the four different experiments. The top performance curve is for the first control study. Note that learning with small obstacles in offline mode is not hard, and the robots perform very well in the online environment. This is due to the fact that the small obstacles make the environment less dense providing the robots sufficient space to navigate. Out of 60 initial robots released in the online environment, 93.3% survived to reach the last goal. With such small obstacles (which is the maximum density examined in the related literature), obstructed perception is not an important issue.

As presented earlier (and in [9]), robots that learned without obstructed perception on larger obstacles had a reasonably high survival rate (78%). The bottom ("No DAEDALUS (large-large)") performance curve shows the effect of obstructed perception (the second control study). Learning with large obstacles in offline mode with obstructed perception is very difficult, and the test results show that out of 60 robots released initially into the online environment only 35% (21 robots) survived to reach the last goal. This is due to the fact that the environments with larger obstacles create large numbers of cul-de-sacs that obstruct perception.

The third control study, where offline training occurs with small obstacles and testing occurs with large obstacles, is surprisingly good (see "No DAEDALUS (small-large)"). Despite an initial drop in performance, performance at the fifth goal is quite acceptable (out of the initial 60 robots, 41.6% (25 robots) survived to reach the final goal). This is a 6.6% improvement over the robots that were trained on larger obstacles. These results run counter to accepted wisdom, which states that it is best to train on the hardest environments that you will encounter. In fact, this example demonstrates that training on simpler problems and applying the knowledge gained to harder problems can potentially provide superior results. Why is this so? As with developmental psychology, one does not train children on hard problems immediately, instead, we train them on easier problems first, in the hopes that they will learn the "basics" (which are important building blocks for solving other, more difficult, problems) more quickly.

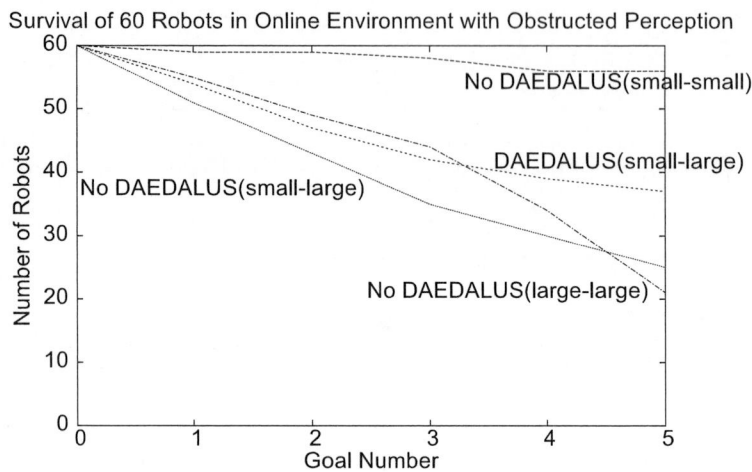

Figure 10. Four different experiments of number of robots surviving - all robots are trained with obstructed perception and tested with and without DAEDALUS. The results are averaged over 100 independent runs.

If we extend the developmental psychology analogy further, we note that we encourage children to experiment and modify their behavior based on changes in the environment. Furthermore, they share the lessons learned. This is precisely what the DAEDALUS system does. The final performance curve in Figure 10 shows the results. With an initial 60 robots, 61.6% or 37 robots survived to reach the last goal. This is a 26.6% improvement over the robots that learned in an environment with the larger obstacles, and a 20% improvement over the robots that learned with small obstacles and tested with the larger obstacles without DAEDALUS. Note that the results are comparable with those achieved without obstructed perception (63%), indicating that obstructed perception is not significantly affecting the performance of DAEDALUS! Hence, these preliminary results are very promising. Although encouraging the robots (or children) to explore and experiment does provide an early drop-off in performance (compared to the "No DAEDALUS (large-large)" curve); the results after three goals are superior. This is a classic example of "exploration" vs "exploitation". Pure exploitation of learned knowledge is good up to a point, but will eventually fail as the problems become more difficult. Exploration provides the key to adapt to these changing environments. DAEDALUS provides just this form of exploration.

5.5.1. Homogeneous DAEDALUS Results

For the DAEDALUS performance curve given above, all robots had the same mutation rate, which was 5%. Hence, each robot had the same rate of exploration. Although the rules for each robot may differ, their mutations rates are identical, and we refer to this system as

"Homogeneous DAEDALUS". However, there are numerous problems with this approach. First, the results may depend quite heavily on choosing the correct mutation rate. How is this mutation rate to be chosen? Second, the best mutation rate may also depend on the environment and should potentially change as the environment changes. How is this to be accomplished?

Since the mutation rate may have a major effect on performance, we decided to explore this effect by conducting several experiments with different mutation rates. Figure 11 shows five independent experiments of Homogeneous DAEDALUS. Five different mutation rates were used: 1%, 3%, 5%, 7%, and 9%. The results are quite striking. Of the five different mutation rates, only 5% and 7% did well (with about 35 robots surviving to the last goal). Recall that the DAEDALUS performance curve shown in Figure 10 resulted from an arbitrarily chosen mutation rate of 5%. As it turns out, we were extremely fortunate in our design decision. For example, with mutation rates of 1%, 3%, and 9%, approximately 20 robots survive to reach the final goal. The performance curve for the 9% mutation rate is especially interesting. Although promising at first, it appears as if the mutation rate is so high that it eventually causes an extremely deleterious mutation to appear. Mutation rates of 1% and 3% are too low to cope with the changed environment.

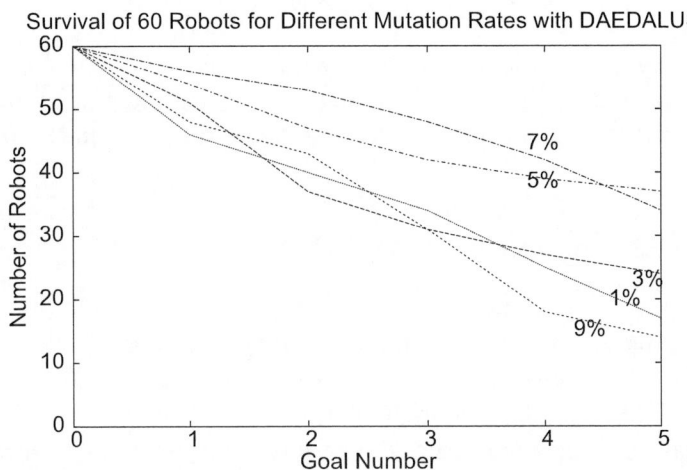

Figure 11. Five different mutation experiments of robots surviving - all robots are trained with obstructed perception and tested with DAEDALUS. The results are averaged over 100 independent runs.

5.5.2. Heterogeneous DAEDALUS Results

In an attempt to address the problem of choosing the correct mutation rate, we divided the robots into five groups of equal size. Each group of 12 robots was assigned a mutation rate of 1%, 3%, 5%, 7%, and 9%, respectively. This mimics the behavior of children that have different "comfort zones" in their rate of exploration. Since different robots have different mutation rates, we refer to this system as "Heterogeneous DAEDALUS". Figure 12 shows the results, in comparison with the three control studies shown in Figure 10. The label "Het.DAEDALUS(small-large)" shows the survivability of robots with pre-assigned mutation rates. Out of the initial 60 robots, 29 or 48% robots survived to reach the final goal. Although this is higher than our second and third control studies, it did not produce results as good as the results achieved with Homogeneous DAEDALUS using a 5% mutation rate (as shown in Figure 11).

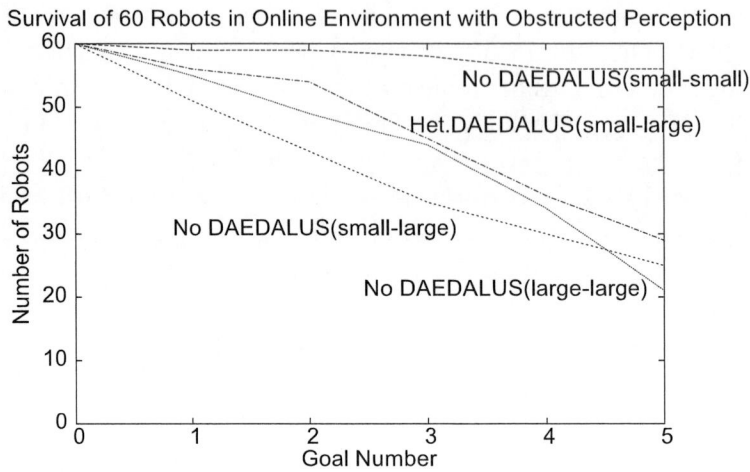

Figure 12. The number of robots surviving with pre-assigned mutation rates. Mutation rates are not exchanged - all robots are trained with obstructed perception and tested with or without DAEDALUS. The results are averaged over 100 independent runs.

5.5.3. Extended Heterogeneous DAEDALUS Results

In an attempt to improve performance, we again borrowed from the analogy of a "swarm" of children learning some task. Not only do they share useful information as to the rules they might use, but they also share meta-information as to the level of exploration that is actually safe! Very bold children might encourage their more timid comrades to explore more than they would initially. On the other hand, if a very bold child has an accident, the rest of the children will become more timid. In "Extended Heterogeneous DAEDALUS", five groups of children are again initialized with mutation rates of 1%, 3%, 5%, 7%, and 9%. However, in this situation, if a robot receives the rules from a neighbor (which, again, occurs if that robot is in trouble), it also receives the neighbor's mutation rate. In this implementation, children in trouble not only change their rules, but their mutation rate. Figure 13 shows the results of this study. The curve labeled with "Ex.Het.DAEDALUS(small-large)" refers to the survivability of robots with pre-assigned mutation rates that also allows the robots to receive a neighbor's mutation rate, if the robot receives the neighbor's rules. The behavior is quite good. On average, 34 robots survive to reach the final goal, which is very close to the optimum value of 37 found by the best Homogeneous DAEDALUS experiment.

5.6. Summary

Traditional approaches to designing multi-agent systems are offline and assume the presence of a global observer. However, this approach will not work in real-time online sys-

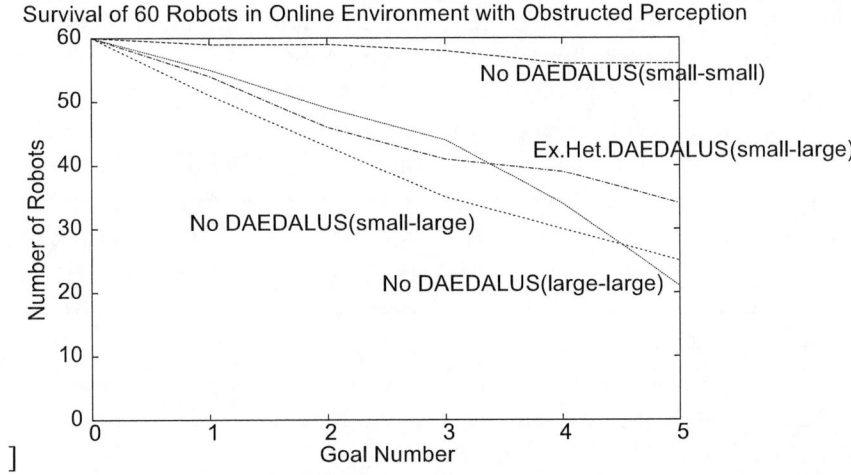

Figure 13. The number of robots surviving with pre-assigned mutation rates. Mutation rates are exchanged - all robots are trained with obstructed perception and tested with or without DAEDALUS. The results are averaged over 100 independent runs.

tems. We presented a novel approach to solving this problem, called DAEDALUS, where we showed how concepts from population genetics could be used with swarms of agents to provide fast online adaptive learning in changing environments.

We addressed the important issue of "obstructed perception" in learning behaviors for swarms of robots that must avoid obstacles while reaching a goal. This issue has been largely absent from the literature. Our obstacle density is also three times higher than the norm, making obstacle avoidance a far more difficult task. Since obstructed perception makes the task far more difficult, DEADALUS had to be extended. Our first extension was to allow different robots to have different rates of exploration, which affects the rate at which they change their behavioral rules. The second extension allows robots to also share their rates of mutation, permitting robots to find the right balance between exploration and exploitation. Results of the extended system are almost as good as the best results we were able to achieve when the exploration rates were controlled by hand. Our framework allows swarms of robots to not only learn and share behavioral rules in changing environments (in real time), but also to learn the proper amount of behavioral exploration that is appropriate.

6. Related Work in Obstacle Avoidance

Most of the swarm robotics literature can be subdivided into *swarm intelligence*, *behavior-based*, *rule-based*, *control-theoretic* and *physics-based* techniques. Swarm intelligence techniques are ethologically motivated and have had excellent success with foraging, task allocation, and division of labor problems [16, 17]. Both behavior-based and rule-based systems [19, 4, 18] have proved quite successful in demonstrating a variety of behaviors in a heuristic manner. Behavior-based and rule-based techniques do not make use of potential fields or forces. Instead, they deal directly with velocity vectors and heuristics for changing those vectors (although the term "potential field" is often used in the behavior-based literature, it refers to a field that differs from the strict Newtonian physics definition). Control-theoretic approaches have also been applied effectively (e.g., [20]). Our approach does not make the assumption of having leaders and followers, as in [21, 22].

One of the earliest physics-based techniques is the *potential fields* (PF) approach (e.g., [24]). Most of the PF literature deals with a small number of robots (typically just one) that navigate through a field of obstacles to get to a target location. The environment, rather than the robots, exert forces. Obstacles exert repulsive forces while goals exert attractive forces. Recently, Howard et al. [25] and Vail and Veloso [26] extended PF to include inter-agent repulsive forces – for the purpose of achieving coverage. Although this work was developed independently of AP, it affirms the feasibility of a physics force-based approach. Another physics-based method is the "Engineered Collective" work by Duncan at the University of New Mexico and Robinett at Sandia National Laboratory. Their technique has been applied to search-and-rescue and other related tasks [27].

The *social potential fields* [28] framework is highly related to AP. Reif and Wang [28] rely on a force-law simulation that is similar to our own, allowing different forces between different robots. Their emphasis is on synthesizing desired formations by designing graphs that have a unique potential energy (PE) embedding. We plan to merge this approach with ours.

In the specific context of obstacle avoidance, the most relevant papers are [4], [3] and [19]. Balch [4] examines the situation of four robots moving in formation through an obstacle field with 2% coverage. In [3], he extends this to an obstacle field of 5% coverage, and also investigates the behavior of 32 robots moving around one medium size obstacle. Fredslund and Matarić [19] examine a maximum of eight robots moving around two wall obstacles. To the best of our knowledge, we are the first to systematically examine larger numbers of robots and obstacles.

The work done in [23] uses an embedded network distributed throughout the environment to approximate the path-planning space and uses the network to compute a navigational path using GNATs when the environment changes. The dynamism of the environment is modeled with an opening and closing door in the experimental setup. However, the embedded network is immobile, whereas our network is completely mobile.

7. Localization via Trilateration

In order to function properly the robots in the swarm must be able to accurately localize their local neighbors and to share information. In this section, we describe our trilateration approach to multi-robot localization, which is fully distributed, inexpensive, scalable, and robust. Our prior research [15] focused on maintaining multi-robot formations indoors using trilateration. As we will show in this chapter, we have now pushed the limits of our trilateration technology by testing formations of robots in an uncontrolled outdoor setting with relatively large inter-robot distances and high speeds.

Our goal is to create a plug-in hardware module to accurately localize neighboring robots, without global information and/or the use of vision systems. Our trilateration approach is not restricted to any particular class of control algorithms [3, 29, 6], and does not preclude the use of other technologies, such as beacons, landmarks, pheromones, vision systems, and GPS.

Two methodologies for robot localization are *triangulation* and *trilateration* [30]. Both compute the location of a point (e.g., a robot) in 2D space. In *triangulation*, the locations of two "base points" are known, as well as the interior angles of a triangle whose vertices comprise the two base points and the object to be localized. The computations are performed using the Law of Sines. In 2D *trilateration*, the locations of three base points are known as well as the distances from each of these three base points to the object to be localized. Looked at visually, 2D trilateration involves finding the location where three circles intersect.

Thus, to locate a remote robot using 2D trilateration the sensing robot must know the locations of three points in its own coordinate system and be able to measure distances from these three points to the remote robot.

7.1. Measuring Distance

Our distance measurement method exploits the fact that sound travels significantly more slowly than light, thereby enabling us to employ a Difference in Time of Arrival technique. The same method is used to determine the distance to a lightning strike by measuring the time between seeing the lightning and hearing the thunder.

To tie this to 2D trilateration, assume that each robot has one radio frequency (RF) transceiver and three ultrasonic acoustic transceivers. The ultrasonic transceivers are the "base points." Suppose robot 2 simultaneously emits an RF pulse and an ultrasonic acoustic pulse. When robot 1 receives the RF pulse (almost instantaneously), a clock on robot 1 starts. When the acoustic pulse is received by each of the three ultrasonic transceivers on robot 1, the elapsed times are computed. These three times are converted to distances, according to the speed of sound. Because the locations of the acoustic transceivers are known, robot 1 is now able to use trilateration to compute the location of robot 2 (precisely, the location of the emitting acoustic transceiver on robot 2). Of the three acoustic transceivers, all three must be capable of receiving, but only one must be capable of transmitting.

Measuring the elapsed times is not difficult. Since the speed of sound is roughly 340.2 meters per second (at standard temperature and pressure), it takes approximately 2.9 ms for sound to travel 1 meter. Times of this magnitude are easily measured using inexpensive electronic hardware.

7.2. Channeling Acoustic Energy into a Plane

Ultrasonic acoustic transducers (also called "transceivers") produce a cone of energy along a line perpendicular to the surface of the transducer. The width of this main lobe (for the inexpensive 40 kHz transducers used in our implementation) is roughly 30°. To produce acoustic energy in a 2D plane would require 12 acoustic transducers in a ring. To get three base points would hence require 36 transducers. This is expensive and is a large power drain. We adopted an alternative approach. Each base point is comprised of one acoustic transducer pointing downward. A parabolic cone is positioned under the transducer, with its tip pointing up toward the transducer (see Figure 15 and Figure 16 later in this chapter). The parabolic cone acts like a lens. When the transducer is placed at the virtual "focal point" the cone "collects" acoustic energy in the horizontal plane, and focuses this energy to the

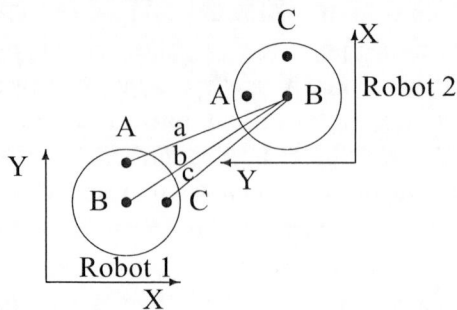

Figure 14. Three base points in an XY coordinate system pattern.

receiving acoustic transceiver. Similarly, a cone also functions in the reverse, reflecting transmitted acoustic energy into the horizontal plane. This works extremely well – the acoustic energy is detectable to a distance of about 2.44 m (8′), which is quite impressive and more than adequate for our needs. Greater range can be obtained with more power (the scaling appears to be very manageable).

7.3. Our Trilateration Approach

Our trilateration approach to localization is illustrated in Figure 14. Assume two robots, shown as circles. An RF transceiver is in the center of each robot. Each robot has three acoustic transducers (also called *base points*), labeled **A**, **B**, and **C**. Note that the robot's local XY coordinate system is aligned with the L-shaped configuration of the three acoustic transceivers. This simplifies the math [15]. Y points to the front of the robot.

In Figure 14, robot 2 simultaneously emits an RF pulse and an acoustic pulse from its transducer **B**. Robot 1 then measures the distances **a**, **b**, and **c**. Without loss of generality, assume that transceiver **B** of robot 1 is located at $(x_{1B}, y_{1B}) = (0,0)$ [31].[2] In other words, let **A** be at $(0,d)$, **B** be at $(0,0)$, and **C** be at $(d,0)$, where d is the distance between **A** and **B**, and between **B** and **C** (see Figure 14). Assume that robot 2 emits from its transducer **B**.

For robot 1 to determine the position of **B** on robot 2 within its own coordinate system, it needs to find the simultaneous solution of three nonlinear equations, the intersecting circles with centers located at **A**, **B** and **C** on robot 1 and respective radii of **a**, **b**, and **c**:

$$(x_{2B} - x_{1A})^2 + (y_{2B} - y_{1A})^2 = a^2 \quad (5)$$
$$(x_{2B} - x_{1B})^2 + (y_{2B} - y_{1B})^2 = b^2 \quad (6)$$
$$(x_{2B} - x_{1C})^2 + (y_{2B} - y_{1C})^2 = c^2 \quad (7)$$

The form of these equations allows for cancellation of the nonlinearity, and simple algebraic manipulation yields the following simultaneous linear equations in the unknowns:

$$\begin{bmatrix} x_{1C} & y_{1C} \\ x_{1A} & y_{1A} \end{bmatrix} \begin{bmatrix} x_{2B} \\ y_{2B} \end{bmatrix} = \begin{bmatrix} (b^2 + x_{1C}^2 + y_{1C}^2 - c^2)/2 \\ (b^2 + x_{1A}^2 + y_{1A}^2 - a^2)/2 \end{bmatrix}$$

---

[2] Subscripts denote the robot number and the acoustic transducer. Thus transducer **A** on robot 1 is located at (x_{1A}, y_{1A}).

Given the L-shaped transducer configuration, we get [31]:

$$x_{2B} = \frac{b^2 - c^2 + d^2}{2d} \qquad y_{2B} = \frac{b^2 - a^2 + d^2}{2d}$$

An interesting benefit of these equations is that they can be simplified even further, if one wants to trilaterate purely in hardware [15].

Analysis of our trilateration framework indicates that, as expected, error is reduced by increasing the "base-line" distance d. Our robots have d equal to 15.24 cm (6″). Error can also be reduced by increasing the clock speed of our trilateration module (although range will decrease correspondingly, due to counter size) [31].

By allowing robots to share coordinate systems, robots can communicate their information arbitrarily far throughout a robotic network. For example, suppose robot 2 can localize robot 3. Robot 1 can localize only robot 2. If robot 2 can also localize robot 1 (a fair assumption), then by passing this information to robot 1, robot 1 can now determine the position of robot 3. Furthermore, robot orientations can also be determined. Naturally, localization errors can compound as the path through the network increases in length, but multiple paths can be used to alleviate this problem to some degree. Heil [31] provides details on these issues.

In addition to localization, our trilateration system can also be used for data exchange. Instead of emitting an RF pulse that contains no information but only performs synchronization, we can also append data to the RF pulse. Simple coordinate transformations allow robot 1 to convert the data from robot 2 (which is in the coordinate frame of robot 2) to its own coordinate frame.

7.4. Trilateration Hardware

Figure 15 and Figure 16 illustrate how our trilateration framework is currently implemented in hardware. Figure 15 shows three acoustic transducers pointing down, with reflective parabolic cones. The acoustic transducers are specially tuned to transmit and receive 40 kHz acoustic signals.

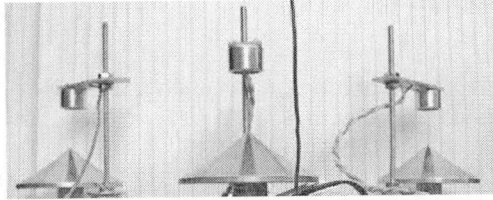

Figure 15. Acoustic transducers and parabolic cones.

Figure 16 (left) shows our in-house acoustic sensor boards (denoted as "XSRF" boards, for *Experimental Sonic Range Finder*). There is one XSRF board for each acoustic transducer. The XSRF board calculates the time difference between receiving the RF signal and the acoustic pulse. Each XSRF contains 7 integrated circuit chips. A MAX362 chip controls whether the board is in transmit or receive mode. When transmitting, a Microchip PIC microprocessor generates a 40 kHz signal. This signal is sent to an amplifier, which then interfaces with the acoustic transducer. This generates the acoustic signal.

Figure 16. The XSRF acoustic sensor printed circuit board (left), and the completed trilateration module (top-down view, right).

In receive mode, a trigger indicates that an RF signal has been heard and that an acoustic signal is arriving. When the RF is received, the PIC starts counting. To enhance the sensitivity of the XSRF board, three stages of amplification occur. Each of the three stages is accomplished with a LMC6032 operational amplifier, providing a gain of roughly 15 at each stage. Between the second and third stage there is a 40 kHz bandpass filter to eliminate out-of-bound noise that can lead to saturation. The signal is passed to two comparators, set at thresholds of ± 2 VDC. When the acoustic energy exceeds either threshold, the PIC finishes counting, indicating the arrival of the acoustic signal.

This timing count provided by each PIC (one for each XSRF) is sent to a MiniDRAGON board[3] powered by a Freescale 68HCS12 microprocessor. The MiniDRAGON performs the trilateration calculations. Figure 16 (right) shows the completed trilateration module from above. The MiniDRAGON is outlined near the center and the three XSRF acoustic sensors are outlined at the bottom.

7.5. Synchronization Protocol

Trilateration involves at least two robots. One transmits the acoustic-RF pulse combination, while the others use these pulses to compute (trilaterate) the coordinates of the transmitting robot. Hence, trilateration is a one-to-many protocol, allowing multiple robots to simultaneously trilaterate and determine the position of the transmitting robot.

The purpose of trilateration is to allow all robots to determine the position of all of their neighbors. For this to be possible, the robots must take turns transmitting. For our current implementation we use a protocol that is similar to a token passing protocol. Each robot has a unique hardware encoded ID. When a robot is transmitting it sends its own ID. As soon as the neighboring robots receive this ID they increment the ID by one and compare it with their own ID. The robot that matches the two IDs is considered to have the token and

---

[3] Produced by Wytec (http://www.evbplus.com/)

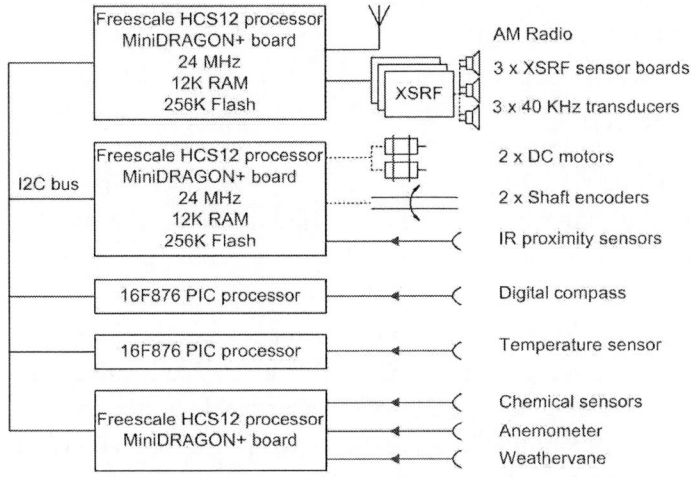

Figure 17. The architecture of the Version 1.0 Maxelbot.

will transmit next. The other robots will continue to trilaterate. Each robot maintains a data structure with the coordinate information, as well as any additional sensor information, of every neighboring robot.

Although this current protocol is distributed, there are a few problems with it. First, it assumes that all robots know how many robots are in the collective. Second, the removal or failure of a robot can cause all robots to pause, as they wait for the transmission of that robot. We are currently working on new protocols to rectify these issues.

7.6. The Maxelbot Robot Platforms

Our University of Wyoming "Maxelbot" (named after the two graduate students who designed and built the robot) is modular. The platform is an MMP5, made by The Machine Lab [4]. A primary MiniDRAGON is used for control. It communicates via an I^2C bus to all other peripherals, allowing us to plug in new peripherals as needed. Figure 17 shows the architecture. The primary MiniDRAGON is the board that drives the motors, and it monitors proximity sensors and shaft encoders. The trilateration module is shown at the top of the diagram. This module controls the RF and acoustic components of trilateration. Additional modules have been built for digital compasses, thermometers, and chemical plume tracing. The PICs provide communication with the I^2C bus.

7.7. Prior Indoor Experiments and Demonstrations

Numerous task-driven formations have been successfully performed with the Maxelbots indoors using trilateration. For details, see [15]. Here, we briefly summarize those results.

One of our research objectives is linear (also called "chain") formations. This type of formation is especially useful for traversing corridor-like environments, such as sewers,

---

[4] See http://www.themachinelab.com/MMP-5.html

underground pipes, or ducts. Our demonstrations show that three Maxelbots, roughly 30.5 cm (12″) apart, will stay in formation as the leader follows a curved trajectory.

A second demonstration is motivated by search and rescue. In particular, a group of three Maxelbots have to pull a box, which contains a doll simulating an infant victim. The desired behavior is as follows. There is one leader and two followers. The followers are 61 cm (24″) behind the leader and are separated by 122 cm (48″) from each other. The lead Maxelbot is not in physical contact with the box, but the box is tethered to the two followers. The Maxelbots succeed in pulling the box in a straight line, despite the unevenness of the load-carrying. The success is due to our trilateration.

The third demonstration/experiment requires maintenance of formations while performing a "chemical plume tracing" (CPT) task [32]. The CPT objective is to locate the source of a hazardous airborne chemical plume by measuring flow properties, such as toxin concentration. This is best performed as a collaborative task. Using ethanol vapors and a Figaro TGS2620 metal oxide chemical sensor mounted on each platform, three Maxelbots succeed in maintaining a triangular formation in a roughly 7.6 m × 7.6 m (25′ × 25′) indoor laboratory environment. Out of 10 trials, the Maxelbot team achieves a 60% success rate with an average search time of just seven minutes. This is competitive with the best CPT results published, despite the fact that the experiments are conducted in a far more unstructured indoor environment than usual, as well as more stringent success criteria than usual. The combination of artificial physics with trilateration not only helps the Maxelbots to stay in their triangular formation, but it also assists in data sharing for the computation of chemical gradients.

The indoor experiments just described are so encouraging that we have decided to push the limits of our technology and venture into highly unstructured outdoor environments. We are unaware of any other rigorous trilateration experiments with such challenging outdoor environments, speeds, and distances as what we describe next.

7.8. New Outdoor Experiments

This section presents an experiment that illustrates the performance of the trilateration system in an uncontrolled outdoor setting. In particular, the Maxelbots are run outside in a region in the center of the University of Wyoming campus called "Prexy's Pasture" (see Figure 18)[5]. Prexy's consists mostly of grass, of average height 5 cm (2″), interspersed with concrete sidewalks, trees, rocks, leaves, and other debris. The grass hits the bottom of the Maxelbot. Although generally flat, the ground slope can change rapidly (within 61 cm or 2′), by up to 20°, at boundaries. Results presented below are averaged over five independent runs, taken over a 20 minute interval.

The control algorithm used during these experiments simply maintains the proper position of robots (with respect to each other) by compensating with speed-ups, slow-downs, and turns. It is not designed to be particularly intelligent. The purpose of these experiments is to validate and test the hardware; the focus is not on the control software.

---

[5]http://www.laramie.willshireltd.com/PrexysPasture.html

Figure 18. Maxelbots outdoors in UW's Prexy's Pasture.

7.9. Accuracy of Diamond Formation Preservation

In this experiment, four Maxelbots are required to maintain a diamond formation. There is a leader and three followers. The leader goes on a curved path, and the followers have to maintain certain XY-coordinates with respect to the leader (see the "Ideal" column in Table 2). The leader is running at 60% power, and the followers are at 80% power with a variable turning speed. The wind speed near the ground ranges from 4 to 9 meters per second (10 to 21 mph).

Table 2 shows the XY-coordinates derived from the trilateration readings, for the three followers. From this table, it can be seen that the mean is very close to the ideal, and the standard deviations are small. Y is within 10% of the desired value, while X is within 5%. In other words, a very good, robust diamond formation is maintained by the trilateration system despite ground disturbances, wind, dust, and relatively high robot speed. The results are averaged over five independent runs. A sequence of snapshots from a typical run is shown in Figure 19.

It is important to note that trilateration will work with any reasonable formation. As mentioned earlier, we are also interested in linear (chain) formations. With our current implementation, the user simply places the robots on the ground in their desired positions. The robots move forward, and trilateration is used to preserve the formation. Figure 20 shows a sequence of snapshots from a typical run with a linear formation. This formation is actually more difficult than the diamond formation, because the robots can potentially block the acoustic signals. However, in practice this does not appear to be an issue – the robots maintain formation extremely well.

Table 2. Accuracy of the three followers' X and Y positions in a diamond formation.

	Ideal cm (inches)	Mean (inches)	Std. dev.
Maxelbot1-X	61 (24)	62.1 (24.5)	1.7 (0.7)
Maxelbot1-Y	61 (24)	54.8 (21.6)	2.7 (1.1)
Maxelbot2-X	-61 (-24)	-64.3 (-25.3)	4.2 (1.7)
Maxelbot2-Y	61 (24)	54.5 (21.5)	3.5 (1.4)
Maxelbot3-X	0 (0)	1.0 (0.4)	5.1 (2.0)
Maxelbot3-Y	122 (48)	111.8 (44.0)	4.6 (1.8)

Figure 19. A sequence of snapshots from a typical run where four Maxelbots maintain a diamond formation.

Figure 20. A sequence of snapshots from a typical run where four Maxelbots maintain a linear formation.

7.10. Summary

We described a robust 2D trilateration framework for the fast, accurate localization of neighboring robots. The framework uses three acoustic transceivers and one RF transceiver. Our framework is designed to be modular, so that it can be used on different robotic platforms, and is not restricted to any particular class of control algorithms. Although we do not rely on GPS, stationary beacons, or environmental landmarks, their use is not precluded. In addition to being robust, our framework is fully distributed, inexpensive, and scalable.

To illustrate the general utility of our framework, we demonstrated the application of our robots in a wide variety of situations. The results from these experiments highlight the surprising accuracy and robustness of our trilateration framework under challenging conditions.

8. Related Work in Localization

Trilateration is a well-known technique for robot localization. Most approaches (including ours) are algebraic, although recently a geometric method was proposed [33]. Many localization techniques, including those involving trilateration, use global coordinates [34];

however ours relies on local coordinates only.

MacArthur [35] presents two different trilateration systems. The first uses three acoustic transducers, but without RF. Localization is based on the differences between distances rather than the distances themselves. The three acoustic transducers are arranged in a line. The second uses two acoustic transducers and RF in a method similar to our own. Unfortunately, both systems can only localize points "in front" of the line, not behind it.

Cricket [36] is another system that makes use of RF and ultrasound for localization. It was developed to be used indoors. Compared to our system, which does not require fixed beacons, the Cricket requires beacons attached to fixed locations in order to function. This is not practical for mobile robot localization in outdoor environments.

Our particular approach was inspired by the CMU *Millibot* project. They also use RF and acoustic transducers for trilateration. However, due to the very small size of their robots, each Millibot can only carry one acoustic transducer (coupled with a right-angle cone, rather than the parabolic cone we use). Hence trilateration is a collaborative endeavor that involves several robots. To perform trilateration, a minimum of three Millibots must be stationary and serve as beacons at any moment in time. The set of three stationary robots changes as the robot team moves. The minimum team size is four robots (and is preferably five). Initialization generally involves having some robots make L-shaped maneuvers, in order to disambiguate the localization [37]. Our approach operates with as few as two robots, and our robots never need to be stationary, due to the presence of three acoustic transducers on each robot.

In terms of functionality, an alternative localization method in robotics is to use line-of-sight IR transceivers. When IR is received, signal strength provides an estimate of distance. The IR signal can also be modulated to provide communication. Multiple IR sensors can be used to provide the bearing to the transmitting robot (e.g., see [38, 39]). We view this method as complementary to our own; however, our method is more appropriate for tasks where greater localization accuracy is required. This is especially important in outdoor situations where water vapor or dust could change the IR opacity of air. Similar issues arise with the use of cameras and omni-directional mirrors/lenses, which require far more computational power and a light source.

9. Conclusion

First, we presented our artificial physics framework, with the use of a "Newtonian" gravitational force law and a generalized Lennard-Jones force law. We summarized how we use evolutionary algorithms to optimize the parameters of the force laws. These force laws were tested within the context of moving robotic swarm formations through obstacle fields to a goal.

Second, we proposed a novel framework called "Distributed Agent Evolution with Dynamic Adaptation to Local Unexpected Scenarios" (DAEDALUS) for engineering multi-agent systems that can be used either offline or online. We showed how concepts from population genetics can be used with swarms of agents to provide fast online adaptive learning in changing environments using DAEDALUS. Our framework allows swarms of robots to not only learn and share behavioral rules in changing environments (in real time), but also to learn the proper amount of behavioral exploration that is appropriate.

Finally, we described a robust 2D trilateration framework for the fast, accurate localization of neighboring robots. To illustrate the general utility of our framework, we demonstrated the application of our robots in a wide variety of situations. For details on this project, see http://www.cs.uwyo.edu/~wspears/maxelbot.

References

[1] Spears, W., Spears, D.: Using Artificial Physics to Control Agents. *IEEE International Conference on Information*, Intelligence, and Systems, (1999)

[2] Hettiarachchi, S., Spears, W.: Moving Swarm Formations Through Obstacle Fields. *International Conference on Artificial Intelligence, CSREA Press* **1** (2005) 97–103

[3] Balch, T., Hybinette, M.: Social Potentials for Scalable Multi-Robot Formations. *IEEE International Conference on Robotics and Automation*, (2000)

[4] Balch, T., and Arkin, R.: Behavior-based Formation Control for Multi-Robot Teams. *IEEE Transactions on Robotics and Automation*, **14** (1998) 1–15

[5] Spears, W., Heil, R., Spears, D., Zarzhitsky D.: Physicomimetics for Mobile Robot Formations. *Proceedings of the Third International Joint Conference on Autonomous Agents and Multi Agent Systems*, (2004) 1528–1529

[6] Spears, W., Spears, D., Hamann, J., Heil, R.: Distributed, Physics-Based Control of Swarm of Vehicles. *Autonomous Robots*. Kluwer, 17 (2004) 137–164

[7] Grefenstette, J.: A System for Learning Control Strategies with Genetic Algorithms. *Proceedings of the Third International Conference on Genetic Algorithms*, Morgan Kaufmann, (1989) 183–190

[8] Wu, A., Schultz, A., Agah, A.: Evolving Control for Distributed Micro Air Vehicles. *Proceedings of the IEEE International Symposium on Computational Intelligence in Robotics and Automation*, IEEE Press, (1999)

[9] Hettiarachchi, S., Spears, W., Green, D., and Kerr, W.: Distributed Agent Evolution with Dynamic Adaptation to Local Unexpected Scenarios. Proceedings of the 2005 Second GSFC/IEEE Workshop on Radical Agent Concepts, Springer, (2006)

[10] Spears, W.: Simple Subpopulation Schemes. *Proceedings of the Evolutionary Programming Conference*, World Scientific, (1994) 296–307

[11] Watson, R., Ficici, S., Pollack, J.: Embodied Evolution: Distributing an Evolutionary Algorithm in a Population of Robots. *Robotics and Autonomous Systems*, Elsevier, 39 (2002) 1–18

[12] Grefenstette, J.: *Credit Assignment in Rule Discovery Systems Based on Genetic Algorithms.* Springer-Verlag, 3 (1988) 225–245

[13] Haeck, N.: Minimum Distance Between a Point and a Line. http://www.simdesign.nl/tips/tip001.html, (2002)

[14] Hettiarachchi, S., Spears, W.: DAEDALUS for Agents with Obstructed Perception. *Proceedings of the 2006 IEEE Mountain Workshop on Adaptive and Learning Systems*, IEEE Press, (2006)

[15] Spears, W., Hamann, J., Maxim, P., Kunkel, T., Heil, R., Zarzhitsky, D., Spears, D., Karlsson, C. Where are you? In Şahin, E., Spears, W., eds.: *Swarm Robotics*, Springer-Verlag, (2006)

[16] Bonabeau, E., Dorigo, M., Theraulaz, G.: *Swarm Intelligence: From Natural to Artificial Systems*. Oxford University Press, Santa Fe Institute Studies in the Sciences of Complexity, (1999)

[17] Hayes, A., Martinoli, A., Goodman, R.: Swarm Robotic Odor Localization. *IEEE/RSJ International Conference on Intelligent Robots and Systems*, (2001)

[18] Schultz, A., Parker, L.: *Multi-Robot Systems: From Swarms to Intelligent Automata.* Kluwer, (2002)

[19] Fredslund, J., Matarić, M.: A General Algorithm for Robot Formations Using Local Sensing and Minimal Communication. *IEEE Transactions on Robotics and Automation*, **18** (2002)

[20] Fax, J., Murray, R.: *Information Flow and Cooperative Control of Vehicle Formations.* IFCA World Congress, (2002)

[21] Desai, J., Ostrowski, J., Kumar, V.: Controlling Formations of Multiple Mobile Robots. *IEEE International Conference on Robotics and Automation*, (1998)

[22] Desai, J., Ostrowski, J., Kumar, V.: Modeling and Control of Formations of Nonholonomic Mobile Robots. *IEEE Transactions on Robotics and Automation*, **17**(2001) 905–908

[23] O'Hara, K. J., Bigio, V. L., Dodson, E. R., Irani, A., Walker, D. B., and Balch, T. R.: Physical Path Planning Using the GNATs. *IEEE International Conference on Robotics and Automation*, Barcelona, Spain, (2005)

[24] Khatib, O.: Real-time Obstacle Avoidance for Manipulators and Mobile Robots. *International Journal of Robotics Research.* MIT Press (1986) 90–98

[25] Howard, A., Matarić, M., Sukhatme G.: Mobile Sensor Network Deployment Using Potential Fields: A Distributed, Scalable Solution to the Area Coverage Problem. *Sixth International Symposium on Distributed Autonomous Robotics Systems*, (2002)

[26] Vail, D., Veloso M.: Multi-robot Dynamic Role Assignment and *Coordination Through Shared Potential Fields.* Multi-Robot Systems, Kluwer, (2003)

[27] Schoenwald, D., Feddema, J., Oppel, F.: Decentralized Control of a Collective of Autonomous Robotic Vehicles. *American Control Conference*, (2001) 2087–2092

[28] Reif, J., Wang, H.: Social potential fields: A Distributed Behavioral Control for Autonomous Robots. *Workshop on the Algorithmic Foundations of Robotics*, (1998)

[29] Fax, J., Murray, R.: Information Flow and Cooperative Control of Vehicle Formations. *IEEE Transactions on Automatic Control,* **49** (2004) 1465–1476

[30] Borenstein, J., Everett, H., Feng, L.: Where am I? *Sensors and Methods for Mobile Robot Positioning*, University of Michigan, (1996)

[31] Heil, R.: A Trilaterative Localization System for Small Mobile Robots in Swarms. Master's thesis, University of Wyoming, Laramie, WY, (2004)

[32] Zarzhitsky, D., Spears, D., Spears, W.: Distributed Robotics Approach to Chemical Plume Tracing. *IEEE/RSJ International Conference on Intelligent Robots and Systems (IROS'05)*, (2005) 4034–4039

[33] Thomas, F., Ros, L.: Revisiting Trilateration for Robot Localization. *IEEE Transactions on Robotics*, **21(1)** (2005) 93–101

[34] Peasgood, M., Clark, C., McPhee, J.: Localization of Multiple Robots with Simple Sensors. *IEEE/RSJ International Conference on Intelligent Robots and Systems (IROS'05)*, (2005) 671–676

[35] MacArthur, D.: *Design and Implementation of an Ultrasonic Position System for Multiple Vehicle Control*. Master's thesis, University of Florida, (2003)

[36] Nissanka, B., P.: The Cricket Indoor Location System. Doctoral thesis, Massachusetts Institute of Technology, Cambridge, MA, (2005)

[37] Navarro-Serment, L., Paredis, C., Khosla, P.: A Beacon System for the Localization of Distributed Robotic Teams. *International Conference on Field and Service Robots, Pittsburgh*, PA, (1999) 232–237

[38] Rothermich, J., Ecemis, I., Gaudiano, P.: Distributed localization and mapping with a robotic swarm. In Şahin, E., Spears, W., eds.: *Swarm Robotics*, Springer-Ve rlag, (2004) 59–71

[39] Payton, D., Estkowski, R., Howard, M.: Pheromone Robotics and the Logic of Virtual Pheromones. In Şahin, E., Spears, W., eds.: *Swarm Robotics*, Springer-Verlag, (2004) 46–58

In: Robotics Research Trends
Editor: Xing P. Guô, pp. 155-191

ISBN 1-60021-997-7
© 2008 Nova Science Publishers, Inc.

Chapter 4

PRELIMINARY-ANNOUNCEMENT FUNCTION OF MOBILE ROBOTS' UPCOMING OPERATION

Takafumi Matsumaru
Shizuoka Univ., Faculty of Engineering
Dept. of Mechanical Engineering

Abstract

We propose approaches and equipment for preliminarily announcing and indicating to people the speed and direction of movement of mobile robots moving on a two-dimensional plane. We introduce the four approaches categorized into (1) announcing the state just after the present and (2) indicating operations from the present to some future time continuously. To realize the approaches, we use omni-directional display (PMR-2), flat-panel display (PMR-6), laser pointer (PMR-1), and projector (PMR-5) for the announcement unit of prototype robots. The four robots were exhibited at the 2005 International Robot Exhibition (iREX05). We had visitors answer questionnaires in a 5-stage evaluation. The projection robot PMR-5 received the highest evaluation score among the four. An examination of differences by gender and age suggested that some people prefer friendly expressions, simple method to inform, and a minimum of information to be presented at one time.

1. Introduction

We propose approaches and equipment for preliminarily announcing and indicating to people the speed and direction of mobile robots moving on a two-dimensional plane. We introduce four prototype robots in which our approaches are realized.

With a dropping birth rate and an aging society, robotics and mechatronics are expected to spread through society to assist, entertain, and otherwise interact with human beings in daily life [1]. Those familiar with the technological potential this involves may see this as a positive thing, but those who are not may find the "invasion of the robots" — "scary", "incompatible", or otherwise unwelcome. This is in part because considerations of affinity remain insufficient. Our purpose is to determine approaches and equipment that will at least let people know when a "robot" is nearby, how fast it's moving and what direction it's headed.

This chapter is organized as follows: Section 2 introduces our research and our approaches with indicating the background and objectives and the relating and previous researches. Section 3 shows the simulation experimentation in which the effect of preliminarily announcing and indicating, the differences according to methods, and the appropriate timing to announce were examined by using a software simulation before the hardware equipments that make the approaches embodiment were designed and manufactured. Section 4 describes the four prototype robots we developed and how they make their upcoming operation known. Section 5 introduces results of a questionnaire we distributed for people to evaluate four prototype robots exhibited at the 2005 International Robot Exhibition (iREX05) in Tokyo in the winter of 2005.

2. Preliminary-Announcement and Indication Function of Robots' Upcoming Operation

2.1. Background

Human beings interact, signaling their own and predicting others' actions and intentions nonverbally through body language, hand gestures, facial expressions, and whole body operations. Most people moving through a crowd, for example, find little trouble plotting a passage through a forest of bodies without bumping into or otherwise upsetting or needlessly distracting others. Human beings hone social and physical skills that make their movement practically second nature based on a sense of "affinity", familiarity, common appearance — sharing what they like to call "common sense".

Robots call up exactly the opposite reaction — disaffinity, unfamiliarity, uncommon appearance — no sharing of the common sense that would make robot movement predictable to people — or human movement predictable to "robots". If I see that you are bent forward, walking fast, and your eyes are on a distant goal, I can predict that getting out of your way is appropriate and I can guess how fast I'll have to move to let you through. If, however, you are a robot, I probably have no idea how to deal with your movement — how fast it might be and what direction you may take. This lack of shared knowledge and common sense between human and artificial organisms is what has made their interaction such a problem — the simplest aspect of which is to avoid the risk of contact and collision.

2.2. Objectives

Research on safety [2,3] has included temporary stop functions when a robot is approached [4, 5], evaluation of resistant value of the pain sense of user at contact [6], detection of contact with objects in a robot arm [7], flexible coating of a robot arm [8], and softening of robot joints by software control [9] or other mechanisms. Such mechanisms are realized, for example, by compressible fluid or nonlinear springs in antagonistic drives [10] or electrical viscous fluid or blade springs in direct drives [11].

We studied functions to notify people of a robot's approach upcoming and intentions before it moves to avoid the risk of unintended contact or collision between people and robots, focusing on mobile robots or transport vehicles moving on a two-dimensional (2D)

Figure 1. Contact and collision avoidance. [12]

plane (**Figure 1**) [12]. Such robots should announce their direction and speed of movement. Such announcements should be simple and immediately understandable enough to be acceptable to people. They must also be easy for people to understand based on common sense.

We studied communication excluding sound and voice, which could transmit information to everyone around, including those not wanting or needing it and disturb the public peace.

The robot assumed here does not avoid contact or collision by announcement and indications alone, but must detect obstacles and avoid them autonomously. We do not assume that the robot comes into contact with those careless in noting robot movement. It is not necessary to inform those unaware of the robot by sound or voice even though safety is ensured, since offering unwanted information could cause more harm than good. The function we studied provides information about upcoming robot action to those responding actively to information. If, for example, a robot and a pedestrian approach from opposite ends of a passage, the robot detects the pedestrian, then (1) stops and waits for the person to pass or (2) continues forward while avoiding touching the person in passing. The pedestrian recognizing the robot's intent then responds to (1) by continuing forward and passing the waiting or to (2) by keeping to the left and passing the robot more safely.

2.3. Relating and Previous Researches

Most research on nonverbal interfacing between human beings and robotic counterparts has been viewed as the transmission of information from the person to the machine to communicate with or operate robotic systems [13]. Much research has involved imaging of body language or gestures [14–16] and facial expressions or glances [17–19], focusing on human action and intentions. Some research has focused on analyzing feelings based on psychological knowledge of human facial expressions [20, 21].

Comparatively little research has concerned the transmission of information from the machine to the user, especially about robotic intent. Indicating a robot's internal state, a screen displays remaining battery charge, internal temperature, etc., in a mobile robot [22]. For communicating upcoming robot movement, projection to common space between the manipulator and the user is proposed [23]. Experiments on industrial robots with LEDs supporting eating by the physically handicapped people have been reported [24, 25]. Toyota Motor Corp. announced "road surface depiction (laser tactile sensing)" in autumn 2004 similar to preliminary announcing its upcoming action [26]. This mainly targets collision avoidance among vehicles, not indicating subsequent movement of the vehicle to pedestrians.

2.4. Proposals

We proposed the four approaches categorized into two types to preliminarily announce and indicate robot speed and direction of movement (**Table 1**) [27–29]. The first type announces the state just after the present and the second type indicates operations from the present to some future time continuously. For the first type we propose the lamp method and the blowout method, and for the second type we propose the light ray method and the projection method.

(a) **Lamp Method**: Lamps on the top of the mobile robot announce the direction of movement by turning the lamp on along which the robot is to move. Blinking speed or colors is used to indicate the speed of movement (**Table 1(a)**).

(b) **Blowout Method**: Blowout is a toy or a party gadget in which blowing air into the cylinder extends it and stopping blowing makes it rewind from top. The blowout put on a turntable is set up on the mobile robot (**Table 1(b)**). Total length and tip direction of the blowout express the speed and direction of movement of mobile robot.

(c) **Light Ray Method**: Movement afterimage of radiant irradiated on running surface expresses the scheduled route (**Table 1(c)**). Period to display the scheduled route is predefined, such as until 3-second-later. Strong point is the "situation on the way", such as straight route or curved route, can be indicated directly.

(d) **Projection Method**: Projection equipment projects a two-dimensional frame on a running surface to indicate both the scheduled route and the state of operation, such as stopping or going backward (**Table 1(d)**). Internal condition of the robot such as remaining battery charge and warnings on worn parts or overheating apparatus is also displayed in the frame.

2.5. Applications

We focused on two applications for announcing robot action (**Figure 2**) [30].

2.5.1. Automobiles

Automobiles are the most familiar example of user/robot coexistence in artificial mechatronics. In passenger cars, the relationship between people and the system has focused mostly to the comfort, maneuverability, and safety of the driver. Considering current transportation condition involving pedestrians, such as narrow roads or roads without walkway, we should take notice further to the relationship between the pedestrian around and the vehicle (**Figure 2(a)**).

Turn indicators (winkers) on automobiles are extremely limited for announcing intended action of any complexity both among drivers and between drivers and pedestrians.

Table 1. Proposed approach. [27–29]

(1) Announcing state just after the present		(2) Indicating continuous operations	
(a) Lamp	(b) Blowout	(c) Light ray	(d) Projection
Several lamps are set on the top. Direction of movement is by turning the lamp on. Speed of movement is by blinking speed or colors.	Blowout put on turntable is set on the top. Speed of movement is by length of blowout. Direction of movement is by tip direction of blowout.	Light ray draws scheduled route on running surface from present to some future time using pan-tilt mechanism. Situation on the way can be indicated.	Not only the scheduled route but also the state of operation, such as stopping or going backward, can be displayed in the projected frame.

The distinction among turning, lane changing, and pulling over to the curb, for example, is not clear using the winker alone. Moreover pedestrians cannot understand which corner the car will turn or wants to turn on intricate system of roads. We focused on replacing the winker with predictive signaling for automobiles.

2.5.2. Robot Vacuum Cleaners

Another suitable application appears to be the robot vacuum cleaner, a home cleaning robot that operates autonomously to clean room floors wall to wall now commercially available and popular.

Robot cleaners are not sophisticated enough to account for the presence of other than immovable object — persons in the same room must watch their steps and cannot anticipate the cleaner's route (**Figure 2(b)**). If the robot cleaner could indicate its upcoming movement, people could avoid contact and collision easily even if he/she wants to pass through nearby the cleaner at work.

3. Simulation Experimentation

The effect of preliminarily announcing and indicating, the differences according to methods, and the appropriate timing to announce were examined by using a software simulation before the hardware equipments that made the approaches embodiment were designed and manufactured [31–33]. The simulation system was programmed by using Visual C++ and OpenGL on Windows PC. The "**chasing task**" was adopted as a method to evaluate quantitatively. First of all, the translational movement to the right and left on a straight line and the rotational movement right-handed/left handed on the spot were examined as one degree of freedom movement. And then, the planar movement on two dimension plane that was the combination of translation and rotation was examined.

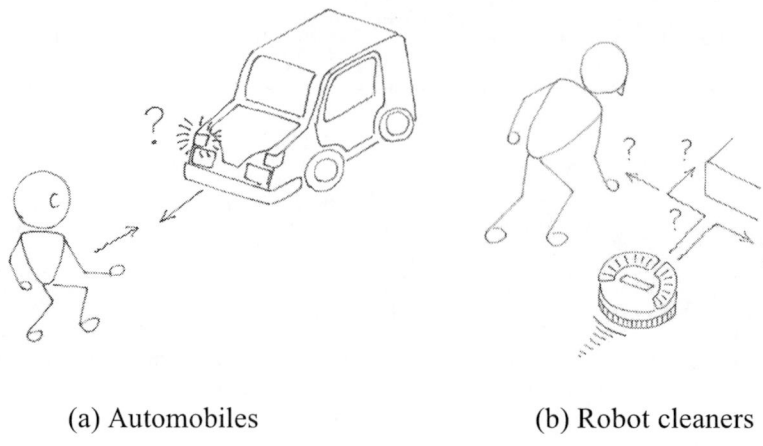

(a) Automobiles   (b) Robot cleaners

Figure 2. Applications. [30]

In planar movement, the mobile robot, $0.4m$ height and $0.4m$ diameter, moves about on a plane while changing the speed and direction of movement at random (**Figure 3**). For subject people, the image from a virtual camera on the operation robot is presented on the computer screen. This image was given as an aspect at $1.6m$ height which is corresponding to the height of human eye in the view looking down at the mobile robot diagonally in front. The subject pursues the mobile robot by controlling the operation robot using a joystick while referring to the preliminarily announcing and indicating the mobile robot's upcoming operation by the equipment set up on the mobile robot. The subject was directed to superpose a virtual mark of the operation robot ($0.5m$ forward and $0.4m$ height from the center of robot) on that of the mobile robot ($0.4m$ height at the center on top of the robot) both in position and in direction as much as possible. Moreover to make the acquired experimental data consistent, we directed the subjects to give priority to matching the position of both marks more than meeting the direction. Movement of mobile robot corresponds to movement of actual robot, and movement of operation robot is relevant to human recognition of preliminary announcement and indication of robot's upcoming movement. Comparative study was made of four types of method (on/off lamp without speed information, lamp with changing colors responding to speed of movement, blowout (telescopic arrow), and light ray (drawing scheduled route)) with six kinds of timing (from $0.5s$ to $3.0s$ before the actual operation in every $0.5s$). The experiment was carried out three times for $60s$ on each condition by each subject.

The result of position error and direction error by ten male subjects in the twenties is shown in **Table 2** and **3**. The following conclusions were obtained from the examination by the simulation experimentations.

- In the preliminary announcement and indication of mobile robot's upcoming operation, not only the direction of movement but also the speed of movement is important information.

- As for a continuous indication (scheduled route drawn by light ray), the mean value of

Preliminary-Announcement Function of Mobile Robots' Upcoming Operation 161

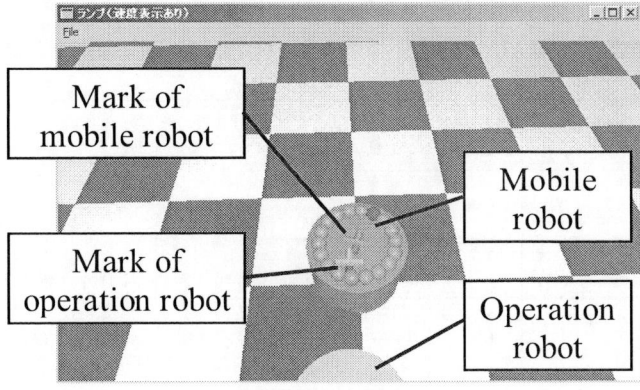

(a) Lamp method

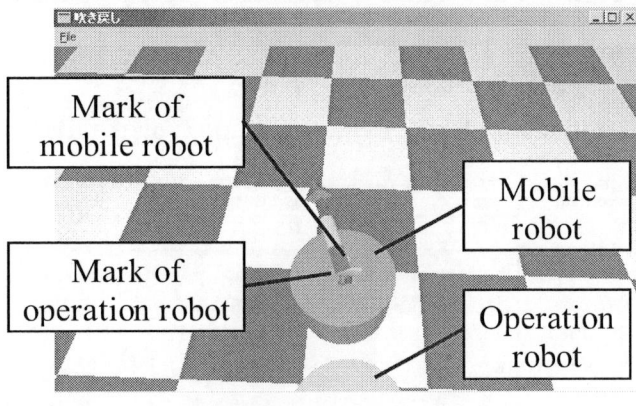

(b) Blowout method

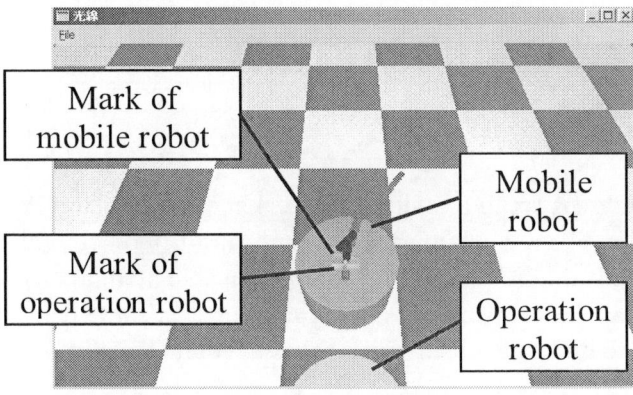

(c) Light ray method

Figure 3. Simulation experimentation. [33]

Table 2. Result of simulation experimentation (distance index). [33]

Method	Timing [s]					
	0.5	1.0	1.5	2.0	2.5	3.0
No	9.12 (1.46)					
Lamp (w/o speed)	8.53 (1.89)	7.95 (1.46)	7.49 (1.17)	7.98 (1.52)	7.81 (1.01)	7.35 (1.25)
Lamp	8.32 (1.52)	7.44 (1.50)	7.18 (0.75)	7.01 (1.07)	7.29 (1.01)	7.35 (1.18)
Blowout	6.54 (1.08)	6.46 (0.70)	6.85 (0.92)	6.70 (0.82)	7.58 (1.16)	7.52 (1.15)
Light ray	7.30 (0.75)	7.18 (1.22)	6.62 (0.83)	6.63 (0.76)	6.20 (0.71)	6.31 (0.65)

unit: cm, upper average, lower: standard deviation

Table 3. Result of simulation experimentation (orientation index). [33]

Method	Timing [s]					
	0.5	1.0	1.5	2.0	2.5	3.0
No	8.33 (1.95)					
Lamp (w/o speed)	8.94 (1.80)	8.47 (1.46)	7.75 (1.52)	8.31 (1.35)	8.62 (1.18)	8.44 (1.82)
Lamp	8.44 (1.57)	7.83 (1.03)	7.31 (0.91)	7.76 (1.71)	8.06 (1.34)	7.93 (1.45)
Blowout	6.12 (0.99)	6.76 (1.28)	7.01 (1.09)	8.37 (1.29)	8.59 (1.46)	8.95 (1.35)
Light ray	7.56 (0.99)	6.62 (1.05)	6.36 (0.85)	6.25 (0.82)	6.37 (1.07)	6.23 (0.85)

unit: deg, upper average, lower: standard deviation

evaluation index is small (smaller is better) and the standard deviation is also small on the whole. For many subjects, a significant difference with other methods was confirmed. This might be the most comprehensible method on announcing and indicating mobile robot's upcoming operation. However some length is necessary on the drawn route to make people around easy to recognize the information.

- In the type of announcing the state just after the present, blowout method is more effective than lamp method. People may understand information easily by change of shape rather than by color variation, and by successive change more than by discrete transition.

- In the type of announcing the state just after the present, it is thought that the most appropriate timing to preliminarily announce is around $1.0s$ to $1.5s$ before the actual operation. If the period between the announcement and the actual movement is too

Table 4. Prototype robots. [30]

(a) PMR-2	(b) PMR-6	(c) PMR-1	(d) PMR-5
D450×W480×H930mm 30kg max180mm/s,22rad/s	D470×W480×H450mm 30kg max180mm/s,22rad/s	D460×W480×H910mm 35kg max180mm/s,22rad/s	D500×W440×H1020mm 35kg max180mm/s,22rad/s

short, it is not possible to synchronize two marks because there is not enough time from recognition to reaction. On the contrary when the period is too long, people tend to react too early or the content of information is not correctly memorable.

4. Prototype Robots

This section explains the four prototype robots we developed to realize the proposed approaches (**Table 4**) [30]. The first type robots announce the state in the near future. Lamp method is embodied using an omni-directional display and blowout method is exteriorized using a flat-panel display. The second type robots indicate continuous operation from the present to some future time. Light ray method is embodied using laser pointers and projection method is exteriorized using a projector.

4.1. PMR-2: Eyeball Robot (Omni-Directional Display)

4.1.1. Outline

In PMR-2, a commercial omni-directional display, magicball(R), is used for announcement (**Table 4(a)**, **Figure 4**) [34–38]. Via $3,000 rpm$ high-speed rotation of 96 LEDs in three lines on a board characters and pictures are displayed in eight colors on side face of spherical body [39]. Messages visible omni-directionally are input on a Windows PC and stored in the magicball memory as 256 by $32 dot$ image frames. A communication command from the PC via RS232-C calls frames from memory and displays them. The frame is refreshed every $0.5s$ due to the required period to read and load the frame.

The mobile unit, common to all four robots, is two-wheel drive driven by DC servomotor with reduction gears and encoders (**Figure 5**). A ball-caster with suspension mechanism is used as trailing wheels. When the program is started, the timer is set and three threads

Figure 4. Overview of PMR-2. [30]

are started: the instruction input, the movement (driving wheel) control, and the announcement display (**Figure 6**). These threads work in $50ms$ at the same time. Robot takes the speed instruction which consist of rotation factor J_x and translation factor J_y every $50ms$. A joystick, *Side Winder Force Feedback 2, Microsoft Co.*, or a game pad, *JC-U912BK, Elecom Co.*, is used as the input device. The movement of robot is controlled with dividing the maximum speed of driving wheels into translation quota and rotation quota (**Figure 7**). Both the distance that the robot should proceed during control cycle ($50ms$) and the rotation angle that each wheel should rotate during the meantime can be calculated. Therefore the target speed of two driving wheels, V_1 and V_2, are decided from the instruction factors, J_x and J_y.

Basic design shown on the display is based on an "**eyeball**" to induce easy familiarization among human beings encountering the robot. Sightline is a clue for people to estimate other's action and intention. We also want to make the robot friendly for people. Targeted movement involves three speeds — high, low, and stop — and four directions of movement — straight, loose turn, tight turn, and on-the-spot rotation. The speed of movement is expressed by the **degree of eye opening** — fully open at high speed, half open at low speed, and closed when stopped (**Figure 8(a)**). The direction of movement is indicated by **eye positioning** — $0deg$ from frontal when going straight, $30deg$ from frontal when making loose turn, $60deg$ from frontal when making tight turn, and $90deg$ from frontal during on-the-spot rotation (**Figure 8(b)**).

Magicball has spherical body so that the frontal that is the criterion to indicate the direction of movement is not understandable. Accordingly we set the fixed frame that covers the viewing area except foreside where eyeballs are displayed. Color of the frame is switched according to the speed of movement in reference to traffic signals — green at high speed, yellow at low speed, and red when stopped. Based on simulation experiments,

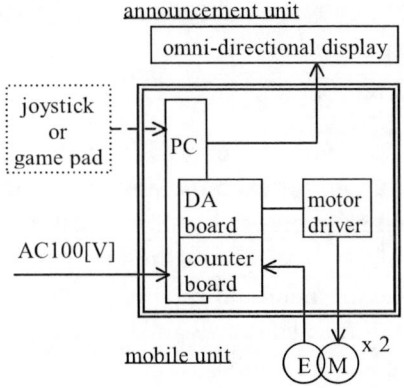

Figure 5. System configuration of PMR-2. [40]

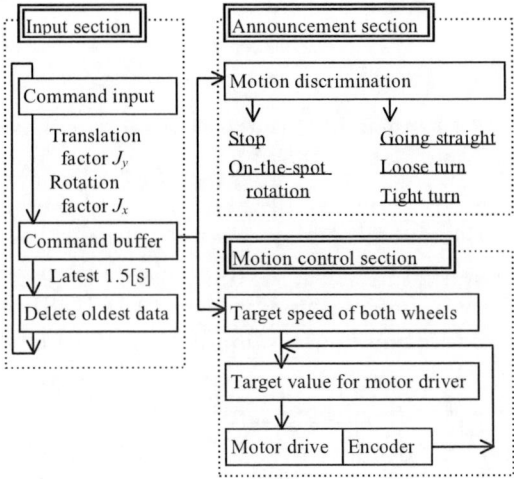

Figure 6. Three sections in program. [40]

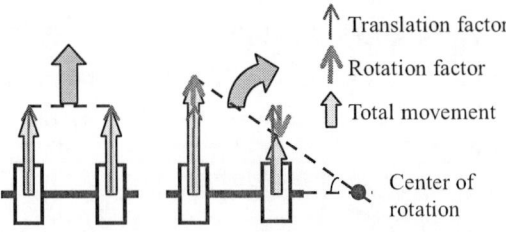

Figure 7. Movement control. [40]

eye behavior corresponding to operation is displayed $1.5s$ before the actual operation [38].

4.1.2. Transition and Continuity

Transition and continuity make the eyeball expression more natural and usual. If the transition is not taken into consideration, for example, in the case that the mobile robot switches the movement from the right-handed loose turn (eye positioning at $30deg$ from frontal on the right) to the left-handed tight turn (eye positioning at $60deg$ from frontal on the left), the eyes are disappeared at the moment and re-appeared at large space immediately. People around might feel it unnatural and sometimes they lose the sight of eyes. To make easy to follow the eye positioning, transition of eye positioning should be considered when changing the eyeball expression (**Figure 9(a),(b)**). Moreover every $30deg$ for eye positioning is too wide viewed in terms of natural feeling. Consequently we decided the eyes are positioned more frequently at every $15deg$ during transition (**Figure 9(c)**).

4.2. PMR-6: Arrow Robot (Flat-Panel Display)

4.2.1. Outline

In PMR-6, a flat-panel display is used instead of some mechanism that imitates actual blowout (telescopic arrow) (**Table 4(b)**, **Figure 10**) [40, 41]. Display is useful to test various composition, shapes, and colors only by changing the content to be displayed, although modification or improvement of mechanism is difficult once it is manufactured. As a multifunctional interface with display, Digital Desk [42] is well known and several trial systems have been reported [43–46]. There are several researches on teleoperation or teaching of robotic systems using multimedia display [47, 48] or PDA (Personal Digital Assistant) [49, 50]. In PMR-6 a commercial liquid crystal display (LCD) is adopted for the announcement unit. It is a 17-inch type — maximum size to be installed on the top of the robot without protrusion — and is selected considering wide viewing angle — $\pm 80deg$.

Basic design shown on the display is made "**arrow**" considering real blowout as telescopic arrow. Arrow is commonly used when indicating direction and comparatively comprehensible as a sign to express movement even if the sign is looked at for the first time. The speed of movement is expressed as the **size (length and width) and color (based on traffic signal) of arrow** — large green arrow at high speed, small yellow arrow at low speed, and red characters when stopped (**Figure 11(a)**). The direction of movement is described with the **curved condition of arrow** — straight when going straight, curved when making loose turn, swerved when making tight turn, and rounded during on-the-spot rotation (**Figure 11(b)**).

4.2.2. Timing to Announce

Arrow or characters are displayed $1.5s$ before the actual operation. This is because the comparison between PMR-6 (flat-panel display, arrow) and PMR-2 (omni-directional display, eyeball) is emphasized on the evaluation about the intelligibility of the preliminary-announcement and indication function. The magicball in PMR-2 takes $0.5s$ to switch the contents to display and it is difficult to change the contents continuously in a short time.

Preliminary-Announcement Function of Mobile Robots' Upcoming Operation 167

straight-fast　　　　straight-slow　　　　　stop
(w. green frame)　　(w. yellow frame)　　(w. red frame)

(a) Speed of movement: degree of eye opening

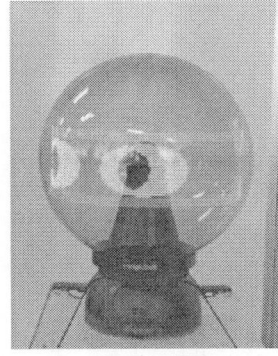  

loose turn　　　　　tight turn　　　　on-the-spot rotation

(b) Direction of movement: eye positioning

Figure 8. Exact description of announcing on PMR-2. [30]

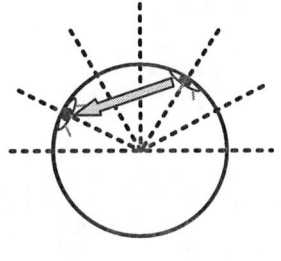 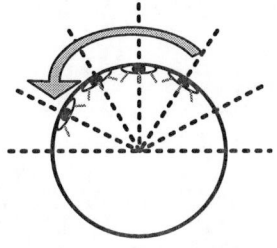 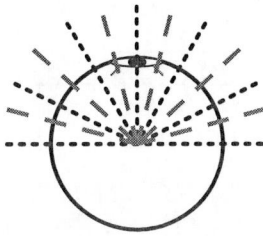

(a) abrupt transition　　(b) smooth transition　　(c) setting for continuity

Figure 9. Transition and continuity (top view). [38]

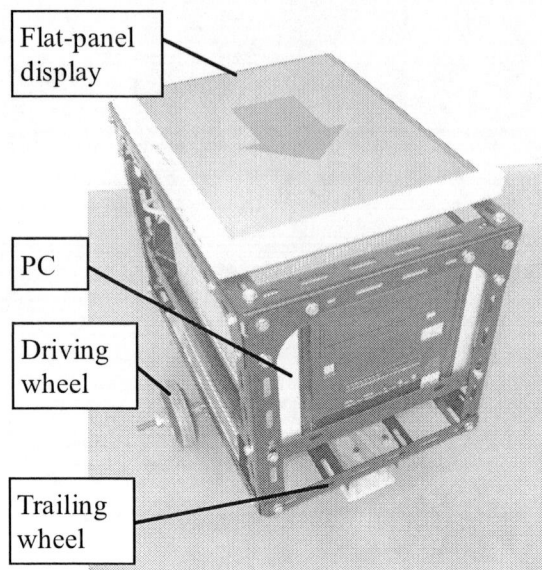

Figure 10. Overview of PMR-6. [40]

Consequently the condition to evaluate is aligned to the discrete expression, $1.5s$ before the actual operation.

4.2.3. Announcement Procedure

Robot takes the speed instruction which consists of rotation factor J_x and translation instruction J_y. Each value is from -1000 to $+1000$. There are two steps to display the upcoming operation.

1) Distinction of state of operation: First the state of operation (the content to be displayed) is distinguished from speed instruction (J_x and J_y) as in the middle row of **Table 5**, regardless whether the robot moves along pre-defined route or it is operated manually in real time.

2) Content to be displayed: The content corresponding to the state of operation which is distinguished in *1)* is displayed as in the right row of **Table 5**.

4.3. PMR-1: Light Ray Robot (Laser Pointer)

4.3.1. Outline

In PMR-1, laser pointers are used as the light source for the announcement unit (**Table 4(c)**, **Figure 12**) [51, 52]. In transferring the intention among human beings using laser pointer, Gesture Cam [53], Geture Laser [54], CTerm [55], Telepointer [56], etc. have been presented to support to transfer the supervisor's intention and instruction to the operator in remote site. In PMR-1 seven laser pointers are bundled to increase amount of radiation so that people can easily recognize the radiant on running surface. Those are set up upward

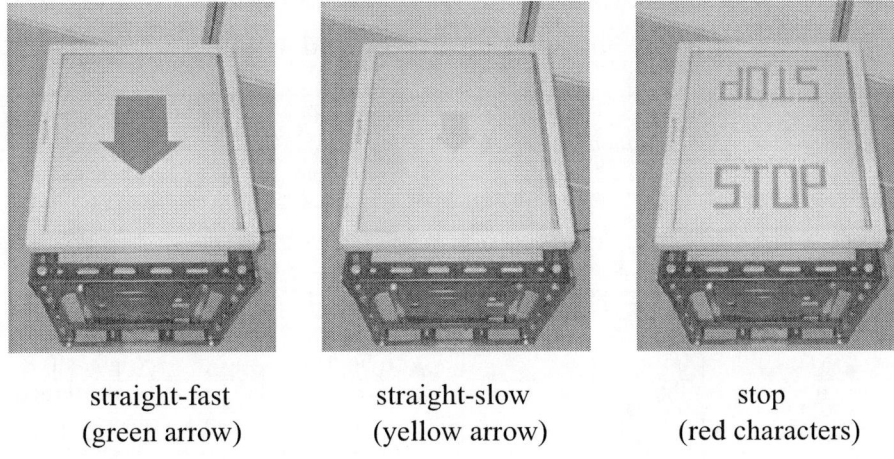

straight-fast 　　straight-slow 　　stop
(green arrow) 　(yellow arrow) 　(red characters)

(a) Speed of movement: size/color of arrow, characters

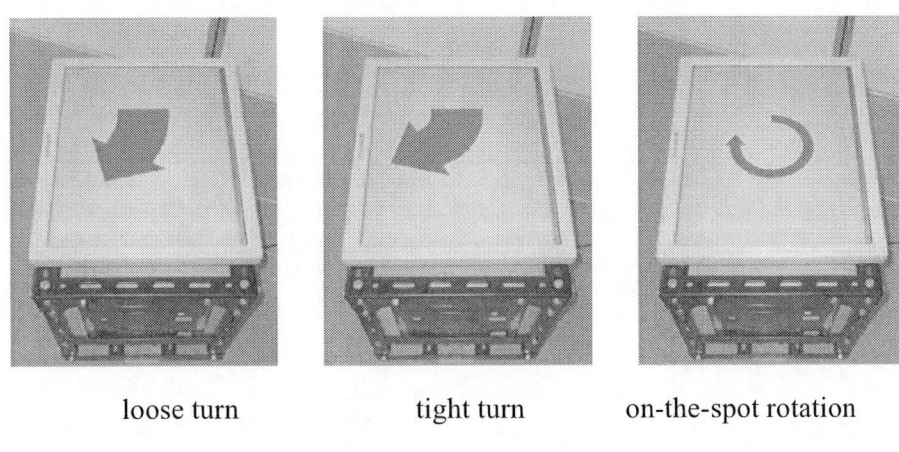

loose turn 　　　tight turn 　　on-the-spot rotation

(b) Direction of movement: curved condition of arrow

Figure 11. Exact description of announcing on PMR-6 (on white background). [40]

just under the mirror and light rays reflected on the mirror are irradiated on running surface. Reciprocating movement of mirror by stepping motors around pan and tilt axes draws the "**schedule route**" as the movement afterimage of radiant. Range of rotation around pan and tilt axes is restricted so that laser lights strictly downwards from the height of the mirror.

Period to display of the scheduled route is predefined. The developed robot draws the route until 3-second-later (**Figure 13(a)**). "Situation on the way" can be presented; for example, when the robot is moving to some point, it can display whether it will go straight directly or it will make a detour to avoid something (**Figure 13(b)**). Accordingly, the speed of movement is expressed as the **length of drawn route** and the direction of movement is shown as the **direction of drawn route** directly. Brightness and shape of the radiant can be controlled with seven laser pointers, for example, to express the change in speed.

Table 5. Operation discrimination. [40]

Input		Operation	Content to be displayed
Translation factor J_y	Rotation factor J_x	(speed) (direction)	(size, color) (orientation)
0	0	Stop	STOP character (red) (for/back)
0	$0 < J_x < \|500\|$	on-the-spot rotation (slow) (right/left)	rounded arrow (thin, yellow) (right/left)
0	$\|500\| < J_x < \|1000\|$	on-the-spot rotation (fast) (right/left)	rounded arrow (thick, green) (right/left)
$0 < J_y < \|500\|$	0	going straight (slow) (fore/back)	straight arrow (thin-short, yellow) (fore/back)
$0 < J_y < \|500\|$	$0 < J_x < \|500\|$	loose turn (slow) (fore/back, right/left)	curved arrow (thin-short, yellow) (fore/back, right/left)
$0 < J_y < \|500\|$	$\|500\| < J_x < \|1000\|$	tight turn (slow) (fore/back, right/left)	swerved arrow (thin-short, yellow) (fore/back, right/left)
$\|500\| < J_y < \|1000\|$	0	going straight (fast) (fore/back)	straight arrow (thick-long, green) (fore/back)
$\|500\| < J_y < \|1000\|$	$0 < J_x < \|500\|$	loose turn (fast) (fore/back, right/left)	curved arrow (thick-long, green) (fore/back, right/left)
$\|500\| < J_y < \|1000\|$	$\|500\| < J_x < \|1000\|$	tight turn (fast) (fore/back, right/left)	swerved arrow (thick-long, green) (fore/back, right/left)

4.3.2. Coordination between Preliminary-Announcement and Movement

The relation among the instruction input, the preliminary-announcement, and the movement control as time goes on is illustrated in **Figure 14**. The execution of instruction will delay for $3s$ after the instruction is input.

The robot takes a speed command every $50ms$ regardless as to whether it moves along the course decided beforehand or it receives the instructions in real time from some input device. The speed commands during the latest $3s$ are always accumulated in the command buffer, and the command value $3s$ ago is used to calculate the target value to control the driving wheels at the time. Cycle time to control the angle position of driving wheels is also $50ms$.

The announcement unit receives the target position of robot for 1-second-later, 2-second-later, and 3-second-later at the beginning of the reciprocating movement of mirror. Then it draws the scheduled route on running surface. However the error between drawn route and robot's actual route will be increased and the round trip error on reciprocating

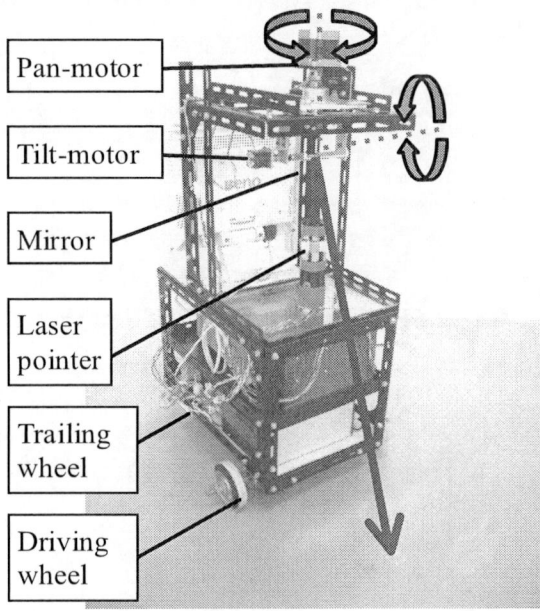

Figure 12. Overview of PMR-1. [51, 52]

movement of radiant will occur, when the robot moves relatively fast, if the target angles of mirror rotation, which is calculated from robot position at 1-second-later, 2-second-later, and 3-second-later, are computed only once at the beginning of the reciprocating movement of mirror. This is because the cycle of the reciprocating movement is set to $2Hz$, relatively slow, considering the visibility of the drawn route. Consequently we decided to recalculate the remaining target angles of mirror rotation whenever some target angle is achieved, considering the robot movement during the period to achieve the target angle (**Figure 15**). For example, at the time the rotation angle of mirror (ω_1, θ_1) for the robot position at 1-second-later (x_1, y_1) is realized, the next target angle (ω'_2, θ'_2) is calculated after computing the relative robot position at 2-second-later (x'_2, y'_2) considering the robot movement during the period to realize the rotation angle (ω_1, θ_1). As explained, the present robot position is referred to whenever it calculates the next target angle of the mirror after realizing the old target angle. When the mirror goes back to the initial position and one reciprocating movement is completed, the announcement unit receives the target robot position at 1-second-later, 2-second-later, and 3-second-later again, then the next reciprocating movement starts to draw a scheduled route.

4.3.3. Experiment on Coordination between Announcement and Movement

In order to check the correspondence of the scheduled route drawn using light ray to the mobile robot's actual trajectory, the experiment on coordination between the announcement and the movement is carried out. It is performed both in the case that the route is given beforehand and in the case that the robot is operated manually in real time.

(1) **Experiment on course given beforehand**: To check whether the announced route

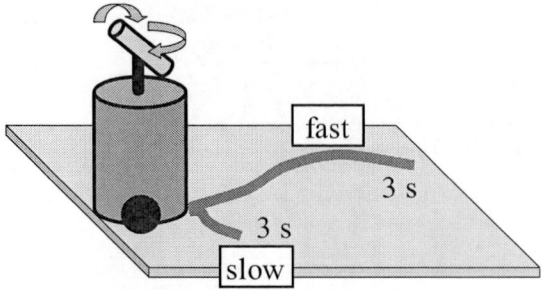

(a) Speed of movement: length of drawn route

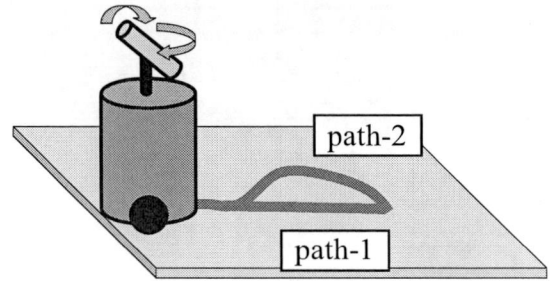

(b) Direction of movement: direction of drawn route

Figure 13. Exact description of announcing on PMR-1. [51, 52]

and the robot trajectory are corresponding we made the robot move along a course including going straight and clockwise/counterclockwise turns at various speed of movement. **Figure 16(a)** shows an experimental result in which actual robot trajectory (square) and the scheduled route drawn by light ray (circle) are given in the world coordinate in which the coordinate origin is at the initial position of the robot. From a static condition the robot (a) goes straight at $140mm/s$ for $3s$, at $200mm/s$ for $3s$, at $140mm/s$ for $4s$, and (b) stops. Then it (c) goes straight in the same way once again and (d) stops. It makes on-the-spot rotation for $90deg$ right-handed to change its orientation. Then the robot makes (e) clockwise turn for $12s$, (f) counterclockwise turn for $12s$, and (g) clockwise turn again for $12s$. All turn is at about $127mm/s$ along the circle of $970mm$ diameter until coming back to the initial position (h). From experimental results we confirmed the correspondence of the announced route to the robot trajectory when the robot moves along a course including going straight and clockwise/counterclockwise turns at various speed of movement.

(2) Experiment in real-time operation: To check whether the announced route and the robot trajectory are corresponding employing the algorithm mentioned in **Figure 14**, we operated the robot manually in real time. **Figure 16(b)** shows an experimental result in which the actual robot trajectory (square) and the scheduled route drawn by light ray (circle) are given in the world coordinate in which the coordinate origin is at the initial position of the robot. Positions of robot every $10s$ are shown in the figure as bordered circles to indicate the robot speed of movement. From experimental results we confirmed the announced route and the robot trajectory are well in agreement also under manual operation in real time.

Consequently we have validated the coordinate algorithm between the announcement

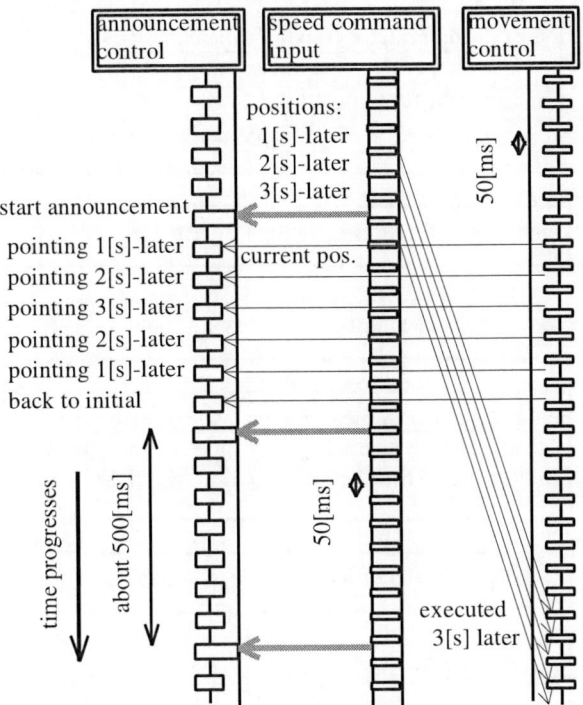

Figure 14. Coordinatoin between announcement and movement. [51, 52]

and the movement.

4.4. PMR-5: Projection Robot (Projector)

4.4.1. Outline

In PMR-5, a liquid-crystal projector is used for announcement (**Table 4(d)**, **Figure 17**) [57, 58]. Many reports have been presented on the real space projection using a projector in the field of augmented reality and computer supported cooperative work, for example, at Procams (IEEE Int. Workshop on Projector-Camera Systems) [59] such as Enhanced-Desk [60] and 3-D Tele-direction Interface [61]. Especially in robotics field, the projection function to common space between manipulator and user [23] and the remote collaboration system with shared view paying attention to both image projection and image capture to improve the efficiency of on-site support for remote worker [62] are proposed. In PMR-5 the projector is set up upward at front of the robot and the frame reflected on the mirror just above the projector is projected on running surface. This structure makes total height of robot, about $1.0m$, lower while the shortest projection distance, $1.2m$, is secured. Size of the projected frame on running surface is 36-inch type ($550mm$ length by $740mm$ width) with the mirror ($140mm$ length by $190mm$ width) at $1.0m$ height.

Main content of the projected frame is also the "**scheduled route**". Schedule route until 3-second-later can be displayed in the frame by the mounted projector and the settled mirror, considering the size of the projected frame and the mobility performance of the

174   Takafumi Matsumaru

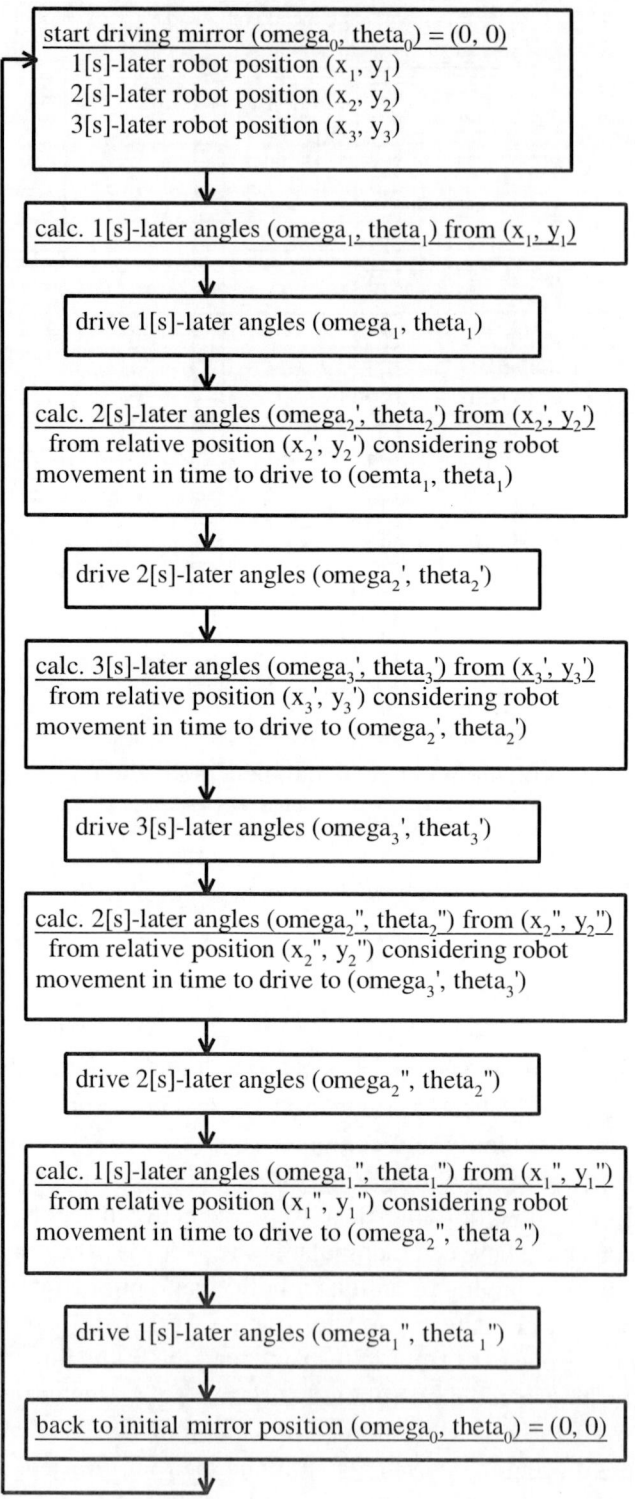

Figure 15. Procedure of drawing route by reciprocating movement of mirror. [51, 52]

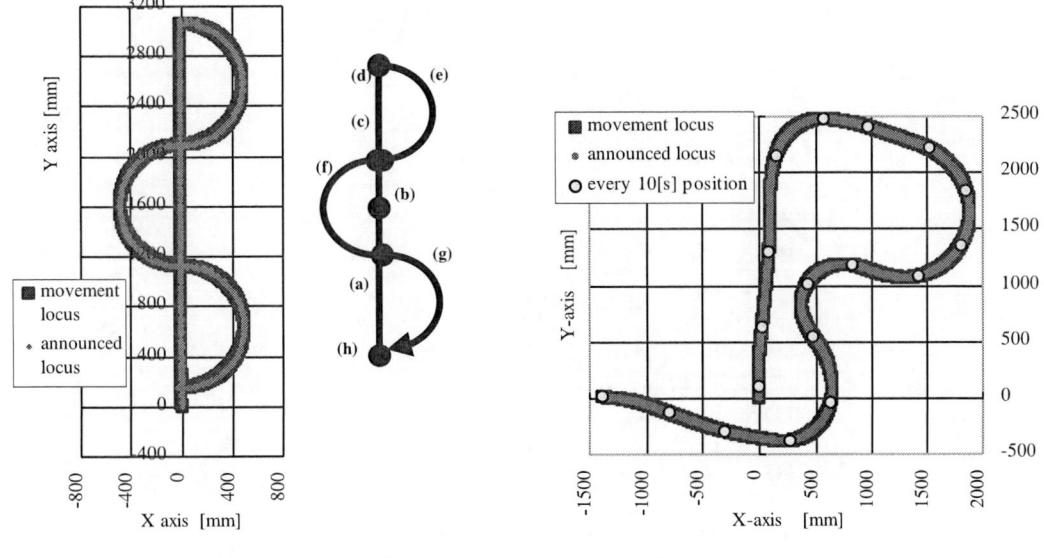

(a) Predefined route (b) Manual operation

Figure 16. Experimental results. [51, 52]

robot. Projected route is shown as **chain of three arrows** in different colors showing the time information of the route — red arrow from the present to 1-second-later, yellow arrow from 1-second-later to 2-second-later, and green arrow from 2-second-later to 3-second-later (**Figure 18(a)**). The arrow can be curved freely and the width and length of the arrow are adjusted depending on the speed of movement (**Figure 18(b)**). "State of operation" can also be seen in the frame (**Figure 18(c)**). For on-the-spot rotation, corresponding sign is displayed at right or left in the frame depending on the direction of movement. Characters for stop or going backward are also displayed in the frame. Dark background is prepared for bright running surface to keep visibility in addition to normal white background (**Figure 19**).

4.4.2. OpenGL Picture Display

The state of operation to display is roughly divided into four kinds — "going forward", "on-the-spot rotation", "going backward", and "stop".

(1) Distinction of state of operation: First the state of operation to display is distinguished from the speed instructions accumulated in the instruction buffer during the latest three seconds (60 arrays).

- There is a instruction in which translation factor is negative. → "Going forward" (because the used joystick is flight simulator type)

- There is a instruction in which translation factor is zero, but rotation factor is not zero. → "On-the-spot rotation"

- There is a instruction in which translation factor is positive. → "Going backward"

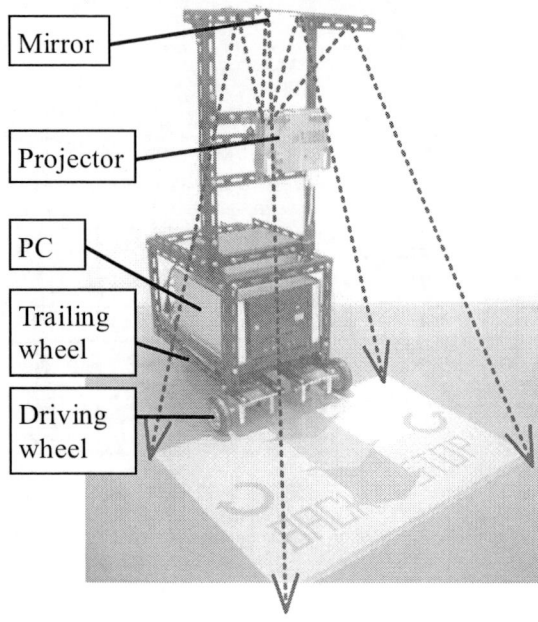

Figure 17. Overview of PMR-5. [57, 58]

- There is a instruction in which both translation factor and rotation factor are zero at the same time. → "Stop"

The state of operation displayed in the projected frame is independent even if two or more states are distinguished simultaneously. So multiple states may be displayed in the frame at the same time, for example, when executing "stop" following "going forward".

(2) Contents and method to display: Only one projector is equipped on the developed robot and can project one frame of a fixed size to predefined position in front of the robot.

a) Going forward: "Going forward" is expressed with the chain of three arrows in which the scheduled route is expressed as the long axis of arrows in different colors showing the time information of the route and the color of each arrow shows the time information, although the scheduled route drawn by laser pointer is a simple line (movement afterimage of red laser light) and the length of the drawn route only shows the average speed during three seconds in PMR-1. The color of three arrows are based on the traffic signal — red arrow from the present to 1-second-later, yellow arrow from 1-second-later to 2-second-later, and green arrow from 2-second-later to 3-second-later — thinking that approaching robot increases risk. Furthermore robot movement at high speed also raises danger, so the width of drawn arrows is adjusted depending on the speed of movement. Consequently not only the length but the width are adjusted according to the speed of movement of the robot.

Sixty polygons are prepared from sixty speed instructions acquired every $50ms$ during the latest three seconds. Those are connected perpendicularly and forms three arrows after transformed so as to look like the correct form. One arrow consists of twenty polygons — one triangular polygon Po_a for the tip of arrowhead, seven quadrangle polygon Po_b for the arrowhead except tip, and twelve quadrangle polygon Po_c for the rod of the arrow

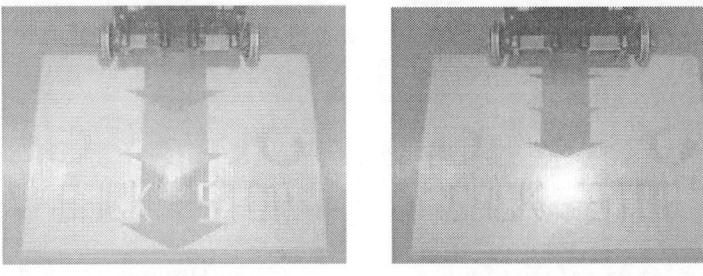

straight-fast   straight-slow

(a) Speed of movement: length/width of chain of arrow

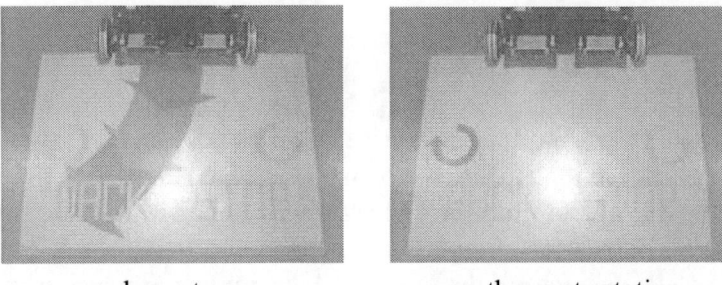

loose turn   on-the-spot rotation

(b) Direction of movement: curved condition of arrow

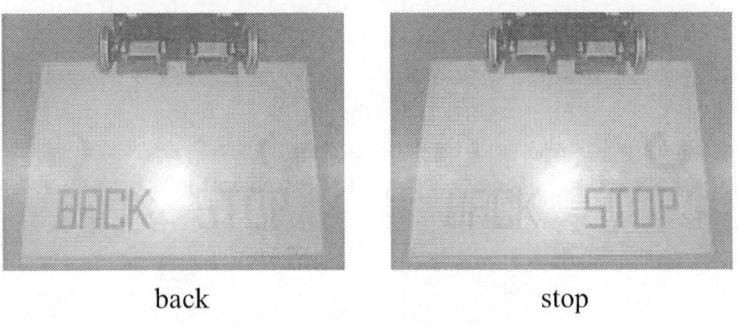

back   stop

(c) State of operation

Figure 18. Exact description on PMR-5 (on white background). [57, 58]

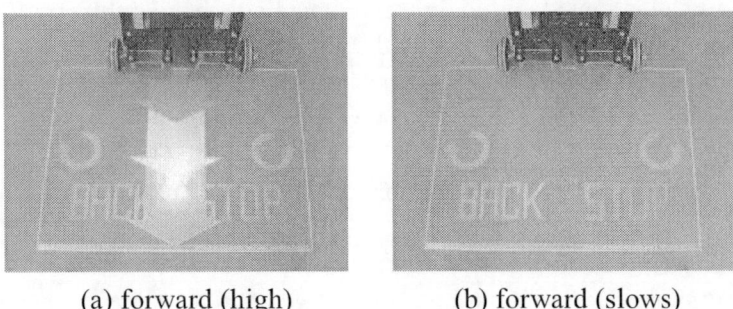

(a) forward (high)   (b) forward (slows)

Figure 19. Projected frame with dark background. [57, 58]

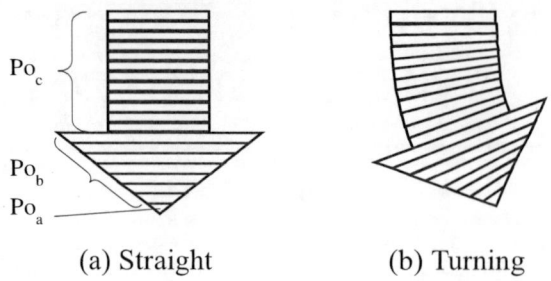

(a) Straight   (b) Turning

Figure 20. Each arrow consists of 20 polygons. [58]

(**Figure 20**).

Three arrows are formed connecting sixty polygons from base to tip using instructions from the oldest to the latest one (**Figure 21**(a)). When a new speed instruction is acquired, the polygon A shaped using the newly acquired instruction is arranged at top of green arrow while sixty polygons are shifted to the base of arrows (toward the robot) (**Figure 21**(b)). The position of polygon A (top of green arrow) shows the position where the robot will reach three seconds later. When the next speed instruction is acquired, the polygon A is shifted once to the base of arrows (toward the robot) and the polygon B which is shaped using the new instruction is arranged at the top of green arrow (**Figure 21**(c)). Thus the polygon A is shifted to the base of arrows (toward the robot) one by one as times goes on. It reaches the base of arrows three seconds later, then the corresponding speed instruction is performed and the robot moves in that manner (**Figure 21**(d)).

b) On-the-spot rotation / Going backward / Stop: In case of "on-the-spot rotation", "going backward", and "stop" the treatment is the same in displaying the state of operation. The corresponding sign for on-the-spot rotation and the characters of "BACK" or "STOP" for going backward and stop respectively are displayed in the frame. Those are always displayed in gray while displayed in yellow if the operation is performed within three seconds, and the color turns in red when the operation is executed at that time.

5. Questionnaire Evaluation

5.1. Purpose and Procedure

Questionnaires were adopted aiming to evaluate understandability of the preliminary-announcement and indication function of the future speed and future direction of movement on the four prototype robots [30]. Especially we pay attention whether the evaluation of understandability will differ depending on gender or age.

The four robots — eyeball robot PMR-2, arrow robot PMR-6, light ray robot PMR-1, and projection robot PMR-5 — were exhibited at the 2005 International Robot Exhibition sponsored by the Japan Robot Association and Nikkan Kogyo Shimbun Ltd., from Nov. 30 to Dec. 03 at Tokyo International Exhibition Hall. Comparatively bright lighting conditions present the same situation in which we had expected robots with preliminary-announcement and indication function to be used at the first onset. After explaining the background and purpose, proposed approaches to announce upcoming operation, robot configurations, etc.,

Preliminary-Announcement Function of Mobile Robots' Upcoming Operation

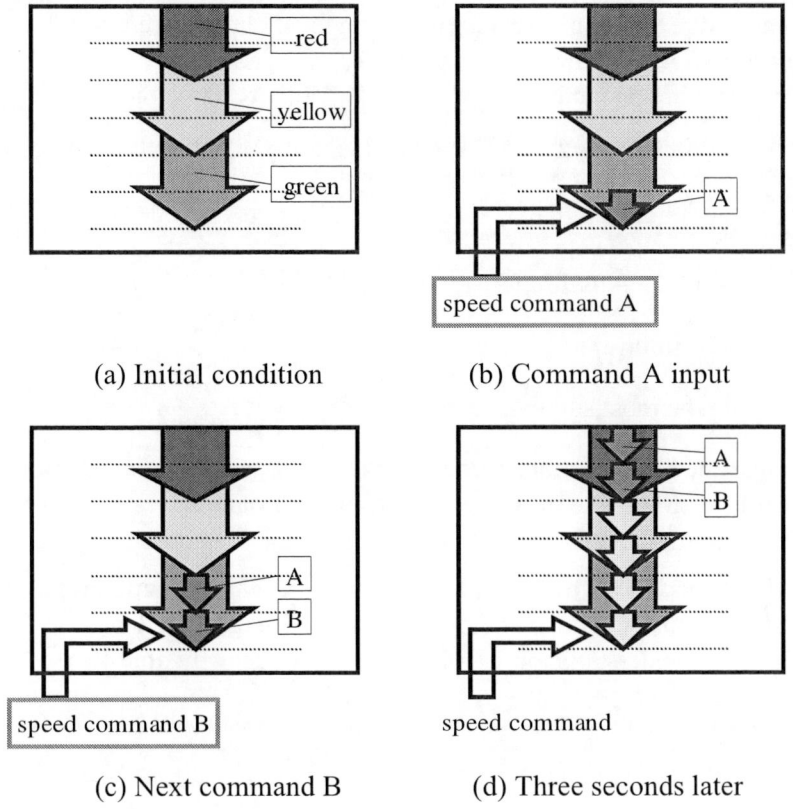

Figure 21. Flow of data. [57, 58]

we asked visitors to fill out questionnaires after watching the four robots move as detailed above. The four robots were operated manually in real time and were followed predefined courses during demonstration/experimentation. Questionnaires were as follows, mainly 5-stage-evaluation on the speed and direction of movement of each robot:

PMR-2 (omni-directional display)

· Please circle the most suitable answer.

1. The future **speed of movement** is previously announced with the **degree of eye opening** shown on the omni-directional display.
   It is (not understandable) 1 - 2 - 3 - 4 - 5 (understandable).

2. The future **direction of movement** is previously announced with the **eye positioning** shown on the omni-directional display..
   It is (not understandable) 1 - 2 - 3 - 4 - 5 (understandable).

· Please fill in your opinions and comments on PMR-2 freely.

PMR-6 (flat-panel display)

· Please circle the most suitable answer.

1. The future **speed of movement** is previously announced with the **size/color of arrow** shown on the flat-panel display.
   It is (not understandable) 1 - 2 - 3 - 4 - 5 (understandable).

2. The future **direction of movement** is previously announced with the **direction of arrow** shown on the flat-panel display.
   It is (not understandable) 1 - 2 - 3 - 4 - 5 (understandable).

· Please fill in your opinions and comments on PMR-6 freely.

PMR-1 (laser pointer)

· Please circle the most suitable answer.

1. The **speed of movement** from now is previously announced with the **length of route** drawn as the movement afterimage of radiant.
   It is (not understandable) 1 - 2 - 3 - 4 - 5 (understandable).

2. The **direction of movement** from now is previously announced with the **direction of route** drawn as the movement afterimage of radiant.
   It is (not understandable) 1 - 2 - 3 - 4 - 5 (understandable).

· Please fill in your opinions and comments on PMR-1 freely.

PMR-5 (projector)

· Please circle the most suitable answer.

1. The **speed of movement** from now is previously announced with the **length of route** shown as chain of three arrows.
   It is (not understandable) 1 - 2 - 3 - 4 - 5 (understandable).

2. The **direction of movement** from now is previously announced with the **direction of route** shown as chain of three arrows.
   It is (not understandable) 1 - 2 - 3 - 4 - 5 (understandable).

· Please fill in your opinions and comments on PMR-5 freely.

We asked respondents to evaluate understandability not relatively among four robots but absolutely in their own sense.

5.2. Results and Discussion

We obtained about 200 replies in four days (**Table 6**). The average and standard deviation of evaluation points are shown in **Table 7**. The respondent's rate in number to the evaluation point is summarized both about the speed of movement and about the direction of movement. **Figure 22** shows overall results and gender-segregated results. **Figure 23** shows age-specific results. Because the number of respondents was small among younger and elderly, evaluation points are calculated on three age groups — younger age (teens and twenties), middle age (thirties and forties), and upper age (fifties and above).

Table 6. Subject persons. [30]

gender	age								total
	10-	20-	30-	40-	50-	60-	70-	no	
male	6	39	43	35	35	16	2	9	185
female	1	6	3	3	0	0	0	3	16
(no)	0	3	4	1	0	0	0	3	11
total	7	48	50	39	35	16	2	15	212

5.2.1. Understandability Comparison among Four Robots

To give official approval whether there is a difference in the population mean of the parent population, the analysis of variance has been done using the data acquired as the evaluation score on each of the four robots. In this analysis of variance, only the data of people who answered about all of the four robots are processed. One-way analysis of variance shows significant different in the level of 5% both on the total average and on each gender group and each age group (**Table 8**). This result shows that the understandability is different among four robots plainly. In any case, the projector robot PMR-5 received the highest evaluation.

5.2.2. Discussion-1

In the first type of the proposed approaches, announcing the state just after the present, we study the result focusing on PMR-6 (flat-plane display).

(1) Comparison of PMR-6 (flat-plane display) with PMR-2 (omni-directional display): Total average of evaluation point about the speed of movement is 3.56 in PMR-6 while it is 2.87 in PMR-2. That about the direction of movement is 3.98 in PMR-6 while it is 3.37 in PMR-2. When the difference of two groups' mean value is given official approval, a significant difference in the level of 5% is admitted both on the speed of movement and on the direction of movement: (speed) $p = 7.18E - 08$, $t = 5.51$, $df = 334$, (direction) $p = 1.14E - 07$, $t = 5.42$, $df = 342$.

The result that the evaluation to PMR-2 is lower than that to PMR-6 leads the following remarks. Eyeball expression on omni-directional display in PMR-2 requires those who look at it of the translation between the sign and the movement and it admits various interpretations. Arrow expression on flat-panel display in PMR-6 is more direct and intuitive for people without translation and interpretation. The blowout method (realized as PMR-6) developed the lamp method (embodied in PMR-2), and consequently the result of the questionnaire evaluation shows the improvement on understandability between PMR-2 and PMR-6 as we aimed.

(2) Gender Difference on PMR-6: The distribution in evaluation point of PMR-6 depending on gender is shown in the second line in **Figure 22**. Females were extremely small number compared with males, and significant difference cannot be confirmed in mean value. However tendency is compared to get some suggestive information. About the speed of movement, females evaluate higher than males on the whole (male: 3.53, female: 3.79) and

Table 7. Evaluation points. [30]

	PMR-2 eyeball robot (omni-directional)		PMR-6 arrow robot (flat-panel)		PMR-6 light ray robot (laser pointer)		PMR-6 projection robot (projector)	
	speed	direction	speed	direction	speed	direction	speed	direction
total	2.87 (1.23)	3.37 (1.13)	3.56 (1.06)	3.98 (0.95)	2.53 (1.09)	2.84 (1.15)	3.91 (1.05)	4.46 (0.79)
male	2.86 (1.21)	3.38 (1.12)	3.53 (1.06)	3.98 (0.96)	2.52 (1.10)	2.84 (1.17)	3.87 (1.05)	4.42 (0.81)
female	3.04 (1.42)	3.35 (1.18)	3.79 (1.16)	4.04 (0.96)	2.91 (1.16)	3.09 (1.16)	4.43 (0.76)	4.79 (0.43)
10-19	3.00 (1.15)	4.14 (1.07)	4.00 (1.55)	4.50 (0.84)	2.29 (1.25)	2.71 (1.25)	4.43 (1.13)	4.57 (0.79)
20-29	2.57 (1.19)	3.34 (1.16)	3.39 (1.07)	3.96 (1.03)	2.55 (1.06)	2.63 (0.92)	4.01 (0.97)	4.50 (0.75)
30-39	2.97 (1.45)	3.17 (1.24)	3.65 (1.13)	3.97 (0.96)	2.10 (0.96)	2.49 (1.02)	3.86 (1.21)	4.52 (0.78)
40-49	2.54 (1.09)	3.15 (1.03)	3.32 (0.83)	3.66 (0.84)	2.77 (1.02)	3.10 (1.14)	3.84 (0.95)	4.47 (0.81)
50-59	3.08 (1.04)	3.59 (1.15)	3.86 (1.03)	4.35 (0.89)	2.85 (1.20)	3.33 (1.32)	4.07 (0.83)	4.38 (0.79)
60-	3.67 (1.11)	3.80 (0.73)	3.50 (1.00)	3.80 (0.82)	3.04 (1.12)	3.21 (1.36)	3.69 (1.11)	4.38 (0.82)

upper: average, lower: standard deviation

the distribution in evaluation point is shifted toward higher. About the direction of motion, the average evaluation point is almost equal (male: 3.98, female: 4.04) and the distribution in evaluation point has similar tendency between males and females. Females' evaluation about the speed of movement is not singular because the same distribution tendency is observed in the direction of movement. Definite reason is not clear but it suggests females accept the **change in size of arrow** and the **change in color based on traffic signal** in response to the speed of movement more favorably than males.

(3) Age Difference on PMR-6: The distribution in evaluation point of PMR-6 according to age is shown in the second line in **Figure 23**. Significant difference by the level of 5% is almost admitted only between middle age and upper age about the direction of movement: $p = 0.056$, $t = 1.94$, $df = 71$. About the speed of movement, the distribution on upper age stands out — it has two peaks. One peak group moderately evaluates it as well as most of middle age. Another peak group evaluates it higher than most of younger age. About the direction of movement, the distribution in evaluation point is almost the same between younger age and upper age, and scores are overall higher than those on middle age. Although middle age severely evaluates rather than other age groups generally, they appreciate the direction indication on PMR-6 relatively higher compared with that on PMR-2 or PMR-1 (light ray by laser pointer).

Those who give high praise to PMR-2 (eyeball expression on omni-directional display) are more on upper age rather than on other age groups. This result indicates some amount

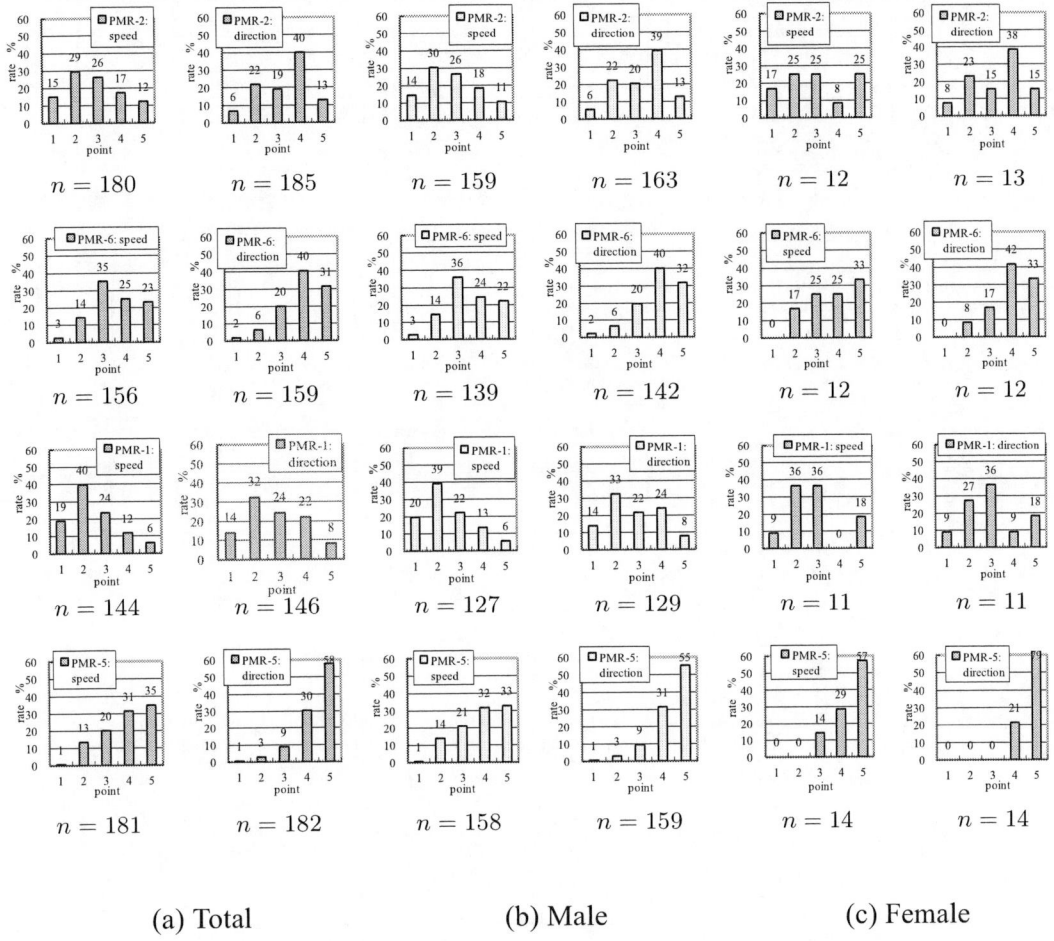

(a) Total (b) Male (c) Female

Figure 22. Overall results and gender-segregated results. [30]

of people think a **friendly expression** to be desirable. Those who have a good opinion on PMR-6 (arrow expression on flat-plane display) are more on upper age and younger age rather than on middle age. This aspect shows some amount of people have favorable impression on a **simple method to inform**.

5.2.3. Discussion-2

In the second type of the proposed approaches, indicating operations from the present to some future time continuously, we study the result focusing on PMR-5 (projector).

(1) Comparison of PMR-5 (projector) with PMR-1 (laser pointer): Total average of evaluation point about the speed of movement is 3.91 in PMR-5 while it is 2.53 in PMR-1. That about the direction of movement is 4.46 in PMR-5 while it is 2.84 in PMR-1. When the difference of two groups' mean value is given official approval, a significant difference in the level of 5% is admitted both on the speed of movement and on the direction of movement: (speed) $p = 1.52E - 25$, $t = 11.48$, $df = 301$, (direction) $p = 1.63E - 34$,

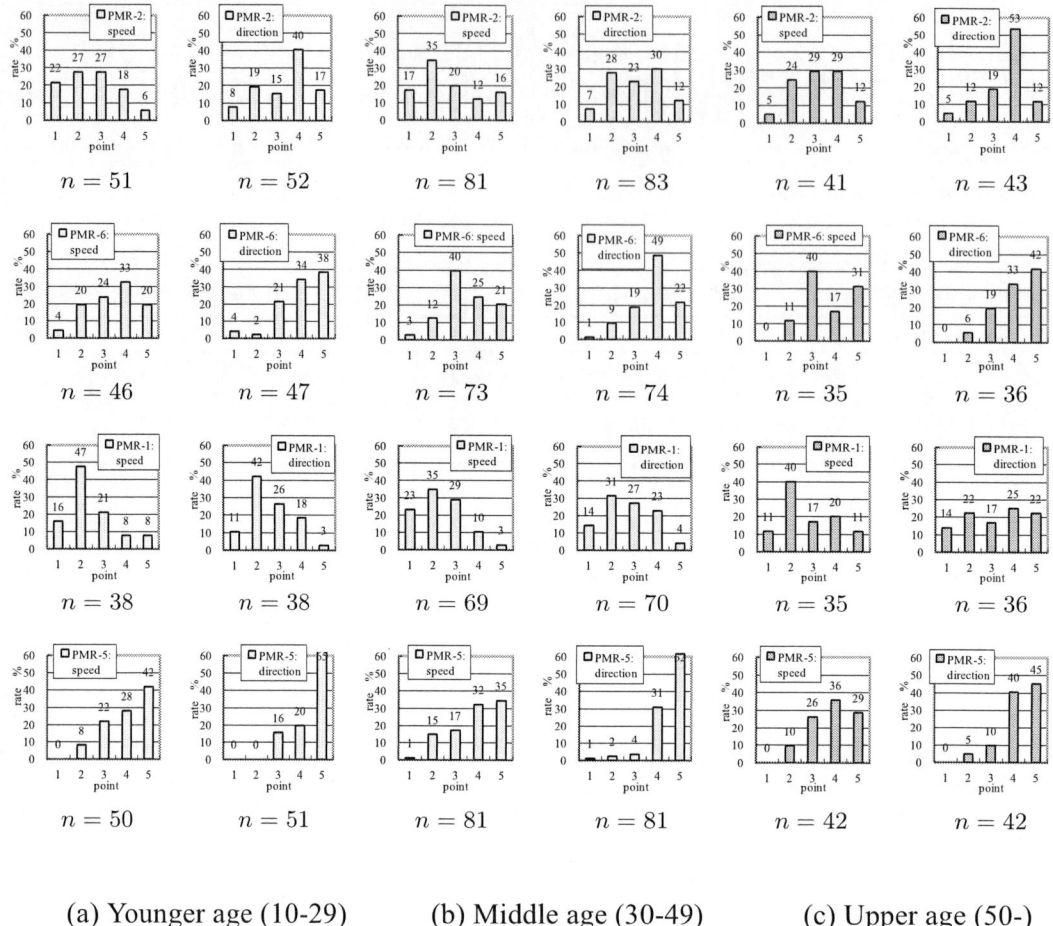

(a) Younger age (10-29) (b) Middle age (30-49) (c) Upper age (50-)

Figure 23. Age-specific results. [30]

$t = 14.40, df = 246$.

The result that the evaluation to PMR-1 is lower than that to PMR-5 leads the following remarks. Scheduled route in PMR-1 by using laser pointers is comparatively not understandable because the irradiated route is shown in single color by red laser light and drawn in simple line. Scheduled route in PMR-5 by using projector is more understandable because the projected route is shown as chain of three arrows in different colors for each period, and the length and width of arrow are adjusted in response to the speed of movement. The projection method (realized as PMR-5) developed the light ray method (embodied in PMR-1), and consequently the result of the questionnaire evaluation shows the improvement on understandability between PMR-1 and PMR-5 as we aimed.

(2) Gender Difference on PMR-5: The distribution in evaluation point of PMR-5 depending on gender is shown in the fourth line in **Figure 22**. Females evaluate higher than males overall both about the direction of movement and about the speed of movement: (speed) male: 3.87, female: 4.43, (direction) male: 4.42, female: 4.79. The distribution in evaluation point about the direction of movement is almost the same between males and

Table 8. Analysis of variance among four robots. [30]

	P-value P	degree of freedom df	F boundary value F
Total	(speed) $2.53E-27$ (direction) $1.93E-38$	3/468	2.62
Male	(speed) $9.80E-23$ (direction) $1.50E-33$	3/412	2.63
Female	(speed) 0.016 (direction) 0.009	3/32	2.90
Younger	(speed) $2.41E-07$ (direction) $4.22E-11$	3/116	2.68
Middle	(speed) $4.80E-15$ (direction) $6.09E-21$	3/228	2.64
Upper	(speed) $7.08E-05$ (direction) $4.32E-07$	3/108	2.69

females. The distribution about the speed of movement is shifted toward higher on females remarkably. It suggests the same result as on PMR-6 (flat-panel display) that females accept the **change in size of arrow** in response to the speed of movement more favorably than males.

(3) Age Difference on PMR-5: The distribution in evaluation point of PMR-5 according to age is shown in the fourth line in **Figure 23**. Significant difference cannot be confirmed in the mean value. People of all ages highly appreciate PMR-5. Roughly speaking there is the tendency that as younger the respondent is as higher the evaluation but it is a slight difference.

Note that the ratio of people who admired PMR-5 is lower and the ratio of people who appreciated PMR-1 is higher in upper age than in other age groups. The comparison result of PMR-5, which can inform various operations in the frame, with PMR-1, which only indicates the scheduled route, suggests that some amount of people prefer **a minimum of information** to be presented at one time especially in upper age.

6. Conclusion and Future Work

We have proposed approaches and equipment for preliminarily announcing and indicating to people the speed and direction of movement of mobile robots moving on a two-dimensional plane. After introducing our approaches categorized into (1) announcing the state just after the present and (2) indicating operations from the present to some future time continuously, we explained the prototype robots we developed — omni-directional display (PMR-2), flat-panel display (PMR-6), laser pointer (PMR-1), and projector (PMR-5) are used for the announcement unit. The four robots were exhibited at the 2005 International Robot Exhibition (iREX05). We had visitors answer questionnaires in a 5-stage evaluation. The projection robot PMR-5 received the highest evaluation score among the four. An examination of differences by gender and age suggested that some people prefer friendly expressions, simple method to inform, and a minimum of information to be presented at one time.

Prototype robots were evaluated during a demonstration after respondents had been explained the purpose and methods. Method to preliminarily announce and indicate the upcoming operation is preferable to be judged and understood from common sense even at first sight. So the evaluation by subject persons without advance knowledge is one of the future works.

Robots with preliminary-announcement function of its upcoming operation are not expected to pose problems in task accomplishment when moving along predefined routes or even if the route is decided on site, like robot cleaners, because the route is announced beforehand and the robot actually moves that way for some time. In manual operation in real time, however, as with automobiles, it will cause problems. The maneuverability in operator will become worse because the time-delay will be always inserted between the instruction and the execution. In that case we have to devise the method that the robot executes the instruction from operator in real time forecasting operator's next operation from operational record and environmental condition [63] and indicating the forecasted operation to surroundings. In addition, future work includes the study on dealing with the case that operator behaves contradictory to the forecasted operation and the case that robot does some autonomous operation suddenly to secure safety like stopping.

We will continue to study to improve the affinity between human being and mechatronic systems to establish a standard for mutual coexistence.

Acknowledgments

We gratefully acknowledge the contribution of the following laboratory staff in this research project: Yu Hoshiba, Yasuhiro Miyata, and Shinji Hiraiwa.

This research was supported in part by the following foundations, to which we are greatly indebted: Suzuki Foundation and Fanuc FA & Robot Foundation (fiscal 2001); Tateishi Science and Technology Foundation and Mitsutoyo Association for Science and Technology (fiscal 2005); and Casio Science Promotion Foundation (fiscal 2006).

References

[1] Research Committee on Human Friendly Robot. Technical Targets of Human Friendly Robots. *J. of the Robotics Society of Japan* **16**(3), 288/294, (1998).

[2] N. Sugimoto. Robot-Safety and Intelligent Fail-Safe. *J. of the Robotics Society of Japan* **2**(2), 158/163, (1984).

[3] N. Sugimoto. Robot and Safety. *J. of the Robotics Society of Japan* **3**(1), 56/59, (1985).

[4] E. Prassler, D. Bank, and B. Kluge. Key Technologies in Robot Assistants: Motion Coordination Between a Human and a Mobile Robot. *Trans. on Control, Automation and Systems Engineering* **4**(1), 56/61, (2002).

[5] J. M. H. Wandosell, and B. Graf. Non-Holonomic Navigation System of a Walking-Aid Robot. *Proc. of the 11th IEEE Int. Workshop on Robot and Human interactive Communication (ROMAN2002)*, 518/523, (2002).

[6] Y. Yamada, and N. Sugimoto. Evaluation of Human Pain Tolerance. *J. of the Robotics Society of Japan* **13**(5), 639/642, (1995).

[7] M. Inaba, Y. Hoshino, and H. Inoue. A Full-Body Tactile Sensor Suit Using Electrically Conductive Fabric. *J. of the Robotics Society of Japan* **16**(1), 80/86, (1998).

[8] T. Morita, Y. Suzuki, T. Kawasaki, and S. Sugano. Anticollision Safety Design and Control Methodology for Human-Symbiotic Robot Manipulator. *J. of the Robotics Society of Japan* **16**(1), 102/109, (1998).

[9] D. Vischer, and O. Khatib. Design and Development of High Performance Torque Controlled Joints. *IEEE Trans. on Robotics and Automation* **11**(4), 537/544, (1995).

[10] K. Koganezawa, M. Yamazaki, and N. Ishikawa. Mechanical Stiffness Control of Tendon-Driven Joints. *J. of the Robotics Society of Japan* **18**(7), 101/108, (2000).

[11] M. Sakaguchi, J. Furusho, G. Zhang, and Z. Wei. Development of ER Actuator and Basic Study on its Force Control System. *J. of the Robotics Society of Japan* **16**(8), 1108/1114, (1998).

[12] T. Matsumaru, T. Kotoku, A. Fujimori, and K. Komoriya. Action Strategy for Remote Operation of Mobile Robot in Human Coexistence Environment. *2000 IEEE Int. Conf. on Industrial Electronics, Control and Instrumentation (IECON-2000)* 1/6, (2000).

[13] K. Mase. Automatic Extraction and Recognition of face and gesture. *J. of the Robotics Society of Japan* **16**(6), 745/748, (1998).

[14] E. Ueda, Y. Matsumoto, M. Imai, and T. Ogasawara. Hand Pose Estimation for Vision Based Human Interface. *Proc. of 10th IEEE Int. Workshop on Robot and Human Communication (ROMAN 2001)*, 473/478, (2001).

[15] L. Bretzner, I. Laptev, and T. Lindeberg. Hand gesture recognition using multi-scale colour features, hierarchical models and particle filtering. *Proc. 5th IEEE Int. Conf. on Automatic Face and Gesture Recognition*, 405/410, (2002).

[16] H. Fei and I. Reid. Dynamic Classifier for Non-rigid Human motion analysis. *British Machine Vision Conf*, **118**, (2004).

[17] Y. Matsumoto, M. Inaba, and H. Inoue. View-Based Approach to Robot Navigation. *J. of the Robotics Society of Japan* **20**(5), 44/52, (2002).

[18] Y. Kuno, N. Shimada, and Y. Shirai. Look where you're going: A robotic wheelchair based on the integration of human and environmental observations. *IEEE Robotics and Automation Magazin* **10**(1), 26/34, (2003).

[19] T. Ohno, and N. Mukawa. A Free-head, Simple Calibration, Gaze Tracking System That Enables Gaze-Based Interaction. *Proc. of Eye Tracking Research & Application Symposium 2004 (ETRA 2004)*, 115/122, (2004).

[20] Y. Liu, K. Schmidt, J.F. Cohn, and R.L. Weaver. Human facial asymmetry for expression-invariant facial identification. *Proc. of the Fifth IEEE Int. Conf. on Automatic Face and Gesture Recognition (FG'02)*, 208/214, (2002).

[21] Y.-L. Tian, T. Kanade, and J. Cohn, Facial expression analysis. in S.Z. Li & A.K. Jain (ed.) *Handbook of face recognition* Springer, (2005).

[22] T. Ogata, and S. Sugano. Emotional Communication between Humans and the Autonomous Robot WAMOEBA-2 (Waseda Amoeba) which has the Emotion Model. *JSME Int. J., Series C* **43**(3), 586/574, (2000).

[23] Y. Wakita, S. Hirai, T. Suehiro, T. Hori, and K. Fujiwara. Information Sharing via Projection Function for Coexistence of Robot and Human. *Autonomous Robots* **10**(3), 267/277, (2001).

[24] Y. Kawakita, R. Ikeura, K. Mizutani. Previous notice method of robotic arm motion for suppressing threat to human. *Japanese J. of Ergonomics* **37**(5), 252/262, (2001).

[25] A. Hagiwara, R. Ikeura, Y. Kawakita, and K. Mizutani. Previous Notice Method of Robotic Arm Motion for Suppressing Threat to Human. *J. of the Robotics Society of Japan* **21**(4), 67/74, (2003).

[26] TOYOTA. Company – News Release: Toyota, Hino, Daihatsu to Jointly Exhibit at 11th ITS World Congress. (September 22, 2004). http://www.toyota.co.jp/en/news/04/0922_2.html

[27] T. Matsumaru, S. Kudo, T. Kusada, K. Iwase, K. Akiyama, and T. Ito. Simulation on Preliminary-Announcement and Display of Mobile Robot's Following Action by Lamp, Party-blowouts, or Beam-light. *IEEE/ASME Int. Conf. on Advanced Intelligent Mechatronics (AIM 2003)*, 771/777, (2003).

[28] T. Matsumaru, H. Endo, and T. Ito. Examination by Software Simulation on Preliminary-Announcement and Display of Mobile Robot's Following Action by Lamp or Blowouts. *2003 IEEE Int. Conf. on Robotics and Automation (2003 IEEE ICRA)*, 362/367, (2003).

[29] T. Matsumaru. The Human-Machine-Information System and the Robotic Virtual System. *J. of SICE*, **43**(2), 116/121, (2004).

[30] T. Matsumaru. Development of four kinds of mobile robot with preliminary-announcement and display function of its forthcoming operation. *J. of Robotics and Mechatronics* **19**(2), 148/159, (2007).

[31] T. Matsumaru, and K. Hagiwara. Method and Effect of Preliminary-Announcement and Display for Translation of Mobile Robot. *Proc. of the 10th Int. Conf. on Advanced Robotics (ICAR 2001)*, 573/578, (2001).

[32] T. Matsumaru, and K. Hagiwara. Preliminary-Announcement and Display for Translation and Rotation of Human-Friendly Mobile Robot. *Proc. of 10th IEEE Int. Workshop on Robot and Human Communication (ROMAN 2001)*, 213/218, (2001).

[33] T. Matsumaru, S. Kudo, H. Endo, and T. Ito. Examination on a Software Simulation of the Method and Effect of Preliminary-announcement and Display of Human-friendly Robot's Following Action. *Trans. of SICE* **40**(2), 189/198, (2004).

[34] T. Matsumaru, K. Iwase, T. Kusada, K. Akiyama, H. Gomi, and T. Ito. Preliminary-Announcement Function of Mobile Robot's Following Motion by using Omni-directional Display. *Proc. of The 11th Int. Conf. on Advanced Robotics (ICAR 2003)* 650/657, (2003).

[35] T. Matsumaru, K. Akiyama, K. Iwase, T. Kusada, H. Gomi, and T. Ito. Eyeball Expression for Preliminary-Announcement of Mobile Robot's Following Motion. *Proc. of The 11th Int. Conf. on Advanced Robotics (ICAR 2003)* 797/803, (2003).

[36] T. Matsumaru, K. Iwase, T. Kusada, K. Akiyama, H. Gomi, and T. Ito. Synchronization of Mobile Robot's Movement and Preliminary-announcement using Omni-directional Display. *IEEE/ASME Int. Conf. on Advanced Intelligent Mechatronics (AIM 2003)* 246/253, (2003).

[37] T. Matsumaru, K. Akiyama, K. Iwase, T. Kusada, H. Gomi, and T. Ito. Robot-to-Human Communication of Mobile Robot's Following Motion using Eyeball Expression on Omni-directional Display. *IEEE/ASME Int. Conf. on Advanced Intelligent Mechatronics (AIM 2003)* 790/796, (2003).

[38] T. Matsumaru, K. Iwase, K. Akiyama, T. Kusada, and T. Ito. Mobile robot with eyeball expression as the preliminary-announcement and display of the robot's following motion. *Autonomous Robots* **18**(2), 231/246, (2005).

[39] Lumino GmbH. Magicball. http://www.magicball.de/

[40] T. Matsumaru. Mobile Robot with Preliminary-announcement and Indication Function of Forthcoming Operation using Flat-panel Display *2007 IEEE Int. Conf. on Robotics and Automation (ICRA'07)* 1774/1781, (2007).

[41] T. Matsumaru, Y. Miyata, Y. Hoshiba, and S. Hiraiwa. Preliminary-announcement and Indication Function of Forthcoming Operation using Flat-panel Display *J. Robotics Society of Japan* (submitted).

[42] P. D. Wellner. The DigitalDesk Calculator: Tactile Manipulation on a Desk Top Display *Proc. ACM Symp. on User Interface Software and Technology (UIST '91)* 27/33, (1991).

[43] H. Koike, Y. Sato, and Y. Kobayashi. Integrating paper and digital information on EnhancedDesk: a method for realtime finger tracking on an augmented desk system *ACM Transactions on Computer-Human Interaction (TOCHI)* **8**(4), 307/322, (2001).

[44] J. Rekimoto. SmartSkin: An Infrastructure for Freehand Manipulation on Interactive Surfaces *Proc. SIGCHI conf. on Human factors in computing systems (CHI '02)* 113/120, (2002).

[45] J. Patten, H. Ishii, J. Hines, and G. Pangaro. Sensetable: A Wireless Object Tracking Platform for Tangible User Interfaces *Proc. of the SIGCHI conference on Human factors in computing systems (CHI '01)*, 253/260, (2001).

[46] B. Piper, C. Ratti, and H. Ishii. Illuminating Clay: A 3-D Tangible Interface for Landscape Analysis *Proc. of the SIGCHI conference on Human factors in computing systems (CHI '02)*, 355/362, (2002).

[47] T. Matsui, and M. Tsukamoto. An Integrated Teleoperation Method for Robots using Multi-Media-Display *J. Robotics Society of Japan* **6**(4), 301/310, (1988).

[48] M. Terashima, and S. Sakane. A Human-Robot Interface Using an Extended Digital Desk Approach *J. Robotics Society of Japan* **16**(8), 1091/1098, (1998).

[49] H. K. Keskinpala, J. A. Adams, and K. Kawamura. PDA-Based Human-Robotic Interface *Proc. of the 2003 IEEE Int. Conf. on Systems, Man, and Cybernetics* 3931/3936, (2003).

[50] T. W. Fong, C. Thorpe, and B. Glass. PdaDriver: A Handheld System for Remote Driving *the 11th Int. Conf. on Advanced Robotics 2003*, 88/93, (2003).

[51] T. Matsumaru, T. Kusada and K. Iwase. Mobile Robot with Preliminary-Announcement Function of Following Motion using Light-ray *The 2006 IEEE/RSJ Int. Conf. on Intelligent Robots and Systems (IROS 2006)*, 1516/1523, (2006).

[52] T. Matsumaru, T. Kusada, and K. Iwase. Development of Mobile Robot with Preliminary-announcement and Display Function of Scheduled Course using Light-ray *J. Robotics Society of Japan* **24**(8), 976/984, (2006).

[53] H. Kuzuoka, T. Kosuge, M. Tanaka. GestureCam: A Video Communication System for Sympathetic Remote Collaboration *Proc. of 5th Conf. on Computer Supported Cooperative Work (CSCW'94)*, 35/43, (1994).

[54] K. Yamazaki, A. Yamazaki, H. Kuzuoka, S. Oyama, H. Kato, H. Suzuki, and H. Miki. GestureLaser and Gesturelaser Car: development of an Embodied Space to Support Remote Instruction *Proc. of The 6th European Conf. on Computer Supported Cooperative Work (ECSCW'99)*, 239/258, (1999).

[55] M. Mikawa, T. Tsujimura, and H. Naruse. Teleoperation and Communication Support System Using CTerm *Proc. The Fourth Asian Fuzzy Systems Symposium (AFSS2000)*, 1072/1075, (2000).

[56] S. Mann. Telepointer: Hands-Free Completely Self Contained Wearable Visual Augmented Reality without Headwear and without any Infrastructural Reliance *Proc. the Fourth International Symposium on Wearable Computers (ISWC'00)*, 177/178, (2000).

[57] T. Matsumaru, T. Kusada and K. Iwase. Mobile Robot with Preliminary-Announcement Function of Following Motion using Light-ray *The 2006 IEEE/RSJ Int. Conf. on Intelligent Robots and Systems (IROS 2006)*, 1516/1523, (2006).

[58] T. Matsumaru, Y. Hoshiba, S. Hiraiwa, Y. Miyata. Development of Mobile Robot with Preliminary-announcement and Display Function of Forthcoming Motion using Projection Equipment *J. Robotics Society of Japan* **25**(3), 410/421, (2007).@

[59] Projector-Camera Systems, http://www.procams.org/

[60] T. Nishi, Y. Sato, and H. Koike. SnapLink: interactive object registration and recognition for augmented desk interface *Proc. IFIP Conf. on Human-Computer Interaction (Interact 2001)* 240/246, (2001).

[61] S. Hiura, K. Tojo and I. Seiji. 3-D Tele-direction Interface using Video Projector *The 30th Int. Conf. on Computer Graphics and Interactive Techniques (SIGGRAPH 2003)* 29-1, (2003).

[62] H. Kawata, T. Machino, S. Iwaki, Y. Nanjo, and K. Shimokura. An Application of Campro-R (Mobile Robot with Camera and Projector) at home - A speculation about structuring information indoors *SICE-ICASE Int. Joint Conf. 2006*, 421/424, (2006).

[63] M. Akamatsu. Establishing Driving Behavior Database and its Application to Active Safety Technologies. *J. of Society of Automotive Engineers of Japan* **57**(12), 34/39, (2003).

In: Robotics Research Trends
Editor: Xing P. Guô, pp. 193-222

ISBN 1-60021-997-7
© 2008 Nova Science Publishers, Inc.

Chapter 5

DEVS AND TIMED AUTOMATA FOR THE DESIGN OF CONTROL SYSTEMS

Norbert Giambiasi[*] *and Hernán P. Dacharry*[†]
LSIS - UMR CNRS 6168
University Paul Cézanne

Abstract

The formal verification of temporal properties is a central issue in the design of real-time control systems, in this context, a multi-formalism framework is proposed using the Timed Automata formalism to describe high-level properties and the DEVS formalism to describe the design-level specification of the control system. The framework introduced lays on a sound mathematical basis allowing the formal verification of timed properties (described with timed automata) of the design specification (given by a DEVS model). Furthermore, the convenience of the framework is illustrated by presenting a case study of a subsystem of an industrial Production Cell generally used in the literature to compare formal methods.

1. Introduction

The design of real-time discrete event control systems is a process that needs dedicated formalisms and adapted tools. These formalisms may differ considerably in their syntax, semantics, and representation. However, there is a common base shared by all discrete event modeling formalisms. First of all, the input/output variables have discrete event trajectories and the state variables have piecewise constant trajectories. Certain formalisms impose explicit set of states (named), in other ones the set of states is implicit (defined by a set of state variables). Furthermore, the state set can be finite or infinite according to the formalism chosen and the system being considered. Finally the time can be represented implicitly (untimed formalisms) or by actual values (timed formalisms).

---

[*]E-mail address: norbert.giambiasi@lsis.org
[†]E-mail address: hernan.dacharry@lsis.org

In addition, some formalisms seem better adapted to the first stages of the design process, others aim towards low-level models with a more accurate representation of the system dynamics. For example, the DEVS (Discrete Event System Specification) formalism [11, 32, 33] is well adapted to represent timed behaviour of real systems. But DEVS is not well suited to high-level specifications of discrete event control systems. This is due to the fact that at the early stages of the design process the models can be nondeterministic and the occurrence times of events are not defined accurately.

The DEVS formalism allows building conceptual discrete event simulation models (deterministic models) with good accuracy by abstracting the approach in which discrete event simulation languages describe the parameters. A DEVS model is a symbolic specification of system semantics and for a given timed input there is only one way to execute a DEVS model. Since DEVS models are deterministic, this can pose an inconvinience for high-level specification of real-time systems. In fact, is for this reason that the DEVS formalism is convenient to introduce the concept of determinism at an early stage in the design level.

In the DEVS formalism one specifies basic atomic models and by connecting together this basic models in an hierarchical manner, one specifies complex coupled models. The DEVS formalism is closed under coupling, this means that a coupled DEVS model can be formally specified as an atomic model, this in particular gives the DEVS formalism sound semantics.

Despite the wide-ranging research in real-time system design [1, 22, 30, 21, 18], and the fact that there exists several computational models and formalism aimed specifically towards the different design stages, we did not found any methodology that properly takes into account the design process as a whole by attacking the problems and errors that originates at the different abstraction levels, simultaneouly with the inconsistencies that arise when using different formalisms and techniques in the various design stages.

We propose the basis for a methodology of design that addresses the problem just mentioned. We propose to use the timed automata formalism for the high-level stage of design of the system specification exploiting the already developed theory of timed automata, focusing in the formal verification of its properties (for example, with model-checking tools). And at the same time using for the low-level phase of design the DEVS formalism developped in the context of system and modeling theory, supporting tools for the simulation and implementation of the system [8].

The main advantage of timed automata formalism in the early phases of design is the existence of model-checking tools and methods already developed and tested for the automatic formal verification of temporal properties of timed automata models. With model-checking tools the properties or requirements of the model are specified by means of a temporal logic formula, and it is checked if the timed automata model conforms the desired property.

On the other hand, the DEVS formalism is convenient for the low-level phase of design since it provides a suitable simulation framework, making possible the validation of the model by simulation. Since DEVS formalism has a sound mathematical semantics it is possible to define a framework to formally verify the behaviour of an implementation (specified with DEVS) with respect to a higher-level specification (using Timed Automata).

This connection between the high-level properties or specification, and the low-level implementation is achieved by formally adapting some common and well developed results from the timed automata theory. The drawback of this approach is that the complexity of the models render the automatic verification of the conformance between the high-level and low-level models unfeasible in the general case.

This is caused by the state explotion problem that frequently appears in the verification of models dealing with a dense time base. Despite this discouraging result, the automatic validation of the conformance relation between an implementation and its specification is still possible, for example, by generating test-cases using as input the high-level specification and applying the test-cases to the low-level model description.

Even if it is not possible the complete automatic formal verification, given the flexibility of the methodology, it still remains suitable for the design and verification. For instance, given a system to be designed, a high-level specification is constructed using a network of timed automata, and the corresponding implementation is expressed with a coupled DEVS model, then it is possible to formally verify the conformance of critical components (atomic DEVS models against timed automata) and the conformance of the whole model exploiting the automatic test case generation.

Since our objective is to formally study the behaviour of DEVS models in terms of the time of event occurrences, and relate the behaviour to what can be described with timed automata or timed transition systems in general, it is necesary to impose some restrictions over the DEVS formalism. In the rest of the chapter when we write about DEVS we will be making reference only to the subclass of DEVS models that have a finite set of input and output values.

Several methodologies for the untimed verification and analysis of DEVS models have been proposed [22, 17, 19], and recently several other [18, 20] attempting to deal and represent timed behaviour. More over in previous work we tackled the problem of timed verification of DEVS models [8, 7].

Beyond presenting our formal methodology and its formal background, we also present a case study to show the advantages of our contribution. The case study chosen is a subsystem of an industrial "Production Cell" [24], presented with the objective of comparing the adequacy of formal methods.

The formal methodology introduced in this chapter, for the design and verification of control systems, was developed as a first step to stimulate the joint cooperation of researchers from two different fields, sytem theory in the case of the DEVS formalism, and computer science in the case of timed automata formalism.

In the next section we give a brief overview, the main research results and the advantages and drawbacks of the two formalisms we propose to use in our methodology to establish a formal background. Over these, in the following section we present the techniques that are needed for the formal verification relating the behaviour of both formalism used. Afterwards, we apply the proposed methodology to a case study taken from the real-time literature, and finally we give our conclusions and remarks along with some directions for future work.

2. Preliminaries

In order to fix the notation we review the concepts, definitions and results introduced in the fields of DEVS models and timed automata.

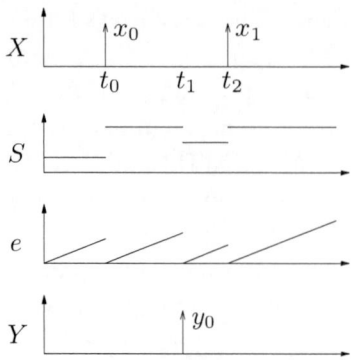

Figure 1. Discrete event segment.

2.1. The DEVS Formalism

The objective of a discete event formalism is to formally describe how to generate new values for variables and when the new values should take effect, thus an event is a change of a variable value that occurs instantaneously. Since a continuous time base is used in the DEVS formalism, the time interval between event occurences is variable and it differentiantes it appart from discrete time approaches.

The DEVS formalism provides a hierarchical modular approach to the specification of complex systems. In the DEVS formalism one specifies basic *atomic* models, atomic models can then be connected to form more complex *coupled* models. These coupled models can itself be used as components of larger coupled models [32, 33] allowing a hierarchical description of complex models using a library.

The atomic DEVS formalism is a mathematical structure for the formal specification of the system's discrete event behaviour.

$$\mathcal{D} = \langle X, Y, S, \delta_{\text{int}}, \delta_{\text{ext}}, \lambda, \text{ta} \rangle$$

where:

X is the set of admissible input events,

Y is the set of admissible ouput events,

S is the set of states,

$\delta_{\text{int}} : S \to S$ is the internal transition function, defines the state changes caused by internal events,

$\delta_{\text{ext}} : Q_\mathcal{D} \times X \to S$ is the external transition function, it determines the state changes produced by external events, $Q_\mathcal{D}$ is the set of total states defined as:

$$Q_\mathcal{D} = \{(s, e) : s \in S, 0 \leq e \leq \text{ta}(s)\}$$

$\lambda : S \to Y$ is the output function, specifies the output values due to the internal transitions, and ta : $S \to \Re_0^+ \cup \{\infty\}$ is the lifetime of the states, it specifies the time interval during which the model will remain in each state if no external event occurs.

A coupled DEVS model is composed by the set of components, the set of input and output ports through which external events are received and send respectively, and the coupling information. The coupling specification consists of:

- the external input coupling which connects the input ports of the coupled model to the input ports of internal components,

- the external output coupling which connects the output ports of internal components to the output ports of the coupled model,

- and the internal coupling which connects the output ports of internal components to input ports of other components.

Since DEVS models are closed under coupling (any coupled DEVS model can be represented as an equivalent atomic DEVS model), in the following sections, we will only consider atomic DEVS models for the formal definitions. The behaviour of an atomic model is described in terms of the basic data items that it produces, in the case of DEVS these are discrete event segments. For an example, see Figure 1, where the input trajectory X is a series of input events occurring at times t_0 and t_2 causing external state transitions, between these event times, at t_1, there is an internal state transition generating the event shown in the output trajectory Y. The state trajectory S shows all these state transitions, the elapsed time trajectory e depicts the time since the last event.

The discrete event segments constitutes the data generated by an atomic model. An abstract simulator for atomic models is depicted in pseudo-code in Figure 2. This abstract simulator is run together with a coordinator which keeps track of the event times, and sends to the simulator the input events when they arrived and the output events at the time indicated by tn. In this way the operational behaviour or semantics of a model is given by its simulator.

One common approach to validate DEVS models in the field of modeling and simulation is to construct an additional model to act as a tester, and simulate the coupled model obtained from connecting the model to be validated to the tester model. For example, in the design of a control, a common approach is to construct a model of the system to be controlled and simulate the coupling of both, the control and the model corresponding to the system.

2.2. Timed Automata

Finite state transition systems are one of the most studied research topic for the specification and verification of systems. These are appropiate for reasoning about qualitative properties, however they are not adequate for dealing with timing properties. Several formalisms have been proposed based on transition system along with a notion of time. One of them known as timed automata [4], it constitutes a simple and natural extension to finite automata.

Since its introduction these models have been thoroughly studied under several aspects. The main studied topics are timed extensions of known problems studied in the classical

```
DEVS-atomic-simulator
    tl   // time of last event
    tn   // time of next internal event
    y    // current output event value
         // of the associated model

    when internal event, at time t:
        y = λ(s)
        send output event (y, t)
        s = δ_int(s)
        tl = t
        tn = tl + ta(s)

    when receive input event (x, t), at
    time t with input event value x:
        e = t - tl
        s = δ_ext(s, e, x)
        tl = t
        tn = tl + ta(s)
end DEVS-atomic-simulator
```

Figure 2. Pseudo-code of a simulator for DEVS atomic models.

automata theory, these includes the decidability of the emptyness problem (is the language accepted the empty one?), reachability (wether some states are reachable from the initial state), universality (does a given automata accepts all possible languages for a given alphabet), inclusion (given two models, is the language accepted by one included in the other?) and equivalence (given two models, are they languages equivalent?) problems.

The operational behaviour of the several timed automata formalisms is specified using timed transition systems [4, 31, 25, 26]. In addition, the timed automata formalisms are not expressive enough to accurately represent the behaviour of DEVS models. Then, we present the formal aspects of the proposed design methodology using timed transition systems, powerfull enough to formally represent the behaviour of both timed automata and DEVS formalisms.

In this section, we recall the basic definitions of timed transition systems. We also recall timed safety automata [14] which are a sub-class of the classical timed automata formalism [4]. Timed safety automata are a natural and simple extension of the classical automata formalism, additionally, timed safety automata is the formalism used in different model-checking tools such as KRONOS [9] and UPPAAL [23].

2.3. Timed Safety Automata

Basicaly a timed safety automaton consists of a finite automaton augmented with a finite set of clock variables, and transitions with clock constraints, aditionally the locations can contain local invariant conditions. The transitions are instantaneous while time can elapse in the vertices or locations, at any instant the value of a clock equals the elapsed time sinc its last reset. A clock can be reset to zero by a transition and the time base for the clocks is the set nonnegative real numbers. Transitions can only take place when the values of the clocks statisfy the respective constraints. An automaton can only remain in a location as long as the local invariants hold. The set of clock constraints is defined inductively by $\phi := x \sim c | x - y \sim c | \neg \phi | \phi_1 \wedge \phi_2$, where x and y are clocks, c is an integer constant, and $\sim \in \{\leq, <, =, >, \geq\}$. The set of all constraints over a set of clocks X is noted $\Phi(X)$

A clock valuation v for a set X of clocks, assigns to each clock a nonnegative real value. For $\phi \in \Re$, $v + \phi$ denotes the clock valuation which maps each $x \in X$ to the value $v(x) + \phi$. For $Y \subseteq X$, $v[Y := 0]$ denotes the clock valuation which assigns the value 0 to each clock $x \in Y$ and equals v for the rest of the clocks.

Then a timed safety automaton $\mathcal{A}$ over an alphabet Σ is defined as a tuple $\langle V, V^0, X, E, I \rangle$, where

- V is a finite set of locations,

- $v_0 \in V$ is the initial location,

- X is a finite set of clocks,

- $E \subseteq V \times \Sigma \times \Phi(X) \times 2^X \times V$ is a set of transitions. A transition $\langle s, a, g, \lambda, s' \rangle$ denotes and edge from location s to location s' on symbol a, the guard g is a clock contraint over X, and $\lambda \subseteq X$ indicates what clocks are reset to 0 with this transition,

- $I : V \to \Phi(X)$ maps invariants to locations.

A timed safety automata generates or accepts timed words (sequences of symbols with their occurrence times).

The semantics of a timed safety automaton $\mathcal{A}$ is defined in terms of a timed transition system over the alphabet $\Sigma \cup \Re$. In the next section we give the definiton of this timed transition system [25].

2.4. Timed Transition Systems

A timed transition system $\mathcal{T}_t$ is an automaton whose alphabet includes $\Re^+$. The transitions corresponding to symbols from $\Re^+$ are referred to as time-passage transitions, while non-time-passage transitions are referred to as discrete transitions. So a timed transition system consists of,

S a possibly infinite set of states,

init an initial state,

Σ a set of discrete actions,

D a set of discrete transitions, noted $s \xrightarrow{x} s'$, where $x \in \Sigma_{\mathcal{T}_t}$ and $s, s' \in S_{\mathcal{T}_t}$, asserting that "from state s the system can instantaneously move to state s' via the occurrence of the event x, and,

T a set of time-passage transitions, $s \xrightarrow{t} s'$, $t \in \Re^+$ and $s, s' \in S_{\mathcal{T}_t}$, asserting that "from state s the system can move to state s' during a positive amount of time t in which no discrete events occurs".

A timed transition system is assumed to satisfy two axioms.

S1 If $s \xrightarrow{t} s'$ and $s' \xrightarrow{t'} s''$, with $t, t' \in \Re^+$, then $s \xrightarrow{t+t'} s''$.

S2 Each time-passage step, $s \xrightarrow{t} s'$, with $t \in \Re^+$, has a trajectory.

Where a trajectory describes the state changes that can occur during time-passage transitions, if I is any closed interval of $\Re_0^+$ beginning with 0 (we are only interested in finite behavior), an I-trajectory is defined as a function, $v : I \to S$ such that:

$$v(t) \xrightarrow{t'-t} v(t') \quad \forall\, t, t' \in I \mid t < t'$$

It is straightforward to described timed safety automaton $\mathcal{A} = \langle V, V^0, X, E, I \rangle$ over an alphabet Σ, in terms of a timed transition system $\mathcal{T}_t$. Let u and v denote clock valuations for the set of clocks X, then the semantics of $\mathcal{A}$ are is the timed transition system $\mathcal{T}_t$, where the states are pairs (s, v) with $s \in V$, and the transtions are defined by:

$(s, v) \xrightarrow{t} (s, v + t)$ if v and $v + t$ satisfies $I(s)$ for $t \in \Re^+$

$(s, v) \xrightarrow{a} (s', v')$ if $\langle s, a, g, \lambda, s' \rangle \in E$, v satisfies g, $v' = v[\lambda := 0]$, and v' satisfies $I(s')$

In order to reason about the behaviour described by timed automata and timed transition systems is necessary to have a formalization of it. This is achieve through the definition of timed execution runs and timed traces.

A timed execution run, describes all the discrete state changes that occur, plus the evolution of the state during time-passage transitions. A timed execution run is defined as a finite alternating sequence $\Upsilon = v_0 a_1 v_1 a_2 v_2 \ldots a_n v_n$, where

1. Each v_i is a trajectory and each $a_i \in \Sigma$.

2. If v_i is not the last trajectory in Υ then $v_i(\sup(I_i)) \xrightarrow{a_{i+1}} v_{i+1}(0)$, where I_i denote the domain of the trajectory v_i. [1]

Note that since a timed execution run is a finite sequence, we implicitly exclude the occurrence of an infinite number of discrete transitions in a finite amount of time, this behaviour is usually refferred to as *Zeno*.

Given a timed execution fragment $\Upsilon = v_0 a_1 v_1 a_2 v_2 \ldots a_n v_n$. For each a_i of Υ we define its time of occurrence t_i as $\sum_{0 \leq j < i}(\sup(I_j))$. Then we define the *trace* associated with the execution fragment Υ, and we note $\text{trace}(\Upsilon)$, as the pair:

$$\text{trace}(\Upsilon) \triangleq (((a_1, t_1), (a_2, t_2), \ldots, (a_n, t_n)), t_n + \sup(I_n))$$

Where the fist component of the trace denote the information of all the discrete transitions that occurred in the execution, and the time when they occurred. The second component corresponds to the total duration of the execution.

---

[1] $\sup(I)$ is the supremum of the interval I

2.5. Verification with Timed Automata

Language inclusion and state reachability constitute the basis of numerous works developed in the area of formal verification. The inclusion problem allows to verify the conformance of the behaviour between an implementation and its specification, while state reachability can guarantee that certain undesirable states are never reached. However, a discouraging result in the timed automata theory is th undecidability of the inclusion problem, in consequence, a lot of research activities has focused in finding a suitable subclass of timed automata in which this problem becomes decidable [15, 3, 16, 12, 2, 28, 6, 10].

Despite the existence of subclasses of timed transition systems for which the inclusion problem is decidable, none of them are expressive enough for dealing with the kind of specification problems at which we aim. Some of this subclasses and their main results are summarized below.

Deterministic Timed Automata [15] This model is less expressive, and it is not suitable to represent input enabled models. Even further the problem of the determinization of a given timed automata is not known to be decidable [29].

Bounded 2-way Deterministic Timed Automata [3] Although they are more expressive than Deterministic Timed Automata, they remain not suitable to represent input enabled models.

Digitization [16] The inclusion problem between an automaton closed under digitization and one closed under inverse digitization is decidable, however checking whether an automaton is closed under inversed digitization is not decidable [27].

Robust Timed Automata [12] The class of models that are possible to express under this formalism is not a proper subset of timed automata, this formalism is based in adding some fuzziness on the acceptance condition. However this model is known not to be determinizable, and since usually the determinization is one of the first steps for solving the inclusion problem, the hope of decidability of this last problem is dampen.

Restrictions on Timed Automata [2, 28] If we restrict to timed automata with only one clock variable, or if the only constant appearing in clock constraints is 0, then the problem of checking if an unrestricted timed automaton is included in a restricted one is decidable, although only for finite words.

Event Clock Automata [6] This subclass consist in associating only one clock variable to each event, and this clocks are reset when the events occur. For this particular subclass the inclusion problem is decidable, although less expressive..

Resource Bounded Automata [10] The number of clocks and granularity of the clocks comprise the resources of the timed automata. Given two timed automata and limited resources, it is shown to be decidable the problem of checking wether there exists a bounded timed automaton such that provides evidence that the first timed automaton is not included in the second.

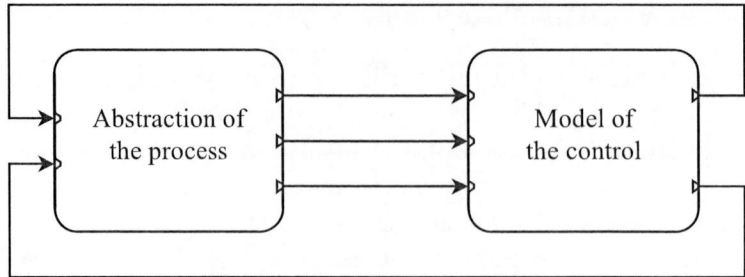

Figure 3. Classic validating approach in DEVS.

Given the undecidability of the inclusion problem, most of the avaiable tools for model-checking timed automata [5, 9, 23] exploit the decidability of the reachability problem to verify properties expressed in a certain temporal logic.

3. Formal Verification for the Design Process

In this section we will start by introducing the design methodology proposed, then in a similar approach as done with timed transition systems we will define the behaviour of DEVS models in terms of timed executions and timed traces. Additionally we will show the correspondence of this formalization with the behaviour implied by a conceptual DEVS simulator.

Afterwards we will proceed to show how to construct a timed transition system from a given DEVS model and we will show that their behaviour is equivalent.

3.1. Design Methodology

The design methodology consists in a classical modular top-down approach. For the high-level specification the timed automata formalism is used, and the low-level implementation is done using the DEVS formalism. The advantage of using timed automata for the high-level specification is that, contrary to DEVS models, it is possible to specify non-deterministic timed behaviour.

While the main advantage of the DEVS formalism is that regardless of the implementation (be it in hardware or software) DEVS enfoces the concept of determinism at the design stage instead of putting it off to the implementation phase. Moreover, it is directly possible to simulate the DEVS models, and several efficient simulators for DEVS models have been developed [13].

The typical approach for validating DEVS control models, consists in building, using the DEVS formalism, an abstraction of the process to be controlled. The model of the process to be controlled will react to the events generated by the control, and it will generate the events to which the control reacts (Figure 3). A coupled model which components are the control model and the model of the process to be controlled is used to run simulations, and the simulation results constitutes the validation of the control.

Our contribution is to introduce, in the design process, formal techniques for reasoning about the conformance of the implementation (DEVS models) and the specification (timed

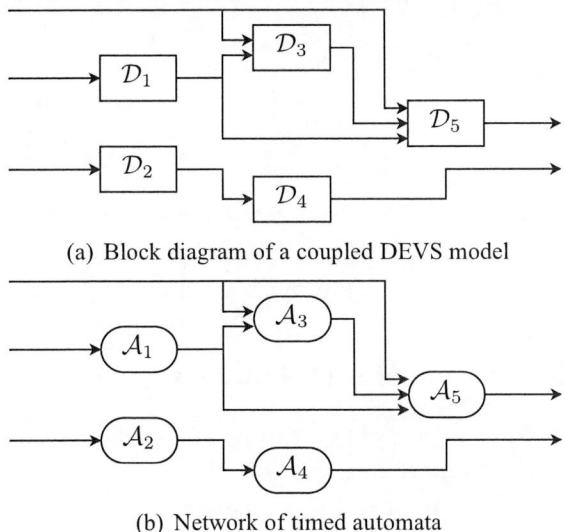

Figure 4. Complex-model verification approach.

automata). This consists basically in defining the semantics of DEVS models in terms of traces of events, in order to compare them to the traces defined by the timed transition systems describing the behaviour of timed automata (in this section when we speak of timed automata we mean any subclasses that can be found in the literature whose behaviour is defined in terms of general timed transition systems).

Once a common semantics for both formalisms is available, we define a *simulation relation*, this provides a sufficient condition for the behaviour conformance between a DEVS model and a timed automata. So the formal verification process consists in finding (or proving the existence) of a simulation relation between the implementation and the specification models.

In the case of complex models where the formal verification of the whole model is not cost-effective, or feasible, using the methodology proposed, is possible to formally verify only some critical subcomponents and validate the behaviour of the full model by testing.

For example, the Figure 4(a) shows a coupled DEVS model, the lines connecting the nodes represents the input, output and internal coupling of the subcomponents. Figure 4(b) gives an overview of a network of timed automata, where each node represents a timed automaton, and the lines connecting them the synchronization relation between the events of the different timed automata.

For instance, if an atomic DEVS subcomponent, lets say $\mathcal{D}_2$, is critical for the safety of the whole model, one can formally verify the behaviour of this component against its corresponding specification given by a timed automaton ($\mathcal{A}_2$). The formal verification and simulation-based validation can be used in a complimentary way. The timed automaton specification can be used to derive test cases to be simulated with the DEVS model, validating in this way the conformance relation.

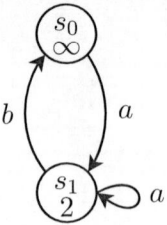

Figure 5. Phase transition of a simple atomic DEVS models.

3.2. Behavioral Comparison of DEVS Models

In DEVS literature [11, 32, 33] the behavior of an atomic DEVS model is given by a conceptual simulator. However formal semantics must be associated to a DEVS model in order to formally reason about its behavior and its properties.

The simulation semantics of an atomic DEVS model is given by the DEVS-simulator as specified by the pseudo-code given in the Figure 2.

We now introduce the formal notion of executions or simulation runs of a DEVS model, and its traces as an alternative way to formally specify the full behavior implied by DEVS models.

Employing these concepts it is possible to formally relate the semantics of DEVS models and timed transition systems. This new description of the behavior of DEVS models is equivalent to the one implied by the conceptual DEVS simulator.

An *execution fragment* of a DEVS model $\mathcal{D}$ is a finite alternating sequence $\Upsilon = \upsilon_0 x_1 \upsilon_1 x_2 \upsilon_2 \ldots x_n \upsilon_n$, where:

time-passage transitions Each υ_i is a function from a real interval $I_i = [0, t_i]$ to the set of total states of $\mathcal{D}$, such that $\forall j, j' \in I_i | j < j'$, if $\upsilon_i(j) = (s, e)$ then $\upsilon_i(j') = (s, e + j' - j)$.

event transitions Each x_i is an input or output event, and if $(s, e) = \upsilon_{i-1}(\sup(I_{i-1}))$, $(s', 0) = \upsilon_i(\inf(I_i))$, one of the following conditions hold:

1. $x_i \in Y_\mathcal{D}$, $\delta_{\text{int}}(s) = s'$, $\text{ta}(s) = e$, and $\lambda(s) = x_i$.
2. $x_i \in X_\mathcal{D}$, $\delta_{\text{ext}}(s, e, x_i) = s'$, and $e < \text{ta}(s)$.

Furthermore we define the set execs($\mathcal{D}$) as the set of all possible executions of the DEVS model $\mathcal{D}$. Given an execution fragment of a DEVS model, $\Upsilon = \upsilon_0 x_1 \upsilon_1 x_2 \ldots x_n \upsilon_n$, then a *trace* of this execution trace(Υ) is a tuple (θ_I, θ_O, t) such that θ_I and θ_O are sequences consisting of all the pairs of events of Υ and their time of occurrence, sorted in chronological order of occurrence, and t is the total time of execution defined as $\sum_{0 \leq j \leq n}(\sup(I_j))$. Formally, the time of occurrence of an event x_i of Υ is equal to $\sum_{0 \leq j < i}(\sup(I_j))$, with I_j the domain of υ_j.

We define the set of all traces of a DEVS model as, traces($\mathcal{D}$) = {trace(Υ)|$\Upsilon \in$ execs($\mathcal{D}$)}.

To ilustrate this definitions we will give a very simple DEVS model as an example. In the Figure 5 is the phase transition diagram of the following atomic DEVS model:

$$\mathcal{D} = \{X, Y, S, \delta_{\text{ext}}, \delta_{\text{int}}, \lambda, \text{ta}\}$$

$$X = \{a\}$$

$$Y = \{b\}$$

$$S = \{s_0, s_1\} \times \Re_0^+ \cup \infty$$

$$\delta_{\text{ext}}(s, \sigma, a, e) = \begin{cases} (s_1, 2) & \text{If } s = s_0 \\ (s_1, \sigma - e) & \text{If } s = s_1 \end{cases}$$

$$\delta_{\text{int}}(s_1, \sigma) = (s_0, \infty)$$

$$\lambda(s, \sigma) = b$$

$$\text{ta}(s, \sigma) = \sigma$$

So the following sequence is a possible execution fragment of the DEVS model $\mathcal{D}$.

$$\Upsilon = v_0 \text{ a } v_1 \text{ b } v_2 \text{ a } v_3 \text{ a } v_4 \text{ b } v_5$$

where

$$\begin{aligned}
v_0(x) &= ((s_0, \infty), x) & I_0 &= [0, 1] \\
v_1(x) &= ((s_1, 2), x) & I_1 &= [0, 2] \\
v_2(x) &= ((s_0, \infty), x) & I_2 &= [0, 1] \\
v_3(x) &= ((s_1, 2), x) & I_3 &= [0, 1.5] \\
v_4(x) &= ((s_1, 0.5), x) & I_4 &= [0, 0.5] \\
v_5(x) &= ((s_0, \infty), x) & I_5 &= [0, 1]
\end{aligned}$$

And the trace of the execution fragment Υ, $\text{trace}(\Upsilon) = (\theta_I, \theta_O, t)$ is

$$\begin{aligned}
\theta_I &= \langle (a, 1), (a, 4), (a, 5.5) \rangle \\
\theta_O &= \langle (b, 3), (b, 6) \rangle \\
t &= 7
\end{aligned}$$

Note that given a DEVS model, $\mathcal{D}$, for a sequence of pairs of input events and their respective time of occurrence, θ_I, a total time of execution t, and a initial state, s_0, there exists a unique execution fragment, $\Upsilon = v_0 x_1 \ldots$, such that $\text{trace}(\Upsilon) = (\theta_I, \theta_O, t)$ and $v_0(0) = (s_0, 0)$. In this sense DEVS models can be said to define deterministic models.

It is simple, as well, to show an equivalence between the formal definition of traces of DEVS models and the behavior exhibited by the pseudo-code of the `DEVS-simulator` (Figure 2).

The initial state for the simulation is the state variable `s` = s_0, the clock variable `t` initialized to `0`, and also the variables, `tl` and `tn`, initialized to `0`.

Each pair of the sequence θ_I, (x_i, t_i), is represented in the simulation algorithm by sending the message `input event (x`$_i$`, t`$_i$`)` at time t_i, which implies that the clock variable `t` has the value t_i when the message is sent. If for all the output event message that the simulator generates `output event (x`$_j$`, t`$_j$`)` we record the output events x_j and its times of occurrence t_j, up to the clock variable `t` reaches the value t, we obtain a sequence identical to θ_O.

3.3. Timed Transition System and DEVS Model

Since our goal is to provide a formal background to compare the behaviour of DEVS models and timed automata in this section we show how to construct a timed transition system that has the same behaviour of a given DEVS model. Later we will use this construction to express properties relating timed transition systems to DEVS models.

Given a DEVS model $\mathcal{D}$ we define its *associated timed transition system* over the alphabet $\Sigma = X_\mathcal{D} \cup Y_\mathcal{D}$, $\text{Taut}(\mathcal{D})$ as a timed transition system $\mathcal{T}_t$, where:

1. the set of states, $S_{\mathcal{T}_t}$ consists of the set of total states of $\mathcal{D}$, $Q_\mathcal{D}$,

2. the initial state, $\text{init}(\mathcal{T}_t)$ is $(s, 0)$, where $s \in S_\mathcal{D}$ and, s is the discrete state defined as the initial state of the DEVS model,

3. the set of discrete transitions, $D_{\mathcal{T}_t}$,

$$D_{\mathcal{T}_t} = \{(s, e) \xrightarrow{x} (s', 0) \mid (\delta_{\text{int}}(s) = s' \wedge \lambda(s) = x \wedge e = \text{ta}(s)) \text{ or } (\delta_{\text{ext}}((s, e), x) = s' \wedge e \leq \text{ta}(s))\}$$

4. and the set of time-passage transitions, $T_{\mathcal{T}_t}$,

$$T_{\mathcal{T}_t} = \{(s, e) \xrightarrow{t} (s, e') \mid (s, e) \in Q_\mathcal{D}, e' = e + t, 0 \leq e + t \leq \text{ta}(s)\}$$

The timed transition system associated with a DEVS model, as defined above, specifies the same set of traces that the DEVS model.

Lemma 1. *If $\mathcal{D}$ is a DEVS model, then* $\text{traces}(\mathcal{D}) = \text{traces}(\text{Taut}(\mathcal{D}))$.

Lemma 1. We can see that for every possible execution fragment of the DEVS model there exists an analogous execution fragment of the associated timed transition system. Suppose $\Upsilon_d = v_0 x_1 v_1 x_2 \ldots x_n v_n$ is any execution fragment of the DEVS model with trace $\text{trace}(\Upsilon_d)$. If $\mathcal{T}_t = \text{Taut}(\mathcal{D})$ then the domain of each function v_i of Υ is contained in $S_{\mathcal{T}_t}$, furthermore each v_i is a trajectory of the timed transition system $\mathcal{T}_t$. So for each event, x_i of the execution fragment Υ_d, by the definition of execution fragment of DEVS models one of the following conditions hold:

1. $x_i \in Y_\mathcal{D}$, $\delta_{\text{int}}(s) = s'$, $\text{ta}(s) = e$, and $\lambda(s) = x_i$.

2. $x_i \in X_\mathcal{D}$, $\delta_{\text{ext}}(s, e, x_i) = s'$, and $e < \text{ta}(s)$.

and so by the definition of Taut (Item 3), there is a discrete transition of $\mathcal{T}_t$, $(s, e) \xrightarrow{x_i} (s', 0)$. Using the definition of execution fragment for timed transition systems Υ is also an execution fragment of the timed transition system $\mathcal{T}_t$, and $\text{trace}(\Upsilon) \in \text{traces}(\mathcal{T}_t)$.

Analogously, if there is an execution fragment of $\text{Taut}(\mathcal{D})$ there is also a correspondent execution fragment in the DEVS model $\mathcal{D}$.

In consequence the sets of traces of a DEVS model and its associated timed transition system are equal. □

There exists several definitions of simulation relations between timed transition systems, this sort of relations are generally used to describe and compare the behaviour of transition systems. They denote preorders between the models, even further if there exists one of this relations between a model $\mathcal{T}_a$ and another $\mathcal{T}_b$, then the set of traces of $\mathcal{T}_a$ is contained in the ones of $\mathcal{T}_b$ [25].

By adapting the results to DEVS and timed transition systems, we provide a method for formally verifying that the behaviour of a DEVS model conforms the behaviour described by a timed transition system. In the next section we give the definition of a *simulation relation* between a DEVS model and a timed transition system, and extend the formal results from timed automata theory.

3.3.1. Simulation Relation

A *simulation relation* from a DEVS model $\mathcal{D}$ to a timed automaton $\mathcal{A}$ is a relation $f \subseteq Q_\mathcal{D} \times S_\mathcal{A}$, such that the following conditions are satisfied:

1. (output event transitions)
   If there is a pair of states $(s, e), (s', e') \in Q_\mathcal{D}$ such that $\delta_{\text{int}}(s) = s'$, $e = \text{ta}(s)$, $e' = 0$ and $u \in f[(s, e)]$ then there exists a state $u' \in f[(s', e')]$ such that $u \xrightarrow{\lambda(s)} u' \in D_\mathcal{A}$.

2. (input event transitions)
   If there is a pair of states $(s, e), (s', e') \in Q_\mathcal{D}$ such that $\delta_{\text{ext}}((s, e), x) = s'$, $e \leq \text{ta}(s)$, $e' = 0$ and $u \in f[(s, e)]$ then there exists a state $u' \in f[(s', e')]$ such that $u \xrightarrow{x} u' \in D_\mathcal{A}$.

3. (time passage steps) If there is a pair of states $(s, e), (s', e') \in Q_\mathcal{D}$ such that $0 \leq e \leq e' \leq \text{ta}(s)$, $s = s'$, and $u \in f[(s, e)]$ then there exists a state $u' \in f[(s', e')]$ such that $u \xrightarrow{e'-e} u' \in T_\mathcal{A}$.

We write $\mathcal{D} \leq_S \mathcal{A}$ if there exists a simulation relation from $\mathcal{D}$ to $\mathcal{A}$.

Theorem 1. *If there exists a forward simulation relation between a DEVS model, $\mathcal{D}$, and a timed automaton, $\mathcal{A}$, then the set of traces of $\mathcal{D}$ is contained in the set of traces of $\mathcal{A}$.*

Proof. Since we can construct a timed transition system that has the same behaviour of $\mathcal{D}$ it is straightforward to extend the classical demonstration [25] to this setting. □

This last result allow us to compare the behaviour of DEVS models and timed transition systems through the search of a simulation relation. In this way it is possible to formally prove that a given DEVS models conforms a certain specification given by a timed transition system.

Regrettably due to the complexity and the dense continuous time base of both formalisms, it is not possible to completely automate the process of searching a simulation relation. Except for a very restricted subclass of timed transition systems and DEVS models, the problem of formally verifying the conformance of an implementation to its respective specification remains undecidable.

4. Case Study Presentation

This case study shows the convenience of the methodology proposed for the design and formal verification of discrete event control systems. A subsystem of the "Production Cell" case study [24], was chosen since it provides a well defined system and it has been used to compare the adequacy of formal methods.

The main goal is to develop the automation software to control a subsystem of a production cell used in a metal processing factory. The subsystem to be controlled consists of travelling crane and a gripper which are used to move metal plates between two conveyor belts.

The model introduced in this section for the travelling crane and gripper is a modular component of a bigger model representing the behaviour of the full "Production Cell". This is possible thanks to the modularity of the formalisms employed in the design methodology.

We start by giving an informal description of the travelling crane and the set of requirements that the control must meet. We formalize these requirements with a timed automaton (in this section when we speak about timed automata, we will be making reference to timed safety automata [14]), and provide its formal specification. In the next design step of the control, we construct a DEVS model representing the travelling crane and the gripper together with a DEVS model implementing the control. Finally, we formally verify the conformance of the DEVS control implementation to the requirement. In addition, the DEVS control implementation can be validated by coupling it with the model of the travelling crane and simulating the coupled model.

4.1. Description of the Travelling Crane

The travelling crane can be classified as a reactive system for which the control must react to changes in the environment. These changes are modelled using a discrete event approach.

The task of the travelling crane consists in picking up metal plates from the deposit belt moving them to the feed belt and unloading them there.

The crane has an electromagnet as a gripper which can perform horizontal and vertical translations. Horizontal mobility serves to cover the horizontal distance between the belts, while vertical mobility is necessary because the belts are placed at different levels. The typical operation is as follow:

- When a photoelectric cell indicates that a metal plate has moved into the unloading area on the deposit belt, the gripper positions itself, through horizontal and vertical translation, over the deposit belt and picks the metal plate.

- The gripper transports the metal plate to the feed belt and unloads it.

Efficiency considerations may lead a designer to move the travelling crane back to the deposit belt at the end of this sequence so that incoming plates can be transported immediately.

In the Figure 6 a sketch of the travelling crane can be seen. The travelling crane can be moved in the horizontal or in the vertical direction, with constant speeds of θ_x and θ_y respectively. The travelling crane is not allowed to move in more than one direction at a

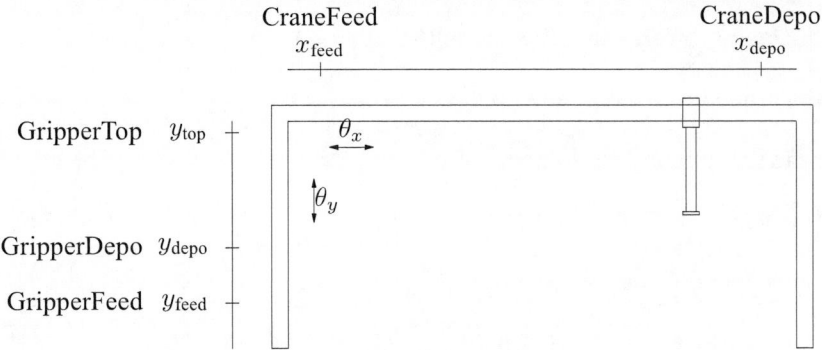

Figure 6. Sketch of the travelling crane.

time, and the movement of the crane is produced by two independent motors, each of which reacts (starts or stops) to events received from the control.

There are sensors in the travelling crane which indicates the position of the travelling crane by sending events when the travelling crane crosses the position of one of this sensors.

These sensors (Figure 6) are located in the following positions:

x_{feed} horizontal position of the feed belt,

x_{depo} horizontal position of the deposit belt,

y_{top} top position of the gripper,

y_{feed} height of the feed belt,

y_{depo} height of the deposit belt.

If the crane and gripper crosses one of this positions and the corresponding motor is not stopped before the crane or gripper moves a certain distance given by two constants x_{buff}, y_{buff}, for the horizontal and vertical motions respectively, an error state is entered, and it stops responding.

The case of the limit y_{depo} is handled differently, the error state is entered only if the gripper, while moving downwards, crosses the position without being stopped, and the travelling crane is horizontally positioned over the deposit belt (to the right of x_{depo}).

The travelling crane generates the following events:

- The crane is over the deposit belt (`CrPosDepo`).

- The crane is over the feed belt (`CrPosFeed`).

- The gripper is in the top position (`GrPosTop`).

- The gripper is at the feed belt level (`GrPosFeed`).

- The gripper is at the deposit belt level (`GrPosDepo`).

And the set of events to which the travelling crane react are:

- Move the crane towards the deposit belt (CrDepo).

- Move the crane towards the feed belt (CrFeed).

- Stop crane movement (CrStop).

- Move the gripper towards the belts (GrDown).

- Move the gripper to the top position (GrUp).

- Stop gripper movement (GrStop).

- Activate the gripper electromagnet (EmAct).

- Deactivate the gripper electromagnet (EmDeact).

Additionally there is an event that can be seen as an external input to the control to indicate that a metal plate has arrived to the unloading area of the deposit belt (Arrived)

There are several safety requirements that the control program should meet [24]. Each safety requirement is a consequence of one of the following principles:

- If the crane is positioned above the feed belt, it may only move towards the deposit belt, and if it is positioned above the deposit belt, it may only move towards the feed belt.

- The gripper of the crane must not be moved downward, if it is positioned above the deposit belt and it is in the position required for picking a metal plate, or if it is positioned above the feed belt and it is in the position required for picking a metal plate.

- The travelling crane is not allowed to knock against a belt laterally (this would happen if the travelling crane moved from the deposit belt to the feed belt without a performing also a vertical upward translation).

- The electromagnet of the crane may only be deactivated, if its magnet is above the feed belt, and sufficiently close to it.

These properties belongs to the class of safety-critical properties, the control software must meet the properties in order to avoid damage to the machines or injury to people.

We formalize the first requirement and to avoid damage to the travelling crane we require to the control to stop the crane motors in less than δ_1 time units after receiving the events of the sensors indicating that the crane has arrived to the limits. It is relevant to note that in the behaviour of the control, the events corresponding to the input (generated by the travelling crane) should always be enabled and can not be constrained by the model specification.

Since this is a high-level property, the behaviour specified by the timed automata describes all possible sequence of events that conforms the requirement. To meet the requirement the set of all possible sequence of events specified by a model of the control should be a subset of the ones described by the timed automaton.

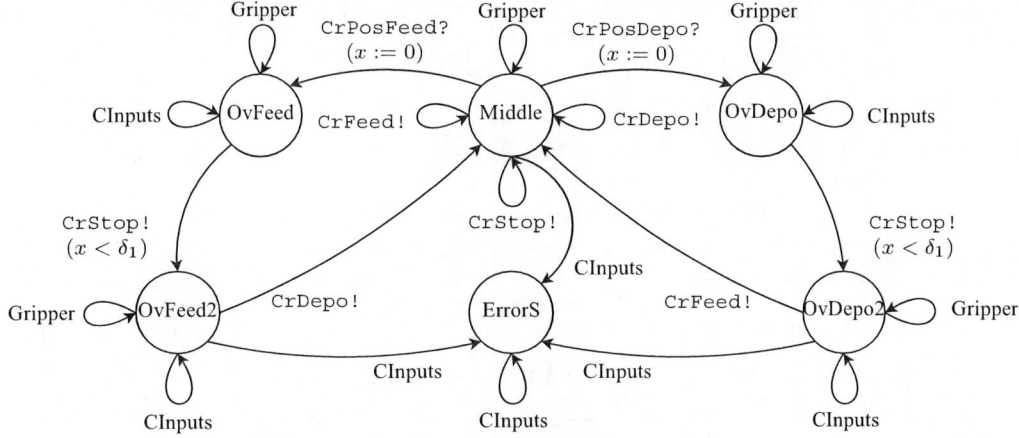

Figure 7. Formal requirement timed automaton.

The timed automaton modeling this requirement is given in the Figure 7. In this figure we abuse of the notation in order to make it more readable. When there are several transitions from the same source node to the same destination and with the same guard we compress all the edges into only one with a label indicating which events corresponds to this transitions. The correspondance of this labels and events is:

Gripper → {GrDown!, GrUp!, GrStop!, GrPosTop?, GrPosFeed?, GrPosDepo?, EmAct!, EmDeact!, Arrived?}

CInputs → {CrPosFeed?, CrPosDepo?, GrPosTop?, GrPosFeed?, GrPosDepo?, Arrived?}

The constant δ_1 and the clock variable x gives a constraint on the maximum amount of time that the motor of the crane can be running without the control passing to an error state after the sensor positioned in the limit of the crane signals that the crane has reached its limit.

The formal definition of the timed automaton of the Figure 7 is given by the timed transition system $\mathcal{R}_1$:

$S_{\mathcal{R}_1} = \{\text{Middle, OvFeed, OvFeed2, OvDepo, OvDepo2, Error}\} \times \Re_0^+$

$D_{\mathcal{R}_1} = \{((\text{Middle}, \sigma), x, (\text{Middle}, \sigma)) | \sigma \geq 0, x \in \text{Gripper} \cup \{\text{CrFeed!, CrStop!, CrDepo!}\}\}$
$\cup \{((\text{Middle}, \sigma), \text{CrPosDepo?}, (\text{OvDepo}, 0))\} \cup \{((\text{OvDepo}, \sigma), \text{CrStop!}, (\text{OvDepo2}, \sigma)) | \sigma < \delta_1\}$
$\cup \{((\text{OvDepo2}, \sigma), \text{CrFeed!}, (\text{Middle}, \sigma))\} \cup \{((\text{Middle}, \sigma), \text{CrPosFeed?}, (\text{OvFeed}, 0))\}$
$\cup \{((\text{OvFeed}, \sigma), \text{CrStop!}, (\text{OvFeed2}, \sigma)) | \sigma < \delta_1\} \cup \{((\text{OvFeed2}, \sigma, \text{CrDepo!}, (\text{Middle}, \sigma))\}$
$\cup \{((\text{Middle}, \sigma), x, (\text{ErrorS}, \sigma)) | x \in \text{CInputs}\} \cup \{((\text{OvFeed2}, \sigma), x, (\text{ErrorS}, \sigma)) | x \in \text{CInputs}\}$
$\cup \{((\text{OvDepo2}, \sigma), x, (\text{ErrorS}, \sigma)) | x \in \text{CInputs}\} \cup \{((\text{ErrorS}, \sigma), x, \text{ErrorS}, \sigma)) | x \in \text{CInputs}\}$
$\cup \{((\text{OvDepo}, \sigma), x, (\text{OvDepo}, \sigma)) | x \in \text{Gripper} \cup \text{CInputs}\}$
$\cup \{((\text{OvFeed}, \sigma), x, (\text{OvFeed}, \sigma)) | x \in \text{Gripper} \cup \text{CInputs}\}$
$\cup \{((\text{OvDepo2}, \sigma), x, (\text{OvDepo2}, \sigma)) | x \in \text{Gripper} \cup \text{CInputs}\}$
$\cup \{((\text{OvFeed2}, \sigma), x, (\text{OvFeed2}, \sigma)) | x \in \text{Gripper} \cup \text{CInputs}\}$

$T_{\mathcal{R}_1} = \begin{cases} \{(s, \sigma), t, (s, \sigma + t)\} & \text{If } s \notin \{ \text{OvDepo, OvFeed} \} \\ \{(s, \sigma), t, (s, \sigma + t)\} & \text{If } s \in \{ \text{OvDepo, OvFeed} \} \text{ and } \sigma + t < \delta_1 \end{cases}$

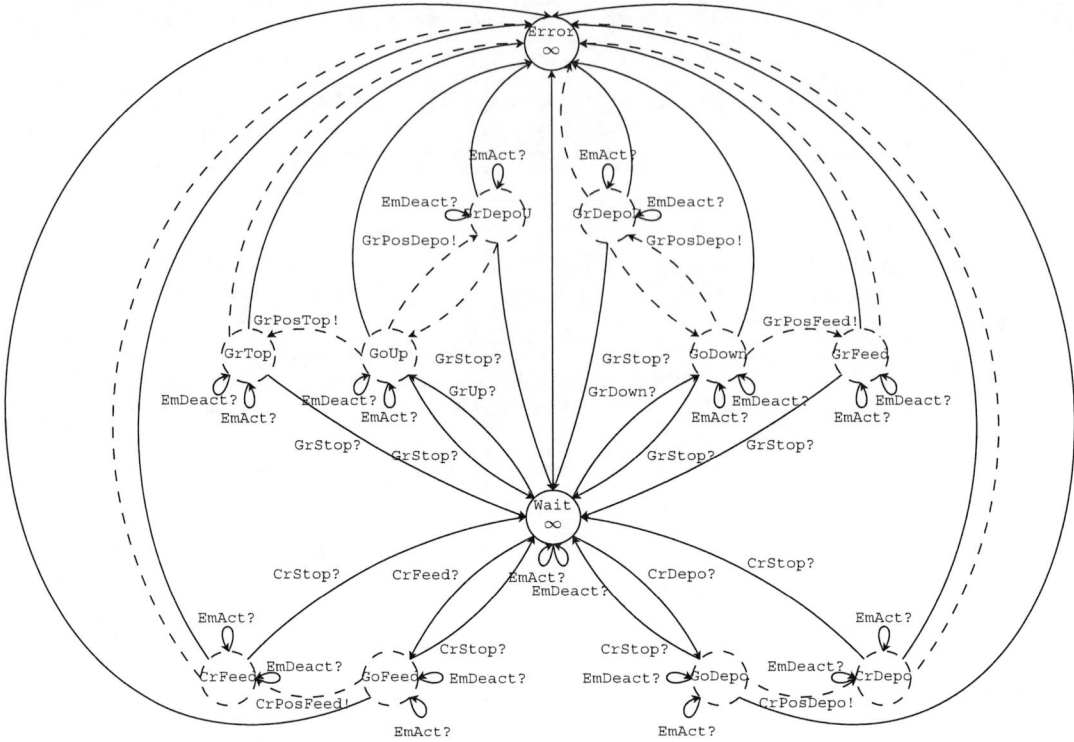

Figure 8. DEVS model of the Travelling Crane.

4.1.1. Travelling Crane Model

We use a phase transition diagram to represent the DEVS model. In this phase transition diagram, the nodes denote a set of discrete phases that constitutes a partition of the state set. The external (responding to input events) and internal (generating output events) transitions between the phases are represented by arrows and dashed arrows respectively.

Usually, the task of creating a formal model implies adopting design decisions to take into account details not mentioned in the original informal description. Here the resulting model is presented without explicitly mentioning and explaining all the choices taken while building the model since it is not in the scope of this case study.

The state space of the travelling crane model consists of a set of phases, a discrete variable denoting the state of the electromagnet (on or off), and three continuous variables, two representing the position in the plane of the gripper and crane, and the other expressing the lifetime of the phase.

A phase transition diagram of the travelling crane model is given in the Figure 8. The complete and formal specification of the DEVS model, $\mathcal{TC}$ is:

$\mathcal{TC} = \{X, Y, S, \delta_{\text{ext}}, \delta_{\text{int}}, \lambda, \text{ta}\}$
$X = \{\texttt{CrFeed?}, \texttt{CrDepo?}, \texttt{CrStop?}, \texttt{GrUp?}, \texttt{GrDown?}, \texttt{GrStop?}, \texttt{EmAct?}, \texttt{EmDeact?}\}$
$Y = \{\texttt{CrPosDepo!}, \texttt{CrPosFeed!}, \texttt{GrPosTop!}, \texttt{GrPosFeed!}, \texttt{GrPosDepo!}, \Lambda\}$
$S = \{\texttt{Wait}, \texttt{Error}, \texttt{GoUp}, \texttt{GoDown}, \texttt{GoDepo}, \texttt{GoFeed}, \texttt{GrTop}, \texttt{GrDepoU}, \texttt{GrDepoD}, \texttt{GrFeed}, \texttt{CrFeed},$
$\quad \texttt{CrDepo}\} \times \{\texttt{EmOn}, \texttt{EmOff}\} \times \Re \times \Re \times \Re_0^+$

$\delta_{\text{int}}(ph, em, x, y, \sigma) =$

$$\begin{cases} (\texttt{GrTop}, em, x, y_{\text{top}}, y_{\text{buff}}/\theta_y) \\ \quad \text{If } ph = \texttt{GoUp} \text{ and } y - \sigma * \theta_y = y_{\text{top}} \\ (\texttt{GrDepoU}, em, x, y_{\text{depo}}, y_{\text{buff}}/\theta_y) \\ \quad \text{If } ph = \texttt{GoUp} \text{ and } y - \sigma * \theta_y = y_{\text{depo}} \\ (\texttt{GoUp}, em, x, y - \sigma * \theta_y, (y - y_{\text{top}})/\theta_y) \\ \quad \text{If } ph = \texttt{GrDepoU} \\ (\texttt{Error}, em, x, y - \sigma * \theta_y, \infty) \\ \quad \text{If } ph = \texttt{GrTop} \\ (\texttt{GrFeed}, em, x, y_{\text{feed}}, y_{\text{buff}}/\theta_y) \\ \quad \text{If } ph = \texttt{GoDown} \text{ and } y + \sigma * \theta_y = y_{\text{feed}} \\ (\texttt{GrDepoD}, em, x, y_{\text{depo}}, y_{\text{buff}}/\theta_y) \\ \quad \text{If } ph = \texttt{GoDown} \text{ and } y + \sigma * \theta_y = y_{\text{depo}} \\ (\texttt{GoDown}, em, x, y + \sigma * \theta_y, (y_{\text{depo}} - y)/\theta_y) \\ \quad \text{If } ph = \texttt{GrDepoD} \text{ and } x > x_{\text{feed}} \\ (\texttt{Error}, em, x, y + \sigma * \theta_y, \infty) \\ \quad \text{If } ph = \texttt{GrDepoD} \text{ and } x \leq x_{\text{feed}} \\ (\texttt{Error}, em, x, y + \sigma * \theta_y, \infty) \\ \quad \text{If } ph = \texttt{GrFeed} \\ (\texttt{CrDepo}, em, x + \sigma * \theta_x, y, x_{\text{buff}}/\theta_x) \\ \quad \text{If } ph = \texttt{GoDepo} \\ (\texttt{Error}, em, x + \sigma * \theta_x, y, \infty) \\ \quad \text{If } ph = \texttt{CrDepo} \\ (\texttt{CrFeed}, em, x - \sigma * \theta_x, y, x_{\text{buff}}/\theta_x) \\ \quad \text{If } ph = \texttt{GoFeed} \\ (\texttt{Error}, em, x - \sigma * \theta_x, y, \infty) \\ \quad \text{If } ph = \texttt{CrFeed} \end{cases}$$

$\delta_{\text{ext}}(ph, em, x, y, \sigma, e, \texttt{GrUp?}) =$

$$\begin{cases} (\texttt{GoUp}, em, x, y, (y - y_{\text{top}})/\theta_y) \\ \quad \text{If } ph = \texttt{Wait} \\ \quad \text{and } y_{\text{top}} < y \leq (y_{\text{depo}} + y_{\text{buff}}) \\ (\texttt{GoUp}, em, x, y, (y - y_{\text{depo}})/\theta_y) \\ \quad \text{If } ph = \texttt{Wait} \\ \quad \text{and } y > y_{\text{depo}} + y_{\text{buff}} \\ (\texttt{Error}, em, x, y, \infty) \\ \quad \text{If not} \end{cases}$$

$\delta_{\text{ext}}(ph, em, x, y, \sigma, e, \texttt{GrDown?}) =$

$$\begin{cases} (\texttt{GoDown}, em, x, y, (y_{\text{depo}} - y)/\theta_y) \\ \quad \text{If } ph = \texttt{Wait} \text{ and } y < y_{\text{depo}} - y_{\text{buff}} \\ (\texttt{GoDown}, em, x, y, (y_{\text{feed}} - y)/\theta_y) \\ \quad \text{If } ph = \texttt{Wait} \\ \quad \text{and } y_{\text{depo}} - y_{\text{buff}} \leq y < y_{\text{feed}} \\ (\texttt{Error}, em, x, y, \infty) \\ \quad \text{If not} \end{cases}$$

$\delta_{\text{ext}}(ph, em, x, y, \sigma, e, \texttt{CrFeed?}) =$

$$\begin{cases} (\texttt{GoFeed}, em, x, y, (x - x_{\text{feed}})/\theta_x) \\ \quad \text{If } ph = \texttt{Wait} \text{ and } x > x_{\text{feed}} \\ (\texttt{Error}, em, x, y, \infty) \\ \quad \text{If not} \end{cases}$$

$\delta_{\text{ext}}(ph, em, x, y, \sigma, e, \texttt{CrDepo?}) =$

$$\begin{cases} (\texttt{GoDepo}, em, x, y, (x_{\text{depo}} - x)/\theta_x) \\ \quad \text{If } ph = \texttt{Wait} \text{ and } x < x_{\text{depo}} \\ (\texttt{Error}, em, x, y, \infty) \\ \quad \text{If not} \end{cases}$$

$\delta_{\text{ext}}(ph, em, x, y, \sigma, e, \texttt{GrStop?}) =$

$$\begin{cases} (\texttt{Wait}, em, x', y', \infty) & \text{If } ph \neq \texttt{Error} \\ (ph, em, x, y, \infty) & \text{If not} \end{cases}$$

$\delta_{\text{ext}}(ph, em, x, y, \sigma, e, \texttt{CrStop?}) =$

$$\begin{cases} (\texttt{Wait}, em, x', y', \infty) & \text{If } ph \neq \texttt{Error} \\ (ph, em, x, y, \infty) & \text{If not} \end{cases}$$

$\delta_{\text{ext}}(ph, em, x, y, \sigma, e, \texttt{EmAct?}) =$

$$\begin{cases} (ph, \texttt{EmOn}, x', y', \sigma - e) & \text{If } ph \neq \texttt{Error} \\ (ph, em, x, y, \sigma) & \text{If not} \end{cases}$$

$\delta_{\text{ext}}(ph, em, x, y, \sigma, e, \texttt{EmDeact?}) =$

$$\begin{cases} (ph, \texttt{EmOff}, x', y', \sigma - e) & \text{If } ph \neq \texttt{Error} \\ (ph, em, x, y, \sigma) & \text{If not} \end{cases}$$

where

$$x' = \begin{cases} x - (\theta_x * e) & \text{If } ph \in \{\texttt{GoFeed}, \texttt{CrFeed}\} \\ x + (\theta_x * e) & \text{If } ph \in \{\texttt{GoDepo}, \texttt{CrDepo}\} \\ x & \text{If not} \end{cases}$$

$$y' = \begin{cases} y + (\theta_y * e) \\ \quad \text{If } ph \in \{\texttt{GoDown}, \texttt{GrDepoD}, \texttt{GrFeed}\} \\ y - (\theta_y * e) \\ \quad \text{If } ph \in \{\texttt{GoUp}, \texttt{GrDepoU}, \texttt{GrTop}\} \\ y \\ \quad \text{If not} \end{cases}$$

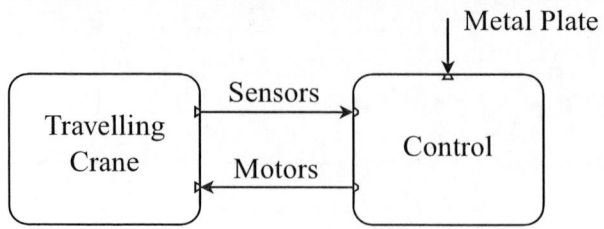

Figure 9. Block diagram, travelling crane and its control.

$\lambda(ph, em, x, y, \sigma) =$

$$\begin{cases} \text{GrPosTop!} & \text{If } ph = \text{GoUp} \\ & \text{and } y - \sigma * \theta_y = y_{\text{top}} \\ \text{GrPosDepo!} & \text{If } ph = \text{GoUp} \\ & \text{and } y - \sigma * \theta_y = y_{\text{depo}} \\ \text{GrPosFeed!} & \text{If } ph = \text{GoDown} \\ & \text{and } y + \sigma * \theta_y = y_{\text{feed}} \\ \text{GrPosDepo!} & \text{If } ph = \text{GoDown} \\ & \text{and } y + \sigma * \theta_y = y_{\text{depo}} \\ \text{CrPosDepo!} & \text{If } ph = \text{GoDepo} \\ \text{CrPosFeed!} & \text{If } ph = \text{GoFeed} \end{cases} \begin{cases} \Lambda & \text{If } ph = \text{GrDepoD} \\ & \text{and } x > x_{\text{feed}} \\ \Lambda & \text{If } ph = \text{GrDepoD} \\ & \text{and } x \le x_{\text{feed}} \\ \Lambda & \text{If } ph = \text{GrFeed} \\ \Lambda & \text{If } ph = \text{CrDepo} \\ \Lambda & \text{If } ph = \text{CrFeed} \\ \Lambda & \text{If } ph = \text{GrDepoU} \\ \Lambda & \text{If } ph = \text{GrTop} \end{cases}$$

$\text{ta}(ph, em, x, y, \sigma) = \sigma$

4.2. Control Model of the Travelling Crane

The control model has two input ports, one through which receives an event to indicate that there is a piece ready to be transported (`Arrived?`), and anotherone through which it receives the events from the sensors of the travelling crane. The output events of the control model deals with the motors and electromagnet of the travelling crane (Figure 9).

Some assumptions over the travelling crane being controlled are made in order to keep the size of the model to a reasonable level. Since some of them are outside the scope of this article they will not be made explicit.

The travelling cranes is assumed to be initially stopped and positioned over the deposit belt, with the gripper in the top position and the electromagnet deactivated. Then, when a piece has been transported, the control returns the travelling crane to this position and waits for a new piece. The motors are supposed to react immediately to the control, although some delays between the outputs generated can be specified through the parameters of the control.

A transport cycle is then composed by the following sequence of actions:

- move the gripper down to the deposit belt position,
- activate the gripper electromagnet,
- move the gripper up to the top position,
- transfer the crane over the feed belt,

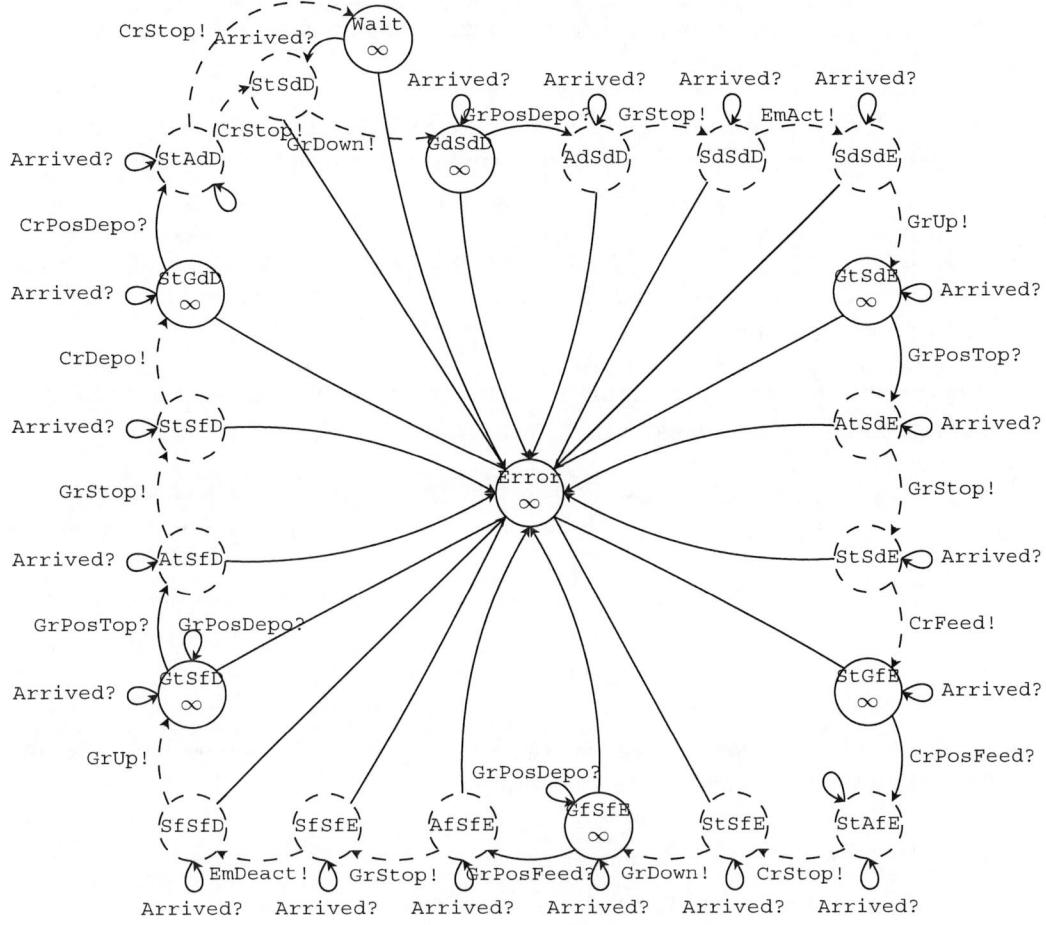

Figure 10. DEVS model of the control.

- move the gripper down to the feed belt position,

- deactivate the gripper electromagnet,

- move the gripper up to the top position, and,

- transfer the crane over the deposit belt.

If one event is received indicating than there is a piece ready while the control is in a transport cycle, it memorizes than there is a piece available in order to restart the transport cycle again.

The DEVS model of the control has three parameters (constants):

σ_{res} the delay response for reacting to an external event (generated by the travelling crane),

σ_{ev} the delay between two consecutive internal events (generated by the controller),

σ_{em} the time to wait after activating the electromagnet.

In the Figure 10 the names of the discrete phases through which the control operates are encoded using five letters,

- two letters to code the state of the gripper (vertical direction), St stopped at the top position, Gd going to the deposit belt position, Af arrived at the feed belt position, etc.

- followed by two letters to code the state of the crane (horizontal direction), Sd stopped at the deposit belt position, Gf going to the feed belt position, Af arrived at the feed belt position, etc.

- and finally one letter representing the state of the electromagnet, E electromagnet activated, and D electromagnet deactivated.

So, for example, a phase named GfSfE represents a phase in which the gripper is going down towards the feed belt while the crane is stopped over the feed belt and the electromagnet is activated.

Figure 10 gives the phase transition diagram of the control. In this figure all the edges without labels that enters to the phase named Error represents external transitions due to an unexpected input from the travelling crane (the sensors).

The state of the DEVS model of the control, $\mathcal{TCC}$, is defined by the phase variable, a state variable to memorize that a piece is available and a continuous state variable to represent the lifetime of the phase.

$\mathcal{TCC} = \{X, Y, S, \delta_{\text{ext}}, \delta_{\text{int}}, \lambda, \text{ta}\}$

$X = \{\text{CrPosDepo?}, \text{CrPosFeed?}, \text{GrPosTop?}, \text{GrPosFeed?}, \text{GrPosDepo?}, \text{Arrived?}, \Lambda\}$

$Y = \{\text{CrFeed!}, \text{CrDepo!}, \text{CrStop!}, \text{GrUp!}, \text{GrDown!}, \text{GrStop!}, \text{EmAct!}, \text{EmDeact!}\}$

$S = \{\text{Wait}, \text{Error}, \text{StSdD}, \text{GdSdD}, \text{AdSdD}, \text{SdSdD}, \text{SdSdE}, \text{GtSdE}, \text{AtSdE}, \text{StSdE},$
  $\text{StGfE}, \text{StAfE}, \text{StSfE}, \text{GfSfE}, \text{AfSfE}, \text{SfSfE}, \text{SfSfD}, \text{GtSfD}, \text{AtSfD}, \text{StSfD},$
  $\text{StGdD}, \text{StAdD}\} \times \{\text{Arr}, \text{Wait}\} \times \Re_0^+$

$\delta_{\text{ext}}(ph, a, \sigma, e, \text{GrPosDepo?}) =$
$\begin{cases} (\text{AdSdD}, a, \sigma_{\text{res}}) & \text{If } ph = \text{GdSdD} \\ (\text{GfSfE}, a, \sigma - e) & \text{If } ph = \text{GfSfE} \\ (\text{GtSfD}, a, \sigma - e) & \text{If } ph = \text{GtSfD} \\ (ph, a, \sigma - e) & \text{If } ph \in \{\text{StAdD}, \text{StAfE}\} \\ (\text{Error}, a, \infty) & \text{If not} \end{cases}$

$\delta_{\text{ext}}(ph, a, \sigma, e, \text{GrPosTop?}) =$
$\begin{cases} (\text{AtSdE}, a, \sigma_{\text{res}}) & \text{If } ph = \text{GtSdE} \\ (\text{AtSfD}, a, \sigma_{\text{res}}) & \text{If } ph = \text{GtSfD} \\ (ph, a, \sigma - e) & \text{If } ph \in \{\text{StAdD}, \text{StAfE}\} \\ (\text{Error}, a, \infty) & \text{If not} \end{cases}$

$\delta_{\text{ext}}(ph, a, \sigma, e, \text{CrPosFeed?}) =$
$\begin{cases} (\text{StAfE}, a, \sigma_{\text{res}}) & \text{If } ph = \text{StGfE} \\ (ph, a, \sigma - e) & \text{If } ph \in \{\text{StAdD}, \text{StAfE}\} \\ (\text{Error}, a, \infty) & \text{If not} \end{cases}$

$\delta_{\text{ext}}(ph, a, \sigma, e, \text{GrPosFeed?}) =$
$\begin{cases} (\text{AfSfE}, a, \sigma_{\text{res}}) & \text{If } ph = \text{GfSfE} \\ (ph, a, \sigma - e) & \text{If } ph \in \{\text{StAdD}, \text{StAfE}\} \\ (\text{Error}, a, \infty) & \text{If not} \end{cases}$

$\delta_{\text{ext}}(ph, a, \sigma, e, \text{CrPosDepo?}) =$
$\begin{cases} (\text{StAdD}, a, \sigma_{\text{res}}) & \text{If } ph = \text{StGdD} \\ (ph, a, \sigma - e) & \text{If } ph \in \{\text{StAdD}, \text{StAfE}\} \\ (\text{Error}, a, \infty) & \text{If not} \end{cases}$

$\delta_{\text{ext}}(ph, a, \sigma, e, \text{Arrived?}) =$
$\begin{cases} (\text{StSdD}, \text{Wait}, \infty) & \text{If } ph = \text{Wait} \\ (ph, \text{Arr}, \sigma - e) & \text{If not} \end{cases}$

$\delta_{\text{ext}}(ph, a, \sigma, e, \Lambda) = (ph, a, \sigma - e)$

$\delta_{\text{int}}(ph, a, \sigma) =$

$$\begin{cases} (\text{GdSdD}, a, \infty) & \text{If } ph = \text{StSdD} \\ (\text{SdSdD}, a, \sigma_{\text{ev}}) & \text{If } ph = \text{AdSdD} \\ (\text{SdSdE}, a, \sigma_{\text{em}}) & \text{If } ph = \text{SdSdD} \\ (\text{GtSdE}, a, \infty) & \text{If } ph = \text{SdSdE} \\ (\text{StSdE}, a, \sigma_{\text{ev}}) & \text{If } ph = \text{AtSdE} \\ (\text{StGfE}, a, \infty) & \text{If } ph = \text{StSdE} \\ (\text{StSfE}, a, \sigma_{\text{ev}}) & \text{If } ph = \text{StAfE} \\ (\text{GfSfE}, a, \infty) & \text{If } ph = \text{StSfE} \end{cases}$$
$\quad\begin{array}{ll} (\text{SfSfE}, a, \sigma_{\text{ev}}) & \text{If } ph = \text{AfSfE} \\ (\text{SfSfD}, a, \sigma_{\text{em}}) & \text{If } ph = \text{SfSfE} \\ (\text{GtSfD}, a, \infty) & \text{If } ph = \text{SfSfD} \\ (\text{StSfD}, a, \sigma_{\text{ev}}) & \text{If } ph = \text{AtSfD} \\ (\text{StGdD}, a, \infty) & \text{If } ph = \text{StSfD} \\ (\text{StSdD}, \text{Wait}, \sigma_{\text{ev}}) & \text{If } ph = \text{StAdD} \text{ and } a = \text{Arr} \\ (\text{Wait}, a, \infty) & \text{If } ph = \text{StAdD} \text{ and } a = \text{Wait} \end{array}$

$\lambda(ph, a, \sigma) =$

$$\begin{cases} \text{GrDown!} & \text{If } ph = \text{StSdD} \\ \text{GrStop!} & \text{If } ph = \text{AdSdD} \\ \text{EmAct!} & \text{If } ph = \text{SdSdD} \\ \text{GrUp!} & \text{If } ph = \text{SdSdE} \\ \text{GrStop!} & \text{If } ph = \text{AtSdE} \\ \text{CrFeed!} & \text{If } ph = \text{StSdE} \\ \text{CrStop!} & \text{If } ph = \text{StAfE} \\ \text{GrDown!} & \text{If } ph = \text{StSfE} \end{cases}$$
$\quad\begin{array}{ll} \text{GrStop!} & \text{If } ph = \text{AfSfE} \\ \text{EmDeact!} & \text{If } ph = \text{SfSfE} \\ \text{GrUp!} & \text{If } ph = \text{SfSfD} \\ \text{GrStop!} & \text{If } ph = \text{AtSfD} \\ \text{CrDepo!} & \text{If } ph = \text{StSfD} \\ \text{CrStop!} & \text{If } ph = \text{StAdD} \text{ and } a = \text{Arr} \\ \text{CrStop!} & \text{If } ph = \text{StAdD} \text{ and } a = \text{Wait} \end{array}$

$\text{ta}(ph, a, \sigma) = \sigma$

4.3. Verification of Compliance

To verify that the behaviour of the DEVS model of the control ($\mathcal{TCC}$, Section 4.2.) conforms the behaviour specified by the timed automaton that gives the safety requirement ($\mathcal{R}_1$) it is only necesary to prove the existence of a simulation relation (Section 3.3.1.).

Given the DEVS model of the control, and the timed automaton representing the formal requirement, we can define a relation

$$f \subseteq Q_{\mathcal{TCC}} \times S_{\mathcal{R}_1}$$

where

$$Q_{\mathcal{TCC}} = \{(s, e) | s \in S_{\mathcal{TCC}}, 0 \leq e \leq ta(s)\}$$

is the set of total states of the DEVS model $\mathcal{TCC}$, and $S_{\mathcal{R}_1}$ the set of states of the timed automaton $\mathcal{R}_1$.

Given $(s, e) \in Q_{\mathcal{TCC}}$, we define the relation f as:

$f = \{((\text{StSdD}, em, \sigma), e), \{(\text{OvDepo2}, x) | x \leq \sigma\}\} \cup$
$\{((\text{GdSdD}, em, \sigma), e), \{(\text{OvDepo2}, x) | x \leq \sigma\}\} \cup \{((\text{AdSdD}, em, \sigma), e), \{(\text{OvDepo2}, x) | x \leq \sigma\}\} \cup$
$\{((\text{SdSdD}, em, \sigma), e), \{(\text{OvDepo2}, x) | x \leq \sigma\}\} \cup \{((\text{SdSdE}, em, \sigma), e), \{(\text{OvDepo2}, x) | x \leq \sigma\}\} \cup$
$\{((\text{GtSdE}, em, \sigma), e), \{(\text{OvDepo2}, x) | x \leq \sigma\}\} \cup \{((\text{AtSdE}, em, \sigma), e), \{(\text{OvDepo2}, x) | x \leq \sigma\}\} \cup$
$\{((\text{StSdE}, em, \sigma), e), \{(\text{OvDepo2}, x) | x \leq \sigma\}\} \cup \{((\text{StGfE}, em, \sigma), e), \{(\text{Middle}, x) | x \leq \sigma\}\} \cup$
$\{((\text{StAfE}, em, \sigma), e), \{(\text{OvFeed}, x) | x \leq \sigma\}\} \cup \{((\text{StSfE}, em, \sigma), e), \{(\text{OvFeed2}, x) | x \leq \sigma\}\} \cup$
$\{((\text{GfSfE}, em, \sigma), e), \{(\text{OvFeed2}, x) | x \leq \sigma\}\} \cup \{((\text{AfSfE}, em, \sigma), e), \{(\text{OvFeed2}, x) | x \leq \sigma\}\} \cup$
$\{((\text{SfSfE}, em, \sigma), e), \{(\text{OvFeed2}, x) | x \leq \sigma\}\} \cup \{((\text{SfSfD}, em, \sigma), e), \{(\text{OvFeed2}, x) | x \leq \sigma\}\} \cup$
$\{((\text{GtSfD}, em, \sigma), e), \{(\text{OvFeed2}, x) | x \leq \sigma\}\} \cup \{((\text{AtSfD}, em, \sigma), e), \{(\text{OvFeed2}, x) | x \leq \sigma\}\} \cup$
$\{((\text{StSfD}, em, \sigma), e), \{(\text{OvFeed2}, x) | x \leq \sigma\}\} \cup \{((\text{StGdD}, em, \sigma), e), \{(\text{Middle}, x) | x \leq \sigma\}\} \cup$
$\{((\text{StAdD}, em, \sigma), e), \{(\text{OvDepo}, x) | x \leq \sigma\}\} \cup \{((\text{Wait}, em, \sigma), e), \{(\text{OvDepo2}, x) | x \leq \sigma\}\} \cup$
$\{((\text{Error}, em, \sigma), e), \{(\text{ErrorS}, x) | x \leq \sigma\}\}$

As this relation, f, was formulated following the constraints given by the definition of simulation relation, it is straightforward to see that it indeed constitutes a simulation relation.

input event transitions For each internal transition of the DEVS model $\mathcal{TCC}$, from a state s to a second s', with $e = ta(s)$, which corresponding output event is $\lambda(s)$. And also for every state of $\mathcal{R}_1$, $u \in f[s,e]$, there exists a state $u' \in f[s',0]$, such that there is a discrete transition in $\mathcal{R}_1$ with the label $\lambda(s)$ from u to u'.

output event transitions Likewise for every external transition of $\mathcal{TCC}$, related to an event x, between the states s and s', with $e \leq ta(s)$ the time since the last transition, and for each $u \in f[s,e]$, there exists $u' \in f[s',0]$ such that there is a discrete transition in $\mathcal{R}_1$ between u and u' with the label x.

time passage steps Finally for every single state s, of $\mathcal{TCC}$, and $u \in f[s,e]$, such that $0 \leq e < ta(s)$, there exists $u' \in f[s,e']$, where $e < e' < ta(s)$, and there is a time passage step in $\mathcal{R}_1$ from u to u' labelled $e' - e$.

It is easy to give superior limits for the time advance function of the DEVS model from the definitions for the relevant states (δ_{ext}),

$$ta(\text{StAfE}, a, \sigma) \leq \sigma_{res}$$
$$ta(\text{StAdD}, a, \sigma) \leq \sigma_{res}$$

And even further, for every total state (s, e), of $\mathcal{TCC}$, corresponding to one of the phases StAfE or StAdD, $e \leq \sigma_{\text{res}}$, so if the constant $\sigma_{\text{res}} \leq \delta_1$ the third condition is met.

By finding this simulation relation between the models, we have verified that the behaviour exhibited by the DEVS model is contained in the behaviour of the timed automaton.

Additionally to having formally verified that the model of the control conforms the required safety property, we have validated its behaviour by running a simulation of a DEVS model constructed by coupling together the model of travelling crane with the model of the control (Figure 9).

5. Conclusion

Generally the methodology for the design of control system within the DEVS formalism is restricted to:

- modeling the system to be controlled,
- modeling of the control for the system,
- validating both models by simulation.

In the extended methodology we propose, we add the posibility of specifying high-level requirements by way of timed automata, and the techniques necessary to verify that a DEVS models conforms a timed automaton specification.

Furthermore, temporal properties of the high-level behaviour can be studied using existing model-checking tools. While at the same time the DEVS models can be promptly simulated.

As can be seen from the case study of the Section 4., the verification of compliance of an atomic DEVS model to a timed automaton of a moderate complexity is not exceptionally tedious. In consequence it proves to be an adequate technique for unit based verfication of complex models.

Despite the possibility of verifying models of a moderate complexity, when dealing with coupled complex models the formal verification becomes impractical due to the exponential increase in complexity implied by the couplage of models. For these cases an automatic verification tool would be handy, regrettably the complete automatization of this technique is proven to be infeasible. Even implementing a general algorithm that given an atomic DEVS model, and a timed automaton, can say if the behaviour of the DEVS model conforms the timed automaton is not possible due to the unbounded state space implied by dense real-time models.

We are currently working in finding a suitable subclass of DEVS models and timed automata, that are expressive enough and admits certain degree of automatization.

We are also working in developping a test case generation technique based on the specification, for testing the conformance relation. In this direction, we are trying to define a fault model that will help identify errors in which the behaviour of the implementation differs from the specification.

Going further, another possible line for future work would be investigating the application of theorem-proovers or similar techniques to assist in the formal verification task.

References

[1] Abadi and Lamport. An old-fashioned recipe for real time. In *REX: Real-Time: Theory in Practice, REX Workshop*, 1991.

[2] Parosh Aziz Abdulla, Johann Deneux, Joël Ouaknine, and James Worrell. Decidability and complexity results for timed automata via channel machines. In *ICALP*, pages 1089–1101, 2005.

[3] Alur and Henzinger. Back to the future: Towards a theory of timed regular languages. In *FOCS: IEEE Symposium on Foundations of Computer Science (FOCS)*, 1992.

[4] R. Alur and D. L. Dill. A theory of timed automata. *Theoretical Computer Science*, **126**:183–235, 1994.

[5] R. Alur and R. P. Kurshan. Timing analysis in COSPAN. In *Proc. Hybrid Systems III*, volume 1066, pages 220–231. LNCS, 1996.

[6] Rajeev Alur, Limor Fix, and Thomas A. Henzinger. Event-clock automata: a determinizable class of timed automata. *Theoretical Computer Science*, **211**(1–2):253–273, 1999.

[7] Hernán Dacharry and Norbert Giambiasi. Formal verification with timed automata and devs models: a case study. In *Proceedings of ASSE'05*. Argentine Society for Computer Science and Operational Research, 2005.

[8] Hernán Dacharry and Norbert Giambiasi. From timed automata to devs models: Formal verification. In *Proceedings of SpringSim'05*. SCS - The Society for Modeling and Simulation International, 2005.

[9] C. Daws, A. Olivero, S. Tripakis, and S. Yovine. The tool KRONOS. In R. Alur, T. A. Henzinger, and E. D. Sontag, editors, *Hybrid Systems III*, volume 1066 of *Lecture Notes in Computer Science*, pages 208–219. Springer-Verlag, 1996.

[10] D'Souza and Madhusudan. Timed control synthesis for external specifications. In *STACS: Annual Symposium on Theoretical Aspects of Computer Science*, 2002.

[11] N. Giambiasi, B. Escude, and S. Ghosh. Gdevs : a generalized discrete event specification for accurate modelling of dynamic systems. *Trans. of S.C.S.I.*, **17-3**:120–134, 2000.

[12] Vineet Gupta, Thomas A. Henzinger, and Radha Jagadeesan. Robust timed automata. In *HART*, pages 331–345, 1997.

[13] A. Hamri and G. Zacharewicz. Lsis-dme: An environment for modeling and simulation of devs specifications. In *AIS-CMS International modeling and simulation multiconference*, pages 55–60, Buenos Aires - Argentina, February 8–10 2007. ISBN 978–2-9520712–6-0.

[14] T. Henzinger, X. Nicollin, J. Sifakis, and S. Yovine. Symbolic model-checking for real-time systems. In *LICS'92*, June 1992.

[15] Thomas A. Henzinger, Peter W. Kopke, Anuj Puri, and Pravin Varaiya. What's decidable about hybrid automata? In *STOC '95: Proceedings of the twenty-seventh annual ACM symposium on Theory of computing*, pages 373–382, New York, NY, USA, 1995. ACM Press.

[16] Thomas A. Henzinger, Zohar Manna, and Amir Pnueli. What good are digital clocks? In *ICALP*, pages 545–558, 1992.

[17] Gyung Pyo Hong and Tag Gon Kim. A framework for verifying discrete event models within a devs-based system development methodology. *Trans. Soc. Comput. Simul. Int.*, **13**(1):19–34, 1996.

[18] Ki Jung Hong and Tag Gon Kim. Timed i/o test sequences for discrete event model verification. In *AIS*, pages 275–284, 2004.

[19] Moon Ho Hwang. Identifying equivalence of devss: Language approach. In *SCSC*, pages 319–324, 2003.

[20] Moon Ho Hwang and Feng Lin. State minimization of sp-devs. In *AIS*, pages 243–252, 2004.

[21] Dilsun Kirli Kaynar, Nancy A. Lynch, Roberto Segala, and Frits W. Vaandrager. Timed I/O automata: A mathematical framework for modeling and analyzing real-time systems. In *RTSS*, pages 166–177, 2003.

[22] T. G. Kim, S. M. Cho, and W. B. Lee. *DEVS framework for systems development: unified specification for logical analysis, performance evaluation and implementation*, pages 131–166. Springer-Verlag New York, Inc., New York, NY, USA, 2001.

[23] K. G. Larsen, P. Pettersson, and W. Yi. UPPAAL in a Nutshell. *Int. Journal on Software Tools for Technology Transfer*, **1**(1–2):134–152, October 1997.

[24] Claus Lewerentz and Thomas Lindner. "production cell": A comparative study in formal specification and verification. In *KORSO Book*, pages 388–416, 1995.

[25] N. A. Lynch and F. W. Vaandrager. Forward and backward simulations – part II: Timing-based systems. Technical report, Laboratory for Computer Science, Massachusetts Institute of Technology, Cambridge, MA, USA, August 1995.

[26] J.S. Ostroff. *Temporal Logic for Real-Time Systems*. John Wiley and Sons, New York, 1989.

[27] Joël Ouaknine and James Worrell. Revisiting digitization, robustness, and decidability for timed automata. In *Proceedings of the eighteenth Annual IEEE Symposium on Logic in Computer Science (LICS'03)*. IEEE Computer Society, 2003.

[28] Joël Ouaknine and James Worrell. On the language inclusion problem for timed automata: Closing a decidability gap. In *Proceedings of the eighteenth Annual IEEE Syposium on Logic in Computer Science (LICS-04)*, Los Alamitos, CA, 2004. IEEE Computer Society.

[29] Anuj Puri. Dynamical properties of timed automata. *Discrete Event Dynamic Systems*, **10**(1-2):87–113, 2000.

[30] J. G. Springintveld, F. W. Vaandrager, and P. R. D'Argenio. Testing timed automata. *TCS: Theoretical Computer Science*, **254**, 2001.

[31] T.A. Henzinger, Z. Manna, and A. Pnueli. Timed transition systems. In *Real Time: Theory in Practice*, pages 226–251, Mook, June 1991. Springer-Verlag.

[32] B. Zeigler, H. Praehofer, and T. G. Kim. *Theory of modelling and simulation, 2d edition*. Academic Press, London, 2000.

[33] Bernard P. Zeigler. *Theory of Modelling and Simulation*. Krieger Publishing Co., Inc., Melbourne, FL, USA, 1984.

Chapter 6

NOISY SURFACE SMOOTHING USING TSALLIS ENTROPY

*Hong Zhou[a], Yonghuai Liu[a]*and Xuejun Ren[b]*
[a]Department of Computer Science
The University of Wales, Aberystwyth
Ceredigion SY23 3DB, UK
[b]School of Engineering
Liverpool John and Moores University
Liverpool L3 3AF, UK

Abstract

3D modelling finds a wide range of applications in robotics research from object recognition and robot localization to path planning and obstacle avoidance. However due to the presence of surface scanning noise and range image registration and fusion errors, the finally reconstructed surfaces are often distorted and thus present obstacles to their applications. In this chapter, we employ the entropy maximization (EntMax) principle in conjunction with the Tsallis entropy from statistical mechanics to optimize the mesh node locations and normals without altering the mesh node connectivity. While the traditional Shannon entropy can only describe extensive systems, the Tsallis entropy can describe a variety of systems: extensive, sub-extensive, and super-extensive. The nodes in the mesh are indeed entangled and interact with each other. The flexibility of the Tsallis entropy in describing different systems is so useful for noisy surface smoothing, since through adjusting the non-extensivity parameter in the Tsallis entropy, it is possible to model various degrees of interaction among neighbouring nodes in the mesh and thus achieve the desired smoothing effect. A comparative study based on real images shows that the proposed algorithm is easy to implement and effectively smoothes the rough surfaces with their geometric details being desirably retained.

Keywords: 3D modelling, scanning noise, reconstruction error, entropy maximization, Tsallis entropy, surface smoothing.

---

*E-mail address: yyl@aber.ac.uk, Tel: +44 1970 621688, Fax: +44 1970 628536. Correspondence author.

1. Introduction

3D modelling finds a wide range of applications in areas such as 3D object recognition [20], simultaneous localization and map (SLAM) building [26] and computer graphics [9]. The quality of reconstructed surface models is of vital importance for reliable path planning and obstacle avoidance in robot navigation and physical realism in model based 3D object recognition and computer graphics. However, the finally reconstructed surfaces are usually rough in the smooth area of real surfaces due mainly to a rapid change of orientations and vertex locations in the reconstructed surfaces, caused by noise introduced in the process of surface scanning and image registration and integration. How the noise from the reconstructed surface models can be effectively filtered, while preserving their desired details, is still an open problem in the 3D imaging and analysis literature. The main challenge for rough surface smoothing lies in two aspects: (1) how to discriminate noise and artefacts from the geometric details, which is usually data dependent, (2) how to balance between surface smoothness and accurate geometry representation, which is usually context dependent. This means that different applications have different requirements of the level of detailed representation of the geometry of the object surface.

1.1. Previous Work

Current smoothing methods can be classified into four main categories, namely Laplacian smoothing, optimization-based smoothing, combining Laplacian and optimization-based smoothing and physically-based methods. Among these methods, Laplacian smoothing is the most commonly used [33, 27, 11, 17, 16, 29, 21, 11, 6]. Laplacian smoothing in its simplest form consists of recursively estimating an optimal position of each node as the average of the nodes connected to it. This technique generally works quite well for flat areas in the mesh. However, it can result in distorted or even inverted elements near concavities in the model.

Optimization-based smoothing techniques [1, 2, 25, 14, 13] measure the quality of the surrounding elements of a node and attempt to optimize it by computing the local gradient of the element quality with respect to the node location. The node is moved in the direction of the increasing gradient until an optimum value is reached. This optimum value is often defined as maximizing the minimum of mesh quality metrics such as the minimum interior angle of triangles and the internal angles at the Steiner points.

Optimization-based smoothing guarantees good quality elements. However, its computational cost is very high when compared with Laplacian smoothing. Therefore, the Laplacian and optimization-based approaches are sometimes combined together, which uses Laplacian smoothing when possible and only uses optimization-based smoothing when local element shape metrics drop below a certain threshold, so that the mesh quality is improved at lower computational cost. Several ways of combination are described in [12, 7, 5].

Based on difference of the smoothing elements, the above three main smoothing approaches can also be classified into two categories. One is to adjust vertex positions so that the overall surface becomes smoother [32, 27, 17] and the other is first to smooth the vertex normals and then to optimize vertex locations based on the smoothed normals [6, 29, 11].

In fact, some other methods do exist that are not truly smoothing techniques, because they require node insertion or deletion when the mesh is not smooth enough. Meanwhile, nodes are relocated based on a simulated physically based attraction or repulsion force. Lohner [24] simulates the force between neighbouring nodes as a system of springs interacting with each other. Shimada, Yamada and Itoh [30] and Bossen and Heckbert [4] regard the nodes as the centre of bubbles that are repositioned to attain equilibrium. With changes in the magnitude and direction of inter-particle forces, different anisotropic characteristics and element sizes can be achieved.

1.2. Tsallis Entropy

While the traditional Shannon entropy: $E_S = -\sum_i w_i \ln w_i$ has been widely used in communications, it can hardly always succeed in describing the probability distribution w_i of the various microstates in statistical mechanics and thermodynamic systems. Thus the Shannon entropy was extended by C. Tsallis in 1988 [34] through introducing a parameter q into the entropy definition: $E_T^q = (1 - \sum_i w_i^q)/(q-1)$ where q is any real positive number except $q = 1$. Actually, $\lim_{q \to 1} E_T^q = E_S$. The generalization is considered to be one of the most viable and applicable candidates for formulating a theory of non-extensive thermodynamics.

One of the most attractive characteristics associated with the Tsallis entropy [34]: $E = (1 - \sum_i w_i^q)/(q-1)$ is the fact that it provides an easy access through the entropy maximization principle [19] to a rich set of probability distribution functions, different from the traditional Gaussian distribution function [15]. Moreover, since the parameter q is introduced, the entropy can be adjusted more freely and thus applicable not only in statistical physics like the analysis of cosmic rays fluxes [35] and hydrodynamic turbulent flows [3], but also in other areas like content based image retrieval [28] and medical image registration [36]. Since q can be predicted in some specific applications, the system can be more accurately analysed and understood [35, 3].

Some interesting differences between the Shannon entropy and the Tsallis entropy can be summarised as:

- While the entropy maximization principle [19] leads the probability distribution of different microstates to follow the exponential law from the Shannon entropy, it leads to a power law from the Tsallis entropy;

- While the Shannon entropy describes only extensive systems in which the entropy of a system is equal to the sum of those of its two independent subsystems A and B: $E_S(A + B) = E_S(A) + E_S(B)$, the Tsallis entropy can describe through changing the non-extensivity index q various systems with and without interaction between their components: $E_T^q(A + B) = E_T^q(A) + E_T^q(B) + (q-1)E_T^q(A)E_T^q(B)$. When $0 < q < 1$, it describes a superextensive system; when $q = 1$, it describes an extensive system; and when $q > 1$, it describes a subextensive system.

1.3. Our Work

While most existing surface smoothing algorithms [33, 27, 11, 17, 16, 29, 21, 11, 6] are developed to mainly deal with scanning noise, in this chapter we develop a novel algorithm that can deal with both scanning noise and image registration and fusion errors. While scanning noise can be modelled as Gaussian white noise, registration and fusion errors depend on specific registration and fusion algorithms and image data. Thus, while the former is relatively easy to handle, the latter is challenging and more powerful smoothing methods have to be developed. To our knowledge, we are not aware of any algorithm specifically developed to handle both scanning noise and image registration and fusion errors in the process of surface smoothing.

Assume that 3D object computer models are represented as triangular meshes, the aim of surface smoothing in this chapter is to optimize the meshes so that the edges of triangles have similar lengths and neighbouring vertices have similar normals. To this end, we construct objective functions to minimize the variation of triangle edge lengths and the normal variation of neighbouring vertices. However, such strategy appears impractical and often leads the surface details to be smoothed out, since it equally treats all the edges of triangles and vertex normals in the mesh without discriminating the details they convey. In the process of surface smoothing, a vertex is usually optimized using its neighbours and different nodes are interwoven and thus influence each other. To characterise this phenomenon, a real number W in the unit interval has to be introduced to represent the extent to which the edges of triangles are of equal length and the normals of neighbouring vertices are similar. Consequently, weighted squared triangle edge length variation and neighbouring vertex normal variation need to be optimized for the optimal positions of mesh nodes without altering the original node connectivity and the optimal normals of vertices. In this case, the noisy surface smoothing is then down to an accurate estimation of the weights (contribution) of the neighbouring vertices to the vertex of interest. To this end, we employ for the estimation of the parameter W the entropy maximization (EntMax) principle [19], since it leads to a least biased estimate W possible on the given information.

To implement the EntMax principle for the inference of the contribution of neighbouring vertices to the vertex of interest, we selected the Tsallis entropy, instead of the traditional Shannon entropy, for the description of the impurity of some measures of vertices of interest in the process of its optimization and update. This is because while the Shannon entropy may not be effective in accurately characterising the uncertainty of the probability distribution of vertex properties, due to the fact that they do not necessarily follow Gaussian distribution, the Tsallis entropy may better model the inter-influence (interaction) of interwoven neighbouring nodes on the node of interest. On the other hand, the Tsallis entropy includes the non-extensivity parameter q which can be used to discriminate noise and artefacts from details of object surface. In other words, q can be used to balance between under-smoothness and over-smoothness, leading the desired level of smoothing to be obtained. Thus, the Tsallis entropy provides an extra degree of freedom for the control of the final desired smoothing quality.

To validate the proposed surface smoothing algorithms, we also implemented three classical algorithms: the signal processing method [32], the original Laplacian method [16], and the mean curvature flow smoothing method [27]. The experimental results based on real images show the advantages of our novel surface smoothing method.

The rest of this chapter is organized as follows: two types of noise are described in Section 2., the mathematical foundation is built and two smoothing methods are developed in Section 3., while the experimental results are presented in Section 4. Finally, some conclusions are drawn in Section 5.

2. Noise Characterization

In the process of constructing accurate 3D object computer models, two main types of noise are often introduced. One is the scanning noise which is introduced due to point sampling of object surface, quantization of measurement, various surface orientations and depth discontinuities, and different reflective properties of object surface. The scanning noise of each point is independent of those of its neighbours and can thus be modelled as Gaussian white noise with a standard deviation of δ.

An example is shown in the first row of Figure 1. The left hand side is a clean surface and the right hand side is a noisy surface subject to Gaussian white noise with a standard deviation of 0.02. It can be clearly seen that the noise uniformly distributes over the surface. When the standard deviation δ of noise is not too large, this kind of noise will not change the total shape significantly, but will distort the details of object surface.

Most existing smoothing methods are developed to mainly deal with scanning noise [33, 27, 11, 17, 16, 29, 21, 11, 6]. However, when a 3D model is reconstructed from several data sets such as range images, the surface reconstruction noise has also been unavoidably introduced, due to the fact that the surface reconstruction process is an ad hoc process with numerous possibilities to fuse the inaccurately registered data and connect neighbouring points into a watertight surface. Accurate reconstruction often requires [18] that the image registration error be an order of magnitude less than the measurement error, the actual automatic registration algorithm [23] at best can only produce an average registration error of 1/3 or 1/2 scanning resolution. Inaccurate image registration and improper data fusion in the detected overlapping area between neighbouring views thus shift points from their actual positions and create artefacts in the reconstructed surfaces. As a matter of fact, reconstruction noise often imposes more severe impact on the quality of reconstructed 3D object models, since such noise usually distorts the shape of object as demonstrated by the bottom row of Figure 1.

As shown in the bottom row of Figure 1, by comparing the more accurate reconstructed surface on the left hand side with the one on the right hand side, it can be observed that while the former generates a limited number of artefacts on the head and tail of the bird surface, the latter did not accurately reconstruct the bird model. While the lower mandible and breast are smooth, the upper mandible, head, wings, tail, and legs are not. The reconstructed bird surface is uneven and more artefacts appear in areas with high curvature. Therefore powerful smoothing algorithms need to be developed to carefully deal with the reconstruction noise.

3. Tsallis Entropy Based Smoothing Algorithm Development

In this section we develop new methods to deal with both scanning noise and reconstruction error. The new methods have ability not only to optimize location of vertices but also to filter noise in vertex normals. To this end, we employ the Tsallis entropy [34] in conjunction with the entropy maximization principle [19] to model the inter-influence (interaction) among the

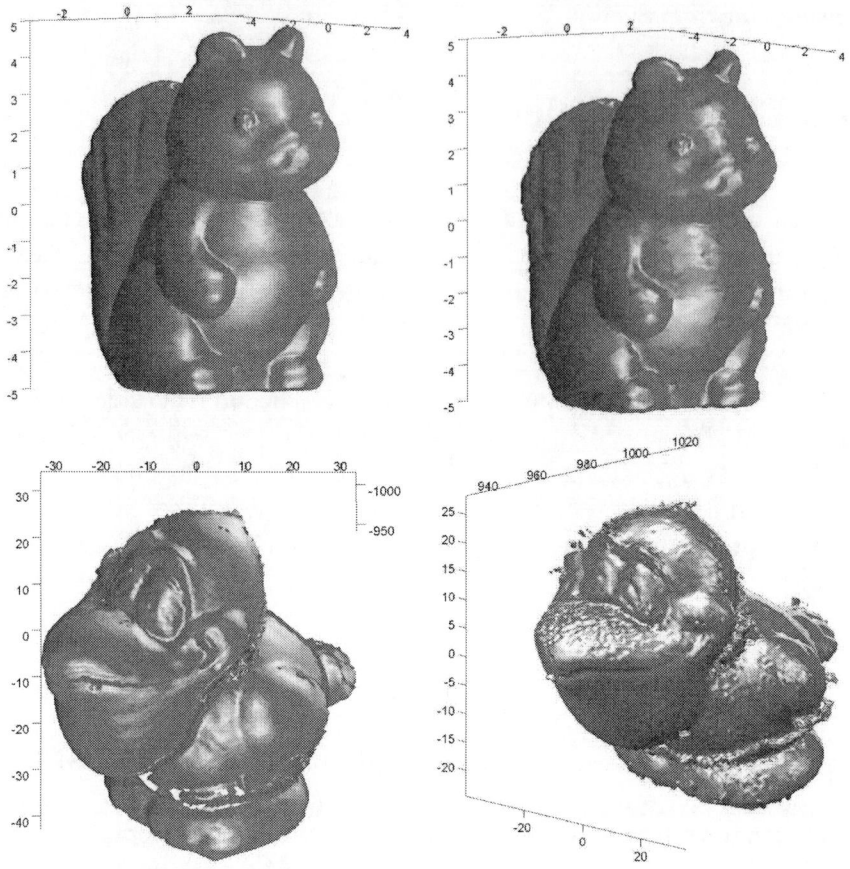

Figure 1. Top left: Original squirrel surface. Top right: Surface corrupted by Gaussian white noise with standard deviation $\delta = 0.02$. Bottom left: More accurately reconstructed surface [38]. Bottom right: Reconstructed model using the volumetric method [10].

interwoven neighbouring vertices in the entangled iterative process of vertex optimization and update.

3.1. Vertex Location Optimization

Firstly, vertex location optimization is considered. If a surface is smooth and optimal, then each triangle should be as similar to equilateral one as possible and their edge lengths should be as equal to each other as possible. Then the edge length variation can be used to characterise the dissimilarity between the existing triangles and optimal ones:

$$\|MaxMinEdge_j\| = \|Max(a_j, b_j, c_j) - Min(a_j, b_j, c_j)\|^2 \tag{1}$$

where a_j, b_j, c_j are the edge lengths of the triangle Tri_j.

However, this strategy is too crude to be practical and it did not take into account the details of object surface represented as varied sizes, shapes, and normals of triangles. While equilateral triangles are desired, scalene triangles are useful to represent

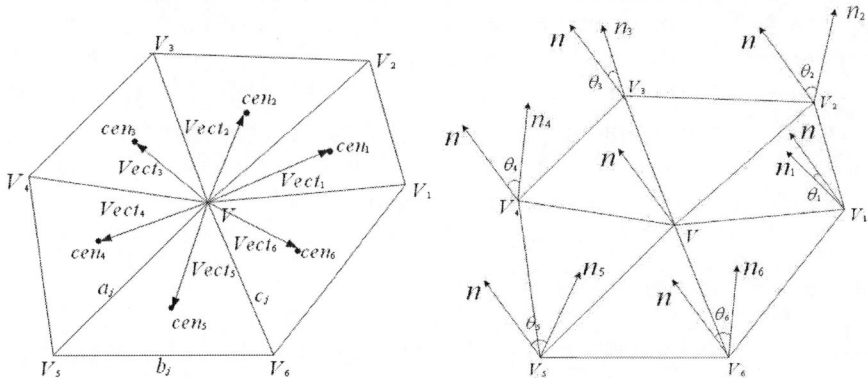

Figure 2. The principle of vertex location smoothing(left) and normal smoothing(right).

the geometrical details of object surface. Thus, a positive real number W_{ij} in the unit interval is introduced to characterise the extent to which the edges of triangles are of equal length. For this purpose, the following objective function is then constructed: $J_{EV} = \min_{W_{ij}} \sum_{i=1}^{M} \sum_{j=1}^{N} W_{ij} \|MaxMinEdge_{ij}\|$ where M is the number of vertices in the mesh and N is the number of incident triangles of the vertex V_i. In order to estimate the parameter W_{ij}, we employ the entropy maximization principle [19] and thus construct a new objective function, consisting of the cost J_{EV} minus the Tsallis entropy of these unknowns W_{ij}, as follows:

$$J = \min_{W_{ij}} \sum_{i=1}^{M} \sum_{j=1}^{N} W_{ij}\|MaxMinEdge_{ij}\| - \frac{1}{\beta} \sum_{i=1}^{M} \sum_{j=1}^{N} \frac{W_{ij} - W_{ij}^q}{q-1} \quad (2)$$

where parameter β is a real positive number and can be regarded as the Lagrange multiplier, controlling the relative contribution of the Tsallis entropy to the constructed objective function.

In order to estimate the unknown W_{ij}, we get the first order derivative of this objective function and set the result equal to zero:

$$\frac{\partial J}{\partial W_{ij}} = \|MaxMinEdge_{ij}\| + \frac{1}{\beta(q-1)}(qW_{ij}^{q-1} - 1) = 0.$$

Thus $W_{ij} = (\frac{1-\beta(q-1)\|MaxMinEdge_{ij}\|}{q})^{\frac{1}{q-1}}$. In order to make sure $1 - \beta(q-1)\|MaxMinEdge_{ij}\| \geq 0$, $0 < q < 1$ is required. In this case, vertices in a triangular mesh are simulated as constituents in a superextensive system.

Then the optimal position of vertex V_i is estimated as:

$$V_{inew} = V_i + \frac{1}{\sum_{j=1}^{N} W_{ij}} \sum_{j=1}^{N} W_{ij} * \vec{Vect}_{ij}, \quad (3)$$

where $\vec{Vect}_{ij} = cen_{ij} - V_i$ and $cen_{i1}, cen_{i2}, cen_{i3} \cdots, cen_{iN}$ are the centroids of incident triangles Tri_{ij} of vertex V_i. Equation 3 shows that the more similar a triangle is to a equilateral one, the smaller the parameter $\|MaxMinEdge_{ij}\|$ is, the larger the weight W_{ij}.

In this case, $\vec{Vect}_{ij}$ from non-equilateral triangles will not make a considerable contribution to the update of vertex location. Such property is useful to prevent details of surface from being smoothed out. Both parameters β and q control the relative contribution of different incident triangles. The principle for vertex location optimization is shown on the left hand side of Figure 2. All vertices are optimized iteratively without changing neighbourhood information until the surface is desirably smoothed.

3.2. Vertex Normal Optimization

The accuracy of surface orientation estimation is very important for the final surface rendering quality. In this section, we follow the same methodology as vertex location optimization described in the last section for vertex normal optimization. The normal of each vertex in the mesh is firstly calculated by averaging the normals of all the triangles weighted by their areas [18] that share the vertex.

If a surface is smooth, then the normal of each vertex should be consistent with those of its neighbours. For each vertex V_i and its neighbouring vertices $V_{i1}, V_{i2}, V_{i3} \cdots, V_{iN}$, let the including angles in degrees between normal vectors N_i at V_i and N_{ij} at V_{ij} be $\|\Delta\theta_{ij}\|$. So for smooth surface, the weighted normal variation $J_{OV} = \sum_{i=1}^{M} \sum_{j=1}^{N} W_{ij} \|\Delta\theta_{ij}\|$ should be minimum where M is the number of vertices in the mesh, N is the number of neighbouring vertices of a vertex and the parameter W_{ij} is in the unit interval indicating the extent to which the normals of neighbouring vertices are desired to be similar.

In order to estimate the parameter W_{ij}, we again employ the EntMax principle and construct the following new objective function that consists of the weighted normal variation J_{OV} minus the Tsallis entropy of these unknowns W_{ij}:

$$J = \min_{W_{ij}} \sum_{i=1}^{M} \sum_{j=1}^{N} W_{ij} \|\Delta\theta_{ij}\| - \frac{1}{\beta} \sum_{i=1}^{M} \sum_{j=1}^{N} \frac{W_{ij} - W_{ij}^q}{q-1} \qquad (4)$$

where parameter β is again a real positive number and can be regarded as the Lagrange multiplier, controlling the relative contribution of the Tsallis entropy to the constructed objective function.

Then we again get the first order derivative of this objective function and set the result equal to zero:

$$\frac{\partial J}{\partial W_{ij}} = \|\Delta\theta_{ij}\| + \frac{1}{\beta(q-1)}(qW_{ij}^{q-1} - 1) = 0.$$

Thus $W_{ij} = (\frac{1-\beta(q-1)\|\Delta\theta_{ij}\|}{q})^{\frac{1}{q-1}}$. Finally, the new normal N_i at vertex V_i is updated as a weighted sum of N_{ij}: $N_{inew} = N_i + \frac{1}{\sum_{j=1}^{N} W_{ij}} \sum_{j=1}^{N} W_{ij} * N_{ij}$ subject to normalization: $N_{inew} = \frac{N_{inew}}{\|N_{inew}\|}$. In order to make sure $1 - \beta(q-1)\|\Delta\theta_{ij}\| \geq 0$, $0 < q < 1$ is required. In this case, vertex normals in a triangular mesh are simulated as constituents in a superextensive system. This update shows that the more similar the normals of neighbouring vertices are to that of the vertex of interest, the larger the weight W_{ij}. In this case, vertex normals that deviate considerably from that of the vertex of interest will not make a significant contribution to the update of the vertex normal being optimized. Such a property is again useful to prevent details of surface from being smoothed out. Both parameters β and

q control the relative contribution of different neighbouring vertex normals. The principle of vertex normal smoothing is illustrated in the right hand side of Figure 2.

After normals of all vertices are optimized, the vertex V_i is updated according to: $V_{inew} = V_i + \frac{1}{\sum_{j=1}^{N} W_{area_{ij}}} \sum_{j=1}^{N} W_{area_{ij}} (\vec{Vect}_{ij} \cdot N_{inew}) N_{inew}$ where $W_{area_{ij}}$ is the area of incident triangles Tri_{ij} of vertex V_i [6].

4. Experimental Results

To detail the advantage and disadvantage of the novel algorithm for smoothing the reconstructed 3D object computer models, the other three classical smoothing methods [16, 32, 27] are also implemented. To compare different algorithms and different noise sources, the following parameters are defined: (1) The mean smoothing error (MSE): the mean distance between optimized and original vertices. The smaller the mean smoothing error, the closer the smoothed surface is to the original one. This measure is useful to characterise the extend to which the details of object surface have been retained. An extreme case is that when a clean model has not been smoothed at all, MSE will be zero. If so, then the noise and artefacts are still there. This implies that smoothing operation has to be carried out. However, it should not change location of vertices arbitrarily. MSE measures the extent to which the location of vertices has been changed after the process of smoothing; (2) The distortion metric [22]: the area of a triangle divided by the sum of the squares of the lengths of its edges and then normalized by a factor $2\sqrt{3}$. The value of the distortion metric is in the unit interval $[0, 1]$. The higher the distortion metric value, the higher the smoothing quality of the triangle; (3) The distribution of interior angles of triangles. The angle distribution shows the global optimal degree of triangles. The closer the interior angles of triangles are to $60°$, the more similar they are to equilateral ones; (4) The computational time; and finally (5) The number of iterations. The experimental results are presented in Figures 3 through 11 and Tables 2 and 3.

4.1. Different Parameters

In order to choose reasonable q and β values in the new method for our experiments, different q and β values were chosen. The experimental results are presented in Figure 3 and Table 1. In the figure, the blue lines represent $\beta = 0.1$, the red lines represent $\beta = 1$. The green line represents $\beta = 0.09$ and the black line represents $\beta = 0.01$ when $q = 0.09$. The magenta line represents $q = 0.01$ and the cyan line represents $q = 0.001$ when $\beta = 0.01$. From Figure 3, it can be observed that when q is large, $q = 0.09$, for example, changing β can only improve the distortion metric slightly but keep MSE almost the same. However, when q is small such as $q = 0.001$, a high distortion metric and a low MSE are obtained. So the novel smoothing method is not sensitive to the set up of parameter β and is effective in retaining more details of surface. Table 1 shows when $q = 0.001$ and $\beta = 0.01$, the highest distortion metric value and the lowest MSE are obtained. Therefore, in the rest of this chapter, we let $\beta = 0.01$ and $q = 0.001$.

To compare the effect of parameters in different filters, the Laplacian filter $W_{ij} = \exp(-\beta * \|Para_{ij}\|)$ [37] and the Gaussian filter $W_{ij} = \exp(-\beta^2 * \|Para_{ij}\|^2)$ [6] are

Table 1. Smoothing results using the new method with parameters q and β taking different values. The 1st, 3rd, 5th rows: Distortion metric value. The 2nd, 4th, 6th rows: Mean smoothing error.

	$\beta = 0.1$	$\beta = 0.01$	$\beta = 0.001$
$q = 0.01$	0.86	0.90	0.90
	1.83	1.76	1.78
$q = 0.001$	0.87	0.90	0.90
	1.82	1.75	1.77
$q = 0.0001$	0.87	0.90	0.90
	1.84	1.76	1.78

also implemented in this chapter. Here $\|Para_{ij}\|$ is a smoothing parameter to be optimized: $\|Para_{ij}\| = \|MaxMinEdge_{ij}\|$. The experimental results are presented in Figures 3 and 4, showing that the distortion metric and MSE for the Gaussian filter and the Laplacian filter change more dramatically with β taking different values than those for the novel smoothing method. Both the Laplacian and Gaussian filters are sensitive to the constant parameter β in the smoothing process. So when β is data dependent and improperly chosen, the smoothing results are not reasonable as shown in the bottom row of Figure 4. However from the top right of Figure 4, it can be observed that not only is the surface smooth but the surface details are also kept. Thus, the Tsallis entropy is realistic in modelling the inter-influence (interaction) among interwoven neighbouring vertices in the entangled iterative process of vertex location and normal optimization. The parameter q provides an extra degree of freedom for the final smoothing quality control.

4.2. Vertex and Normal Smoothing

To compare the impact of surface orientation on 3D rendering, a single lobster surface rendered from its range image with noisy vertices and normals is used. The experimental results are presented in Figure 5, showing that even the vertices are still corrupted by Gaussian white noise, after the new method has smoothed the noisy normals, the rendered surface is much smoother. Next, the noisy vertices are updated using the method [6] that first optimizes vertex normals and then locations. It is clear that the surface with smoothed normals and noisy vertices is very similar to that with smoothed normal and smoothed vertices. Not only do both of them look smoother but the details of surface are also kept. The bottom right figure is smoothed by firstly optimizing vertex location and then recalculating the surface orientation. The resulting surface is smooth but some details such as the abdomen of the lobster becomes blurred. This shows that details of a surface can be determined more easily and effectively by using surface normals rather than by using vertex locations, thus the surface normals impose a greater impact on the model's perceived quality.

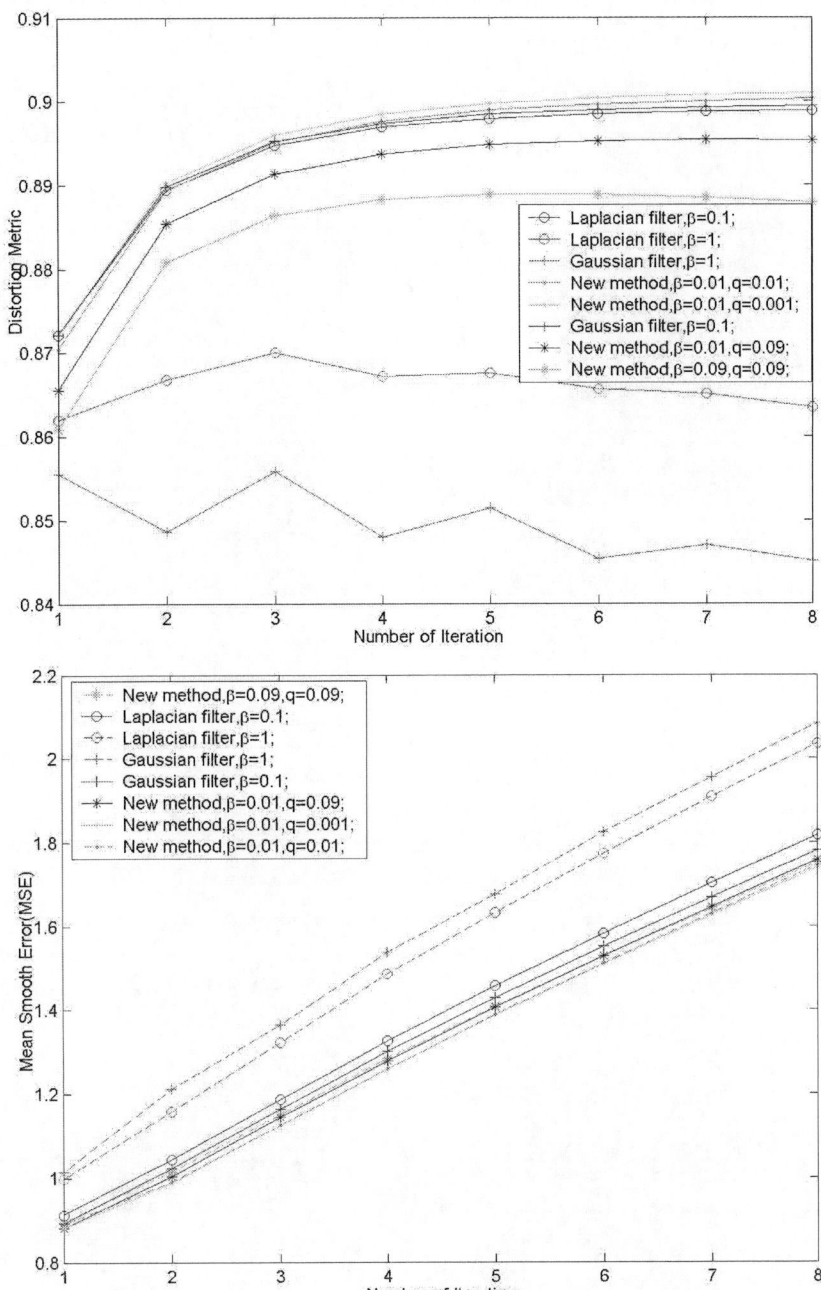

Figure 3. Experimental results of different filters with different parameters taking different values. Top: distortion metric. Bottom: MSE.

4.3. Different Smoothing Algorithms

In this section, we compare different algorithms using images with different resolutions corrupted by different levels of Gaussian white noise. The experimental results are presented in Figures 6 through 9 and Table 2.

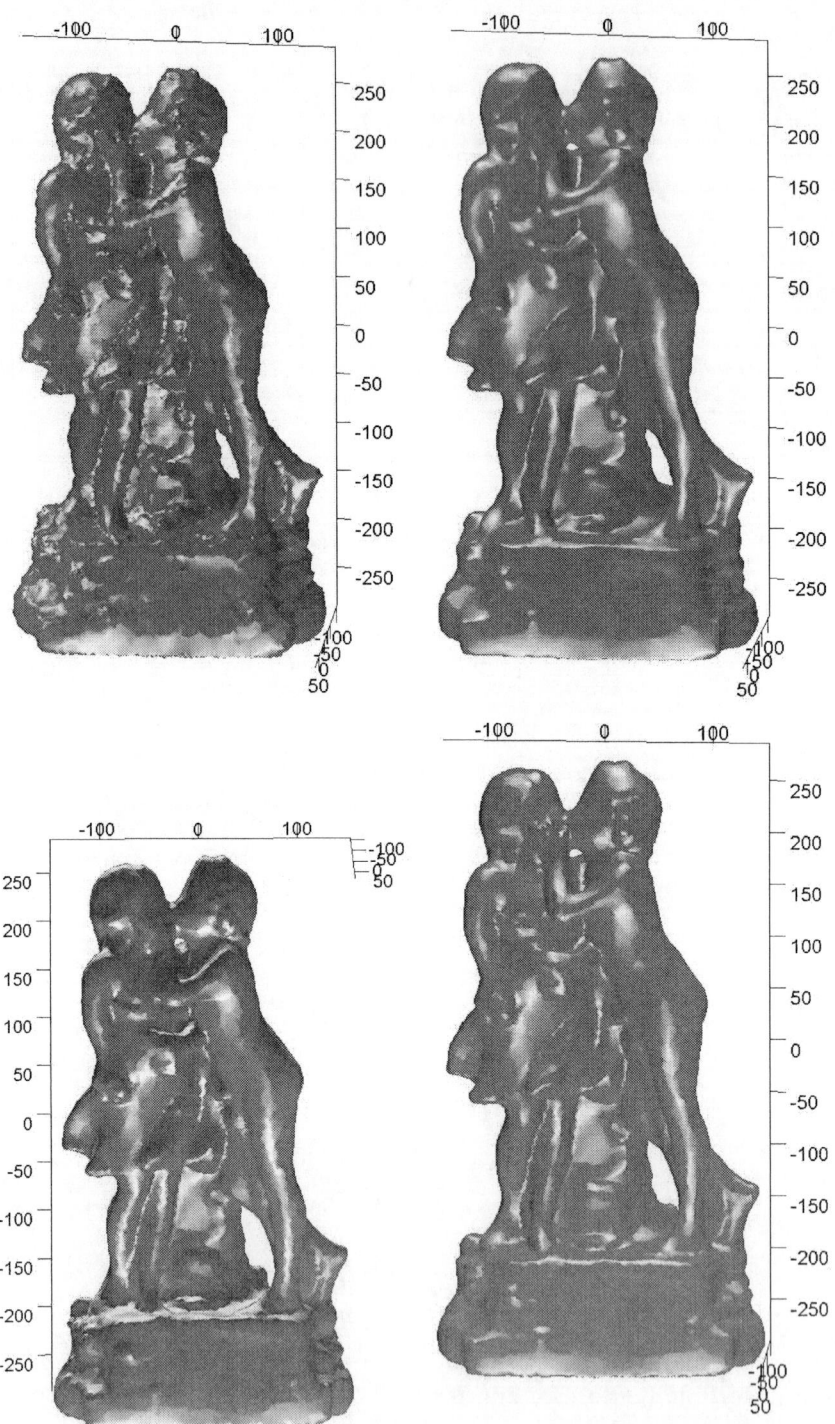

Figure 4. Noisy surface smoothing results using different algorithms. Top left: Noisy surface corrupted by Gaussian white noise with a standard deviation of $\delta = 0.9$. Top right: The novel smoothing method. Bottom left: Laplacian smoothing method [37]. Bottom right: Gaussian smoothing method [6].

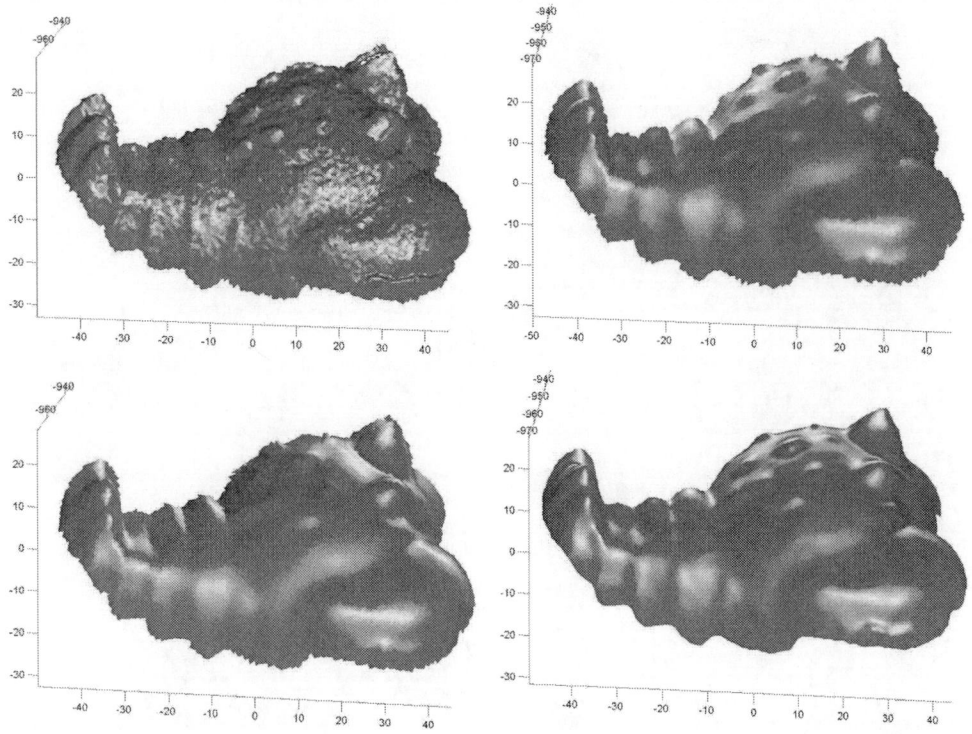

Figure 5. Noisy surface smoothing results with different parameters being optimized. Top left: Noisy lobster due to Gaussian white noise with a standard deviation of $\delta = 0.2$. Top right: Smoothing only vertex normals. Bottom left: Smoothing first vertex normals and then locations. Bottom right: Smoothing first vertex locations and then recalculating their normals.

In Figure 6, the foot surface was corrupted by Gaussian white noise with a standard deviation $\delta = 0.9$. The surface by using the signal processing smoothing method [32] is still non-smooth after 15 iterations. However our new method and the original Laplacian method smooth the noisy surface efficiently. Since the resolution of this surface is as high as $2.33mm$, there are enough sampled points for the representation of the details of surface. The distance between the neighbouring nodes is quite small. So the optimal node can be reasonably modelled as a weighted average of the neighbouring nodes, rendering the surface smoother. When the surface depth does not change rapidly, as is the case for the instep of the foot surface, the new method produces similar smoothing results to the original Laplacian method [16]. But the distortion metric, 0.96, produced by the new method is still larger than that, 0.94, produced by the original Laplacian method. Since the signal processing smoothing method [32] employed a low-pass filter and a high-pass filter alternately, some high frequency noise still remains and the instep of the foot surface is still not smooth after as many as 15 iterations. Although this method can protect shrinkage of surface caused by node moving, it usually happens only to surface boundaries. So it could not shrink considerably when the surface is a closed manifold, as is the case for the foot surface. Meanwhile, since the interior angles of triangles on the surface do not change rapidly,

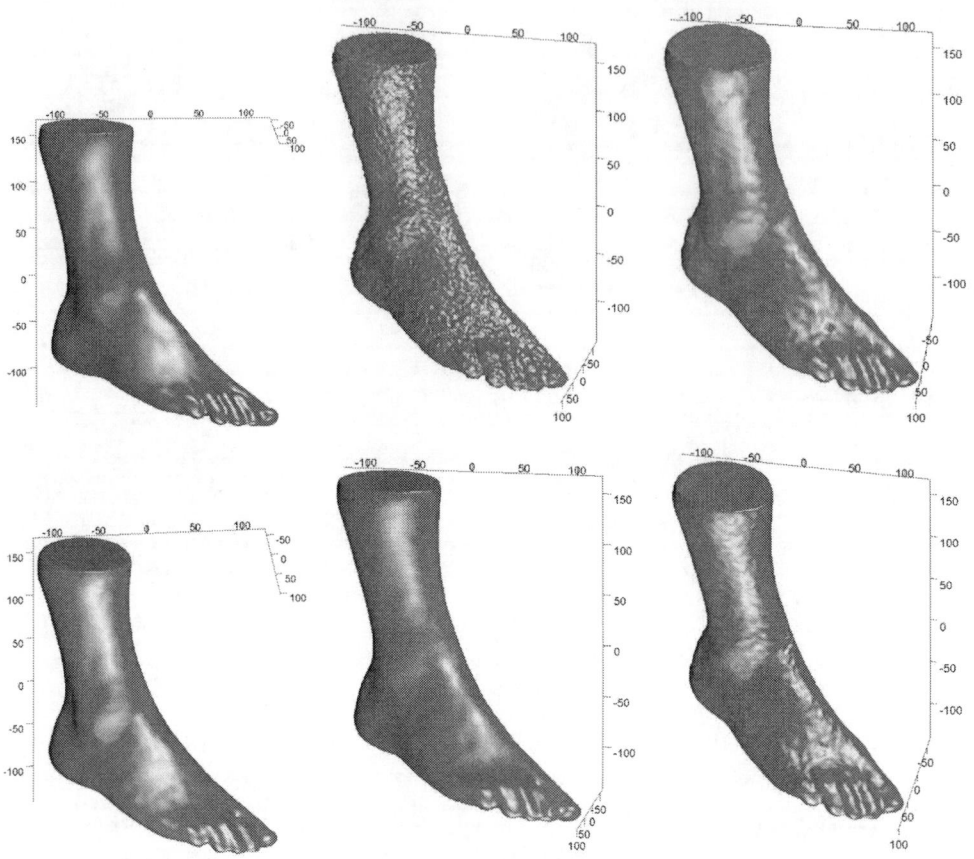

Figure 6. Noisy surface smoothing results using different algorithms. From top left to bottom right: First: Clean foot surface. Second: Noisy foot surface subject to Gaussian white noise with a standard deviation of $\delta = 0.9$. Third: Signal processing smoothing method [32]. Fourth: The new method. Fifth: Original Laplacian method [16]. Sixth: Mean curvature flow smoothing method [27].

the mean curvature flow method [27] is also not efficient for smoothing even after a large iteration number of 10.

A low resolution head surface $R = 4.8mm$ and a high resolution lion surface $R = 0.26mm$ with more details were used to further test smoothing effectiveness of different methods. From Figure 7 and Table 2, it can be seen that the new method and the signal processing method yield best smoothing results. However, the latter needs much more iterations (12) than the former (2) and it required 3 times the computational time the former required. This means that the former is significantly more efficient than the latter. The original Laplacian method produces a smoother surface than the new method but at a cost of some details, such as the eyes, mouth and hair on the head surface and the mouth on the lion surface. This conclusion has been confirmed by Table 2. While the original Laplacian method produces the highest distortion metric values of 0.87 and 0.89 for the head and lion surfaces respectively, it also produces the largest smoothing errors of $1.35mm$ and $0.13mm$. This means that although the triangles become more similar to the equilateral

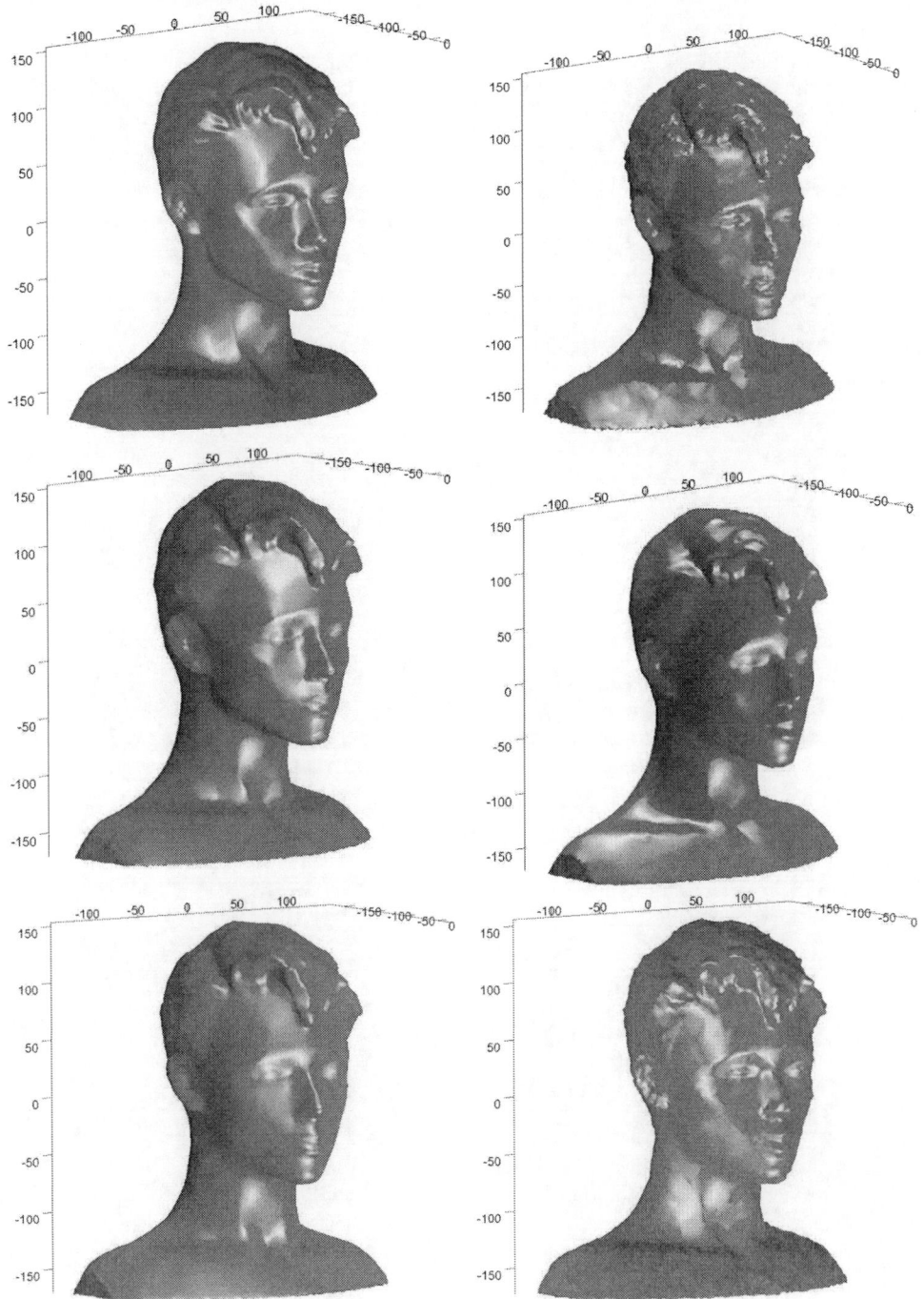

Figure 7. Noisy surface smoothing results using different algorithms. From top left to bottom right: First: Clean head surface. Second: Noisy head surface subject to Gaussian white noise with a standard deviation of $\delta = 0.9$. Third: Signal processing smoothing method [32]. Fouth: the new method. Fifth: Original Laplacian method [16]. Sixth: Mean curvature flow smoothing method [27].

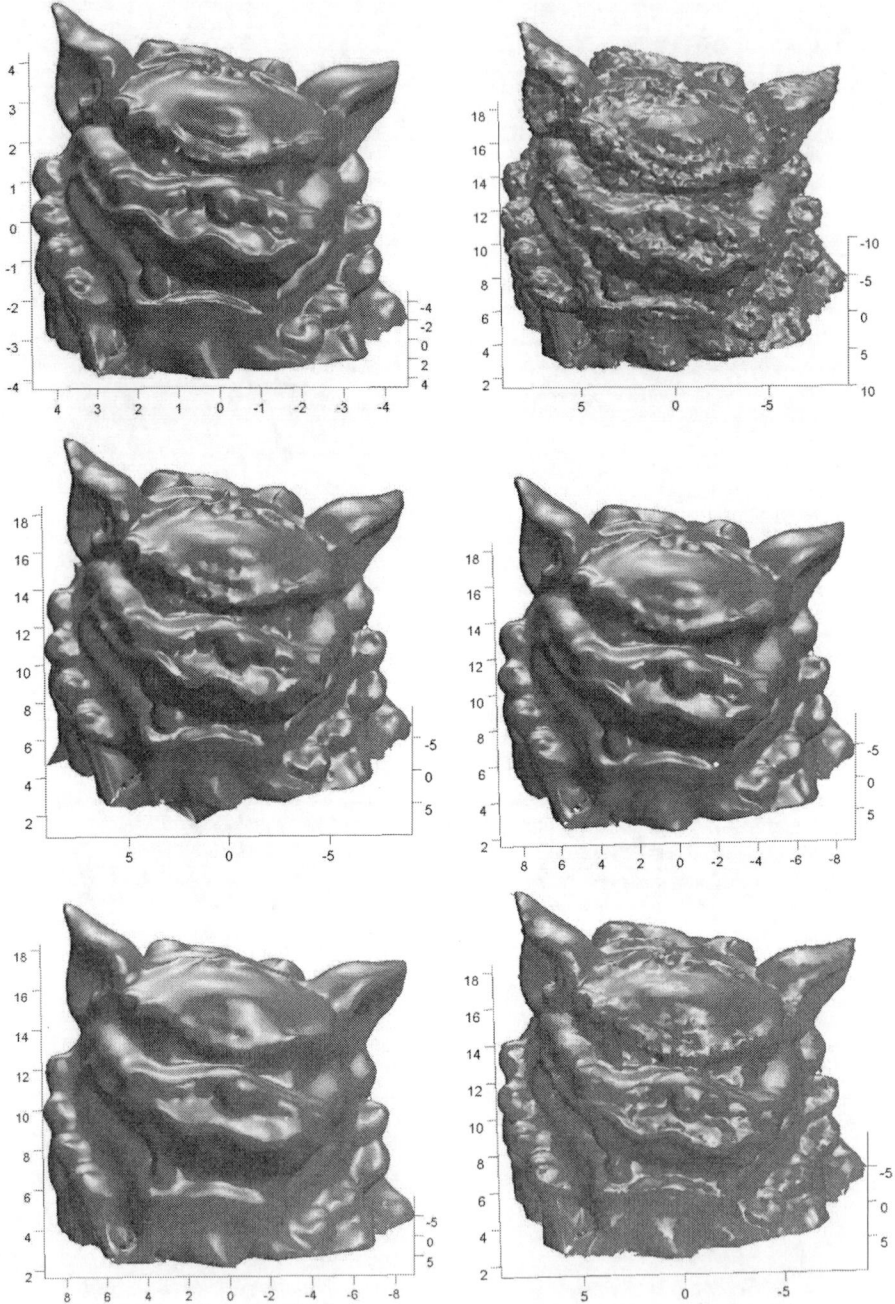

Figure 8. Noisy surface smoothing results using different algorithms. From top left to bottom right: First: Clean guard lion surface. Second: Noisy guard lion surface subject to Gaussian white noise with a standard deviation of $\delta = 0.05$. Third: Signal processing smoothing method [32]. Fouth: The new method. Fifth: Original Laplacian method [16]. Sixth: Mean curvature flow smoothing method [27].

ones and the quality of the surface is improved, some vertices have to shift a long distance from their original locations, resulting in some details being lost. However, the new method produces not only the most optimal mesh but also the smallest smoothing error. Although the mean curvature flow method keeps the details of the head surface, it blurred the details about the mouth of the lion surface. On the other hand, it requires the longest time for computation.

By comparing the angle distributions of the foot, head and lion surfaces in Figure 9, it can be clearly seen that the new method yielded the best results, since it results in the interior angle distribution of triangles being almost Gaussian with peak around 60°, as desired. In summary, Figures 6 through 9 and Table 2 show that the new method yields the highest distortion metric and lowest MSE and thus the most optimal surface without losing main details in these three surfaces with varied complexities of geometry.

Table 2. Scanning noise smoothing results using different algorithms and images. MSE: Mean smoothing error. DistMetric: Distortion metric. The 2nd, 7th, 12th rows: The signal processing method [32]. The 3rd, 8th, 13th rows: The new algorithm. The 4th, 9th, 14th rows: The original Laplacian method [16]. The 5th, 10th, 15th rows: The mean curvature flow method [27].

Image	Face	Vertex	Resolution(mm)	MSE(mm)	DistMetric	Iteration	Time(s)
				1.26	0.73		
				0.57	0.90	15	422
Foot	59690	25845	2.33	0.43	0.96	6	293
				0.43	0.94	5	204
				0.47	0.85	10	2000
				1.26	0.73		
				1.13	0.78	12	92
Head	20222	10113	4.80	1.09	0.87	2	32
				1.35	0.87	2	31
				1.20	0.74	4	284
				0.10	0.70		
				0.08	0.82	15	523
Lion	50000	24930	0.26	0.08	0.89	3	201
				0.13	0.89	5	236
				0.08	0.73	4	942

4.4. Scanning Noise and Reconstruction Error

After we have tested different algorithms for smoothing scanning noise, in this section, we test them in handling the scanning noise as well as reconstruction error. The mesh based reconstruction algorithm [31] and the volumetric reconstruction algorithm [10] were implemented to produce the noisy reconstructed teletubby and dinosaur surfaces with range images registered using the automatic registration algorithm [23]. The noisy reconstructed teletubby and dinosaur surfaces are shown in the left upper corners of Figures 10 and 11 respectively. The telebubby surface was reconstructed from 20 range images and contains

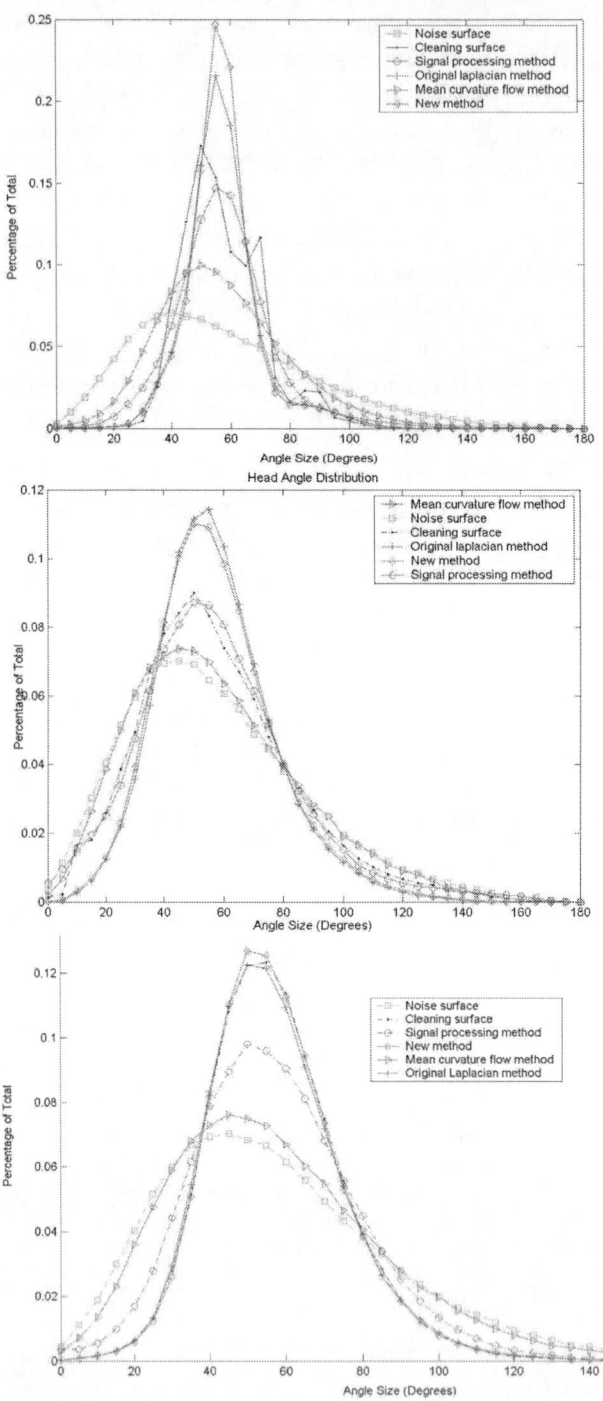

Figure 9. The distribution of interior angles of triangles in different surfaces. Top: Foot. Middle: Head. Bottom: Lion.

33478 faces and 17243 vertices. The dinosaur surface was reconstructed from 8 range images and contains 88444 faces and 43965 vertices. The mesh based integration method [31] reconstructs the surface with the original meshes from different range images. So if the registration error is large, as is the case for the registration of the teletubby images, the connected meshes will depart away from each other, the final reconstructed surface will not be smooth, as illustrated by the chest and arms of the teletubby. For the volumetric method [10], since the voxel size is difficult to choose and data sampling is non-uniform, many artefacts appear on the reconstructed surface, as illustrated by the profile of the dinosaur.

The experimental results for smoothing the reconstructed teletubby and dinosaur surfaces are presented in Figures 10 and 11 and Table 3. In the table, MSE was not computed due to the fact that these surfaces were reconstructed from multiple registered range images, the ground truth of original points is not available. Figures 10 and 11 and Table 3 show the new method performs best with the distortion metric values of two surfaces increased from 0.81 and 0.75 to 0.92 and 0.90 within 5 and 4 iterations respectively. Although the original Laplacian method [16] also produced high distortion metric values 0.91 and 0.90 after 4 iterations, it blurred the details of the eyes and chest of the teletubby and the profile of the dinosaur. After 7 and 5 iterations, however, the signal processing method and the mean curvature flow method [27] did not produce smooth teletubby and dinosaur surfaces with the distortion metric values increased only to 0.86, 0.80 and 0.82, 0.77 respectively.

It can also be seen from Figure 12 that the angle distribution of noisy surfaces marked with blue line do not distribute smoothly with two peaks around 45° and 85° respectively. However after smoothing using the original Laplacian method [16] and the new method the interior angles of most triangles have been driven closer to 60° and their distributions are smoother. This means that most triangles approach equilateral ones. While the signal processing method [32] also produces an optimal distribution of interior angles of triangles with a lower peak, the mean curvature flow method [27] hardly changes the angle distribution.

On the other hand, although the values of parameters of interest from the original Laplacian method and the new method are nearly the same in Table 3 and Figures 10 through 12. However, more details about the eyes of the teletubby and the profile of the dinosaur are kept by the new method but not by the original Laplacian method. This is because the Tsallis entropy is effective in modelling the entangled optimization of the vertices of interest using their interwoven neighbours. Furthermore, the Laplacian method is sensitive to the parameter β and its assumption that triangle edge length variations and vertex normal variations follow Gaussian distribution is often violated. Thus it can hardly distinguish noise and artefacts from details and consequently often smooth out details. The mean curvature flow method [27] is the most computationally expensive, since it tries to minimize the area of triangular meshes and averages the curvature of surface [31]. While the finally reconstructed teletubby and dinosaur surfaces include many ill-shaped triangles with small areas and the curvature does not change rapidly, it is not effective for smoothing.

Comparing Figure 9 with Figure 12, it can be seen that surfaces corrupted by Gaussian white noise often lead to smooth interior angle distributions, while the reconstructed surfaces often lead to interior angle distributions with peaks deviating from 60°. This is because inaccurate registration and integration errors result in ill-shaped triangles being created, driving the peaks of the interior angle distributions away from the optimal 60°.

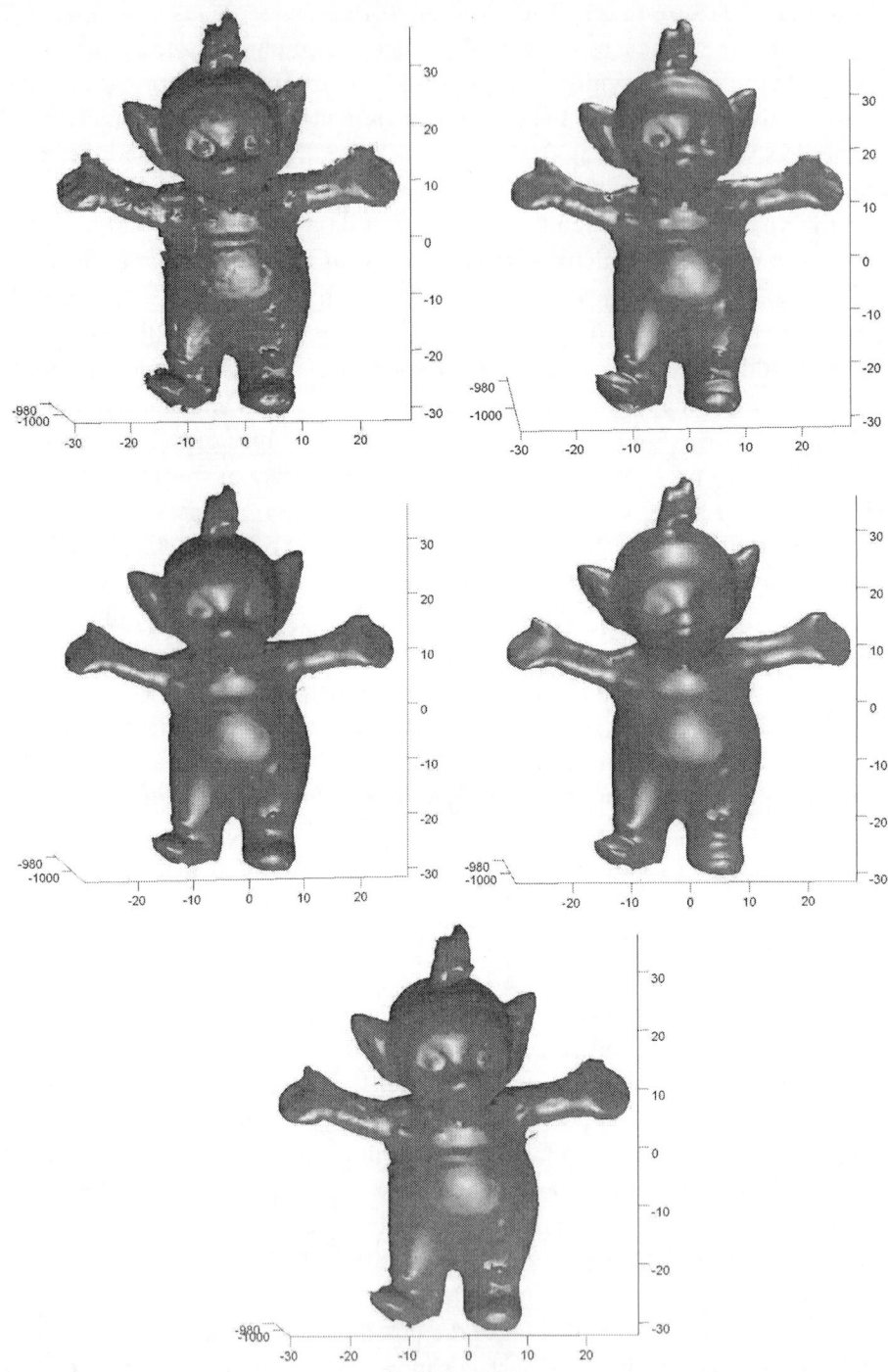

Figure 10. Reconstructed surface smoothing results with different algorithms. From top left to bottom: First: Reconstructed teletubby surface with 20 range images [31]. Second: Signal processing smoothing method [32]. Third: the new method. Fourth: Original Laplacian method [16]. Fifth: Mean curvature flow smoothing method [27].

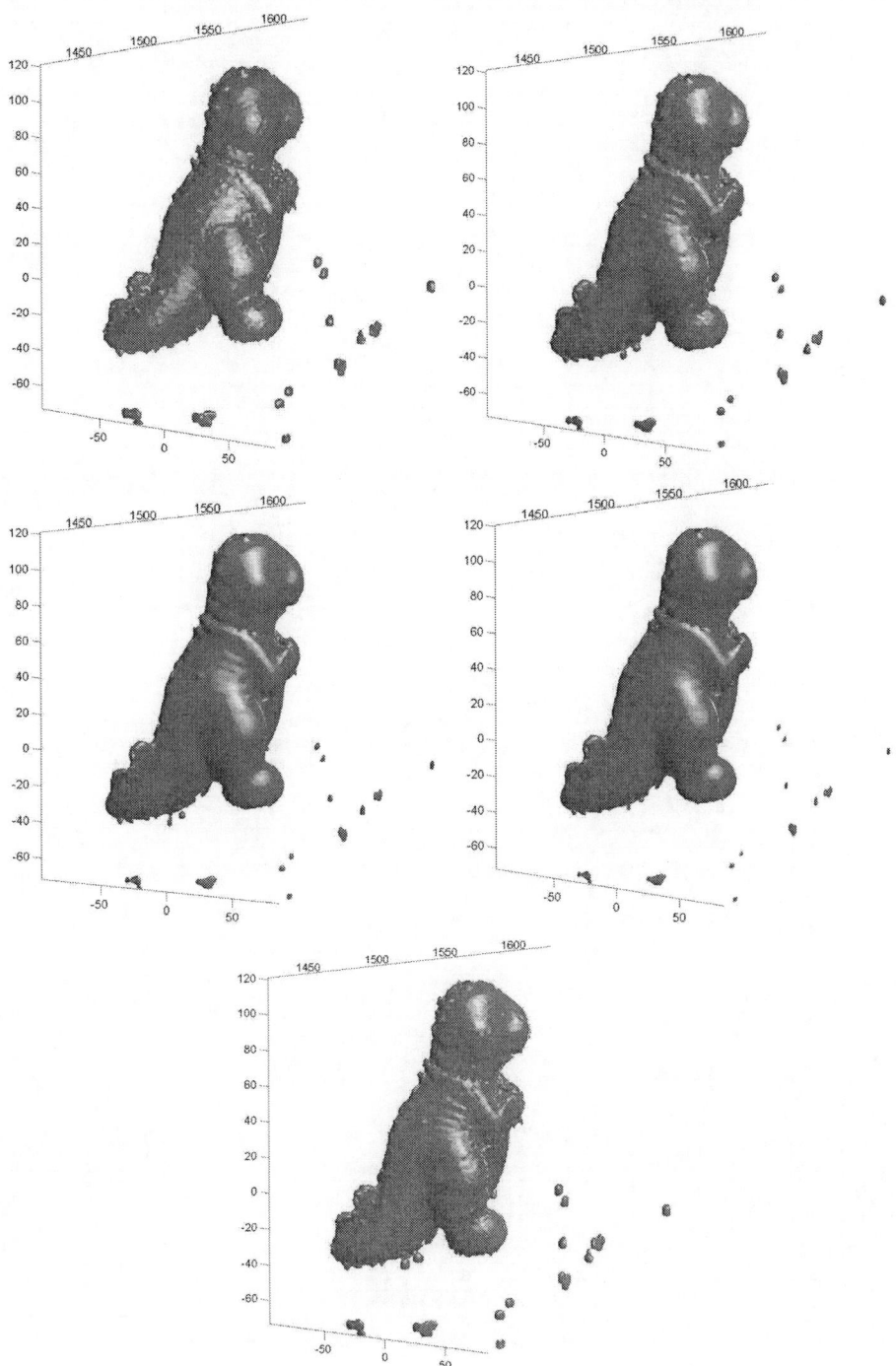

Figure 11. Reconstructed surface smoothing results with different algorithms. From top left to bottom: First: Reconstructed dinosaur surface with 8 range images [10]. Second: Signal processing smoothing method [32]. Third: The new method. Fourth: Original Laplacian method [16]. Fifth: Mean curvature flow smoothing method [27].

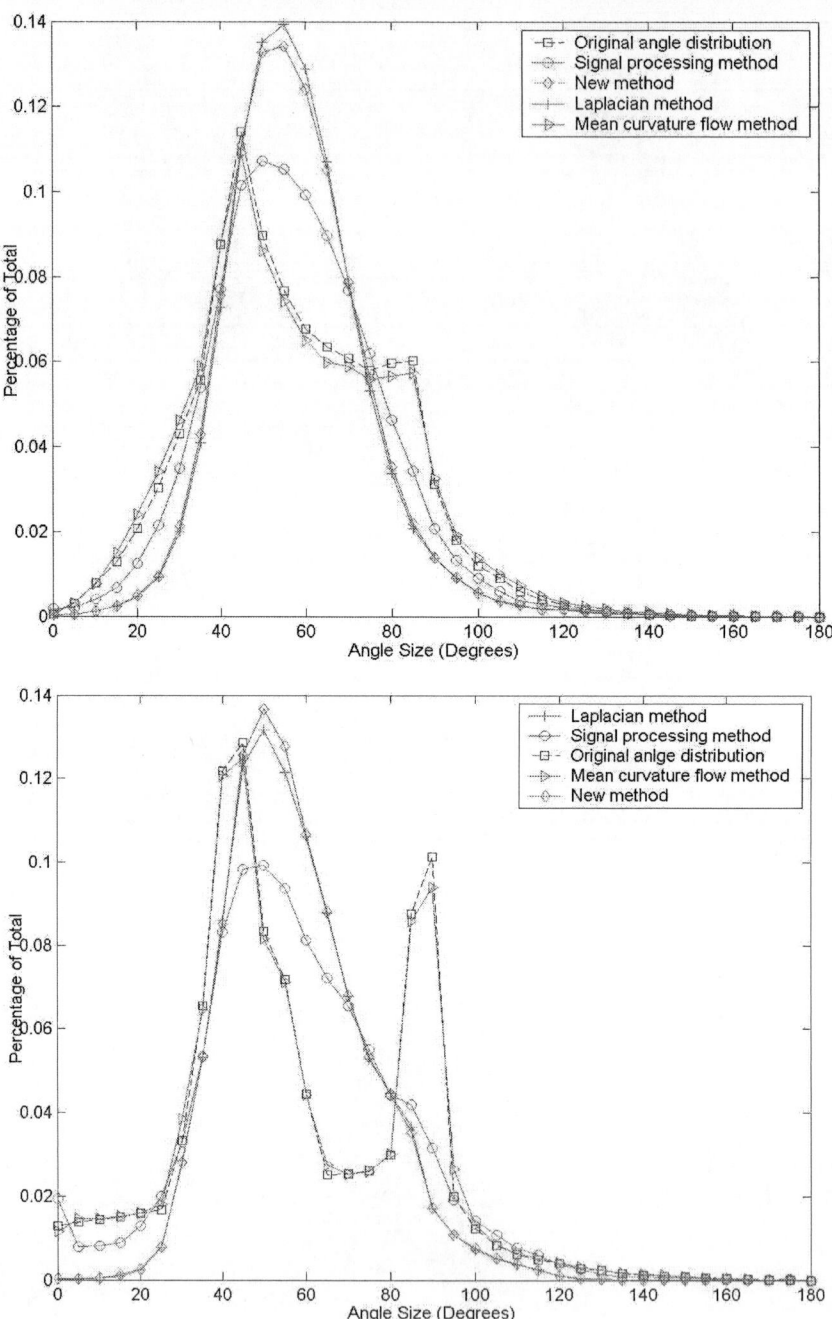

Figure 12. The distribution of interior angles of triangles in the reconstructed surfaces smoothed using different smoothing methods. Top: Teletubby surface reconstructed using the mesh based reconstruction method [31] from 20 range images. Bottom: Dinosaur surface reconstructed using the volumetric method [10] from 8 range images.

Table 3. Scanning noise and reconstruction error smoothing results using different algorithms and images. DistMetric: Distortion metric. NOI: Number of images. The 2nd, 7th rows: The signal processing method [32]. The 3rd, 8th rows: The new algorithm. The 4th, 9th rows: The original Laplacian method [16]. The 5th, 10th rows: The mean curvature flow method [27].

Object	Face	Vertex	Resolution(mm)	NOI	DistMetric	Iteration	Time(s)
Teletubby	33478	17243	0.68	20	0.81		
					0.86	7	179
					0.92	5	102
					0.91	4	106
					0.82	5	800
Dinosaur	88444	43965	1.40	8	0.75		
					0.80	10	1000
					0.90	4	689
					0.90	4	661
					0.77	5	1106

5. Conclusion and Further Work

In this chapter, we proposed a novel method for smoothing reconstructed surfaces corrupted by both scanning noise and reconstruction error due to inaccurate image registration and integration. In order to smooth noisy surfaces, we are desired to construct meshes with triangles as similar to equilateral ones as possible and normals of neighbouring vertices as similar to each other as possible. Thus, we construct an objective function to minimize the weighted length variations of edges of triangles to optimize the vertex locations and the weighted normal variations of neighbouring vertices to optimize the vertex normals.

In the smoothing process, a vertex is generally optimized using its neighbours. This is an iterative process. When the vertices in the mesh vary significantly in the sense of location and orientation, the Shannon entropy is not effective for the estimation of the contribution of the neighbouring vertices to the vertex of interest for its optimization and update. Therefore, the Tsallis entropy was used instead in conjunction with the EntMax principle to accurately estimate the contribution.

The proposed algorithm has an advantage of easy implementation. On the other hand, since the optmization and update of a node considers only its neighbours, it has a linear computational complexity with regard to the number of nodes on the surface. While the new method is as efficient as the original Laplacian method in dealing with high resolution surface or the flat area in a surface, it outperforms the original Laplacian method in the sense of retaining details of a surface when the surface resolution is low or the surface contains a rapid change of depth. Comparing the novel method with the signal processing method [32] and the mean curvature flow method [27], the former is more computationally efficient and effective in producing high quality meshes with a better balance between details and smoothness. Experimental results based on real images have shown that the novel smoothing method yields on the whole the best smoothing results in the sense of MSE, distortion metric, interior angle distribution, computational time, and the number of iterations for smoothing. This is because the EntMax principle in conjunction with the Tsallis entropy are expressive in modelling the inter-influence (interaction) among interwoven neighbouring vertices in the process of surface smoothing and distinguishing details from noise and artefacts in the surface being smoothed. The Tsallis entropy also provides an extra degree of freedom for the final smoothing quality control.

In this chapter, triangular edge length variation and vertex normal variation were chosen to be optimized. However, other parameters such as polygonal angles, surface curvatures and polygonal areas can also be optimized using our entropy optimization method. Thus, we are planning to apply various measures for the optimization of vertex locations and normals. Research is under way and the results will be reported in the future.

Acknowledgement

We would like to express our great gratitude to the following websites that provide free access to the images and data used in this chapter:

http://sampl.ece.ohio-state.edu/database.htm for bird, lobster, teletubby and dinosaur,

http://www.mpi-sb.mpg.de/shin/Research/PubPolyData/PubPolyData.html for squirrel and lion-dog, and

http://fastscan3d.com/download/samples/vrml/ for head, foot and sculpture.

References

[1] N. Amenta, M. Bern, and D. Eppstein. Optimal point placement for mesh smoothing. *Proc. ACM-SIAM Symp. on Discrete Algorithms*, 1997, pp. 528-537.

[2] R.E. Bank and R.K. Smith. Mesh smoothing using a posteriori error estimates. *SIAM J. Numerical Analysis* **34**(1997) 979-997.

[3] C. Beck. Application of generalized thermostatistics to fully developed turbulence. *Physica A* **277**(2000) 115-123.

[4] F.J. Bossen and P.S. Heckbert. A pliant method for anisotropic mesh generation. *Proc. 5th IMR*, 1996, pp. 63-76.

[5] S.A. Canann, J.R. Tristano, and M.L. Staten. An approach to combined laplacian and optimization-based smoothing for triangular quadrilateral and quad-dominant meshes. *Proc. 7th IMR*, 1998, pp. 479-494.

[6] C.Y. Chen, K.Y. Cheng. A sharpness dependent filter for mesh smoothing. *Computer Aided Geometric Design*, **22**(2005) 376-391.

[7] Z. Chen, J.R. Tristano, W. Kwok. Combined Laplacian and optimization-based smoothing for quadratic mixed surface meshes. *Proc. IMR*, 2003, pp. 847-857.

[8] H.L. de Cougny, M.S. Shephard and M.K. Georges. Explicit node point smoothing within the octree mesh generator. Troy-NY, Report 10-1990, *SCOREC*, Rensselaer Polytechnic Institute.

[9] B. Curless and M. Levoy. A volumetric method for building complex models from range images. *Proc. SIGGRAPH*, 1996, pp. 303-312.

[10] C. Dorai, G. Wang. Registration and integration of multiple object views for 3D model construction. *IEEE Trans. PAMI* **20**(1998) 83-89.

[11] S. Fleishman, I. Drori, D. Cohen. Bilateral mesh denoising. *Proc. ACM SIGGRAPH*, 2003, pp. 950-953.

[12] L. Freitag. On combining Laplacian and optimization-based mesh smoothing techniques. *AMD Trends in Unstructured Mesh Generation, ASME* **220**(1997) 37-43.

[13] L. Freitag and P. Plassmann. Local optimization-based simplicial mesh untangling and improvement. *Internat. J. Numer. Methods Engrg* **49**(2000) 109-125.

[14] L. Freitag, P. Knupp. Tetrahedral mesh improvement via optimization of the element condition number. *Internat. J. Numer. Mehtods Engrg* **53**(2002) 1377-1391.

[15] A.N. Gorban. Family of additive entropy functions out of thermodynamic limit. *Physical Review E* **67**(2003) 016104.

[16] L.R. Herrman. Laplacian-isoparametric grid generation scheme. *J. Engineering Mechanics* **102**(1976) 749-756.

[17] K. Hildebrandt, K. Polthier. Anisotropic filtering of non-linear surface features. *Proc. EUROGRAPHICS*, 2004, pp. 391-400.

[18] A. Hilton, A.J. Stoddart, J. Illingworth and T. Windeatt. Implicit surface-based geometric fusion. *CVIU* **69**(1998) 273-291.

[19] E.T. Jaynes. Information theory and statistical mechanics. *The Physical Review* **106**(1957) 620-630.

[20] A. Johnson and M. Hebert. Using spin images for efficient object recognition in cluttered 3D scenes. *IEEE Trans. PAMI* **21**(1999) 433-449.

[21] T.R. Jones, F. Durand, M. Desbrun. Non-iterative, feature-preserving mesh smoothing. *Proc. SIGGRAPH*, 2003, pp. 943-949.

[22] C.K. Lee, S.H. Lo. A new scheme for the generation of a graded quadrilateral mesh. *Computers and Structures* **52**(1994) 847-857.

[23] Y. Liu, L. Li, and B. Wei. 3D shape matching using collinearity constraint. *Proc. ICRA*, 2004, pp. 2285-2290.

[24] R. Lohner, K. Morgan, O.C. Zienkiewicz. Adaptive grid refinement for the compressible Euler equations. *Accuracy Estimates and Adaptive refinements in Finite Element Computations*, John Wiley and Sons Ltd, 1986, pp. 281-297.

[25] A.A. Mezentsev. A Generalized Graph-Theoretic Mesh Optimization Model. *Proc. IMR*, 2004, pp. 255-264.

[26] A. Nuchter, H. Surmann, S. Thrun. 6D SLAM with an application in autonomous mine mapping. *Proc. ICRA*, 2004, pp. 1998-2003.

[27] K. Polthier. Polyhedral surfaces of constant mean curvature. *Habilitationsschrift Technische University Berlin*, 2002, pp. 1-212.

[28] P.S. Rodrigues, G.A. Giraldi, A. de A. Araujo. Using Tsallis entropy into a Bayesian network for CBIR. *Proc. ICIP*, 2005, pp. 1028-1031.

[29] Y. Shen, K.E. Barner. Fuzzy vector median-based surface smoothing. *IEEE Trans. Viz. Comp. Graphics* **10**(2004) 252-265.

[30] K. Shimada, A. Yamada and T. Itoh. Anisotropic triangular meshing of parametric surfaces via close packing of ellipsoidal bubbles. *Proc. 6th IMR*, 1997, pp. 375-390.

[31] Y. Sun, C. Dumont. Mesh-based integration of range and color images. *Proc. SPIE*, vol. 4051, 2000, pp. 110-117.

[32] G. Taubin. A signal Processing approach for fair surface design. *Proc. SIGGRAPH*, 1995, pp. 351-358.

[33] G. Taubin. Geometric signal processing on polygonal meshes. *Proc. Eurographics: State of the Art Report (STAR)*, Interlaken, Switzerland, 2000.

[34] C. Tsallis. Possible generalization of Boltzmann-Gibbs statistics. *J. Statistical Physics* **52**(1988) 479-487.

[35] C. Tsallis, J.C. Anjos, E.P. Borges. Fluxes of cosmic rays: a delicately balanced stationary state. *Physics Letters A* **310**(2003) 372-376.

[36] M.P. Wachowiak, R. Smolikova, and T.M. Peters. Multiresolution biomedical image registration using generalised information measures. *Proc. MICCAI*, 2003, *LNCS* **2879**, pp. 846-853.

[37] H. Zhou, Y. Liu, and L. Li. Incremental mesh-based integration of registered range images: robust to registration error and scanning noise. *Proc. ACCV, LNCS* **3851**, 2006, pp. 958-968.

[38] H. Zhou, Y. Liu. Flag guided integration of multiple registered range images. *Proc. ICPR*, 2006, pp. 17-20.

[39] T. Zhou, K. Shimada. An angle-based approach to two-dimenshinal mesh smoothing. *Proc. IMR*, 2000, pp.373-384.

Chapter 7

AN EMBEDDED REAL-TIME CONTROL ARCHITECTURE FOR UNMANNED MARINE VEHICLES

Gabriele Bruzzone[*] *and Massimo Caccia*[†]
Consiglio Nazionale delle Ricerche
Istituto di Studi sui Sistemi Intelligenti per l'Automazione
Via De Marini, 6 - 16149 Genova, Italy

Abstract

An innovative embedded real-time control architecture has been designed, implemented and applied to the field of marine robotics. Based on recent enhancements both in computing power of commercially available off-the-shelf (COTS) boards and in performance of free software, the proposed platform is able to manage the different scheduling requirements of the modules constituting advanced intelligent control architectures. The architecture is being satisfactorily used for developing the family of marine robotic vehicles of CNR-ISSIA. In particular, discussion focuses on the application case of the Charlie unmanned surface vehicle (USV) presenting experimental results in typical operating conditions.

1. Introduction

In the last years the increasing growth of computer computational power and flash memory card capacity, as well as the ever-increasing enhancement and diffusion of the open source GNU/Linux [1][2] operating system, have allowed a large number of applications to run satisfactorily on systems consisting of standard GNU/Linux and commercially available off-the-shelf (COTS) hardware with no need of special purpose, and often expensive, hardware and proprietary real-time operating systems (OSs) such as VxWorks [3], LynxOS [4], QNX [5], etc.
In particular, the benefits of a COTS hardware and free software based platform in terms

---
[*]E-mail address: gabriele.bruzzone@ge.issia.cnr.it
[†]E-mail address: max@ge.issia.cnr.it, Phone: +39-010-6475-657/612, Fax: +39-0106475600

of portability, flexibility, modularity, and availability of device drivers, has awoke a great interest in the field of robotic research applications. An example is given by the adoption of GNU/Linux as operating system for the last generation of Remotely Operated Vehicles (ROVs) developed by the Woods Hole Oceanographic Institution and John Hopkins University [6]. In this context, this chapter presents a careful evaluation of how a native Unix-like kernel such as Linux can be made real-time, of its advantages and limitations, and of its timing performance, in order to help robotics and manufacturing application developers to choose the *right* embedded real-time operating system, in terms of reliability, performance and costs. An application to the family of unmanned marine vehicles (UMVs) developed by CNR-ISSIA is then presented, focusing on the implementation of the intelligent control architecture of the Charlie unmanned surface vehicle (USV).

The chapter is organised as it follows. Section 2. discusses the hardware and software state-of-the-art in the field of embedded real-time control systems (section 2.1.), explaining in detail the proposed embedded real-time infrastructure (section 2.2.). The specific application to the field of marine robotics is dealt with in section 3.. After a brief overview of unmanned marine vehicles (section 3.1.1.) and recent advances in control architectures (section 3.1.2.), Charlie, the prototype USV used for the tests, is described (section 3.3.), as well as system integration (section 3.3.2.) and experimental results (section 3.3.3.).

2. Embedded Real-Time Systems

2.1. State-of-the-Art

In spite of its historical development as an operating system for the desktop/server environment, GNU/Linux has become quite an attractive choice in the field of embedded OSs [7]. As a matter of fact, thanks to its basic features of reliability, scalability, portability, open source specifications and standards, support and large programmer base, and free of charge availability [8], GNU/Linux provides rapid application development with a reduced time to market. Its main limitation is that, being Linux a Unix-like kernel designed to guarantee a fair sharing of resources among many users and applications, it does not satisfy real-time performance requirements, which are fundamental in the embedded operating system world.

In order to obtain real-time capabilities, many solutions have been proposed and are currently available both commercial and free, basically following two different approaches:

1. introducing a new software layer (essentially a real-time kernel called micro/nano kernel) between Linux, which runs as a low priority process of the new real-time kernel, and the hardware, as proposed by RTLinux [9], RTAI [10] and more recently by ADEOS [11] and Xenomai [12];

2. applying a set of patches to a standard Linux to make it a real-time kernel, as initially proposed by the KURT project [13] and TymeSys Linux [14], and become a standard configuration option for Linux starting from release 2.5.4-pre6.

The result is that GNU/Linux has become a practical option in robotics research and industry, as demonstrated by projects as MCA (Modular Controller Architecture), where a modular, network transparent and real-time capable C/C++ framework for controlling robots and other kind of hardware was developed [15], COMEDI (COntrol and MEasurement Device Interface), where open-source drivers, tools, and libraries for data acquisition are provided [16], and OROCOS (Open RObot COntrol Software), whose aim is to develop a general-purpose and open robot control software package [17]. Moreover, as discussed for example in [18], the trend of next generation manufacturing and factory automation systems is based on building flexible open systems using open source OSs and tools as those provided by GNU/Linux, that can easily, quickly and cost-effectively be upgraded or expanded to meet the ever-changing production requirements. OSACA (Open System Architecture for Controls within Automation systems) [19], OMAC (Open Modular Architecture Control) [20], OSEC (Japan's Open System Environment Consortium) [21] and OSADL (Open Source Automation Development Lab) [22] are a few examples of efforts in this direction. For up to date lists of industrial automation and robotics projects carried out using GNU/Linux the reader can refer to the Real Time Linux Foundation web site [23].

On the other hand, in these last years the dramatic increase in hardware performance allowed the development of systems with very strict timing requirements using standard GNU/Linux running on COTS hardware. This is, for instance, the case of the data acquisition system for nuclear physics developed at the National Super-conducting Cyclotron Laboratory of the Michigan State University [24], where a precision of the order of a few microseconds was obtained by using the readout computer as an instrument rather than a general-purpose computer and utilizing a few amortization techniques of software and systems overheads. In addition, an off-the-shelf GNU/Linux distribution, without the help of particular software optimization techniques, was sufficient to guarantee accurate updates of the 10-Hertz controller of the tele-operated underwater robotic vehicle Jason II developed by the Woods Hole Oceanographic Institute [6].

The research presented in the following is situated in this last trend, i.e. trying to utilize GNU/Linux running on COTS hardware to implement an innovative embedded real-time control architecture for UMVs.

2.2. System Infrastructure

2.2.1. Linux and Real-Time

Nowadays Linux has proven to be a very stable and high-quality kernel for several application types. However, due to the fact that GNU/Linux was originally designed as a Unix-like operating system, its kernel does not guarantee real-time performances, i.e. a worst case interrupt response time and a deterministic execution time. In other words, a real-time operating system must be able to quickly preempt any task that is currently executing when an interrupt occurs. Linux is not able to do that basically because of the so-called *scheduler latency* problem, i.e. the fact that the delay between the occurrence of an interrupt and the running of the thread that is in charge of serve it can be, in particular critical situations, very long (of the order of tens of milliseconds). This problem is essentially caused by the no preemptibility of threads when running kernel code and by the presence of long critical code sections in the kernel that can not be interrupted. In the last years two different approaches have been proposed to solve the scheduler latency problem and providing Linux with real-time features.

The former solution, initially proposed and developed by RTLinux [25] at the New Mexico Institute of Technology in 1996, consists in inserting a highly efficient code layer called a micro-kernel, between the hardware and Linux. The micro-kernel takes care of all the real-time functionalities including interrupts, scheduling, and high-resolution timing, and the standard GNU/Linux kernel is run as a background task. A similar solution was adopted by RTAI (Real Time Application Interface, also known as Linux-RT) [10] an open-source project started in 2000 at Dipartimento di Ingegneria Aerospaziale - Politecnico di Milano. More recently, in 2002, a different code layer, called nano-kernel, was proposed in the framework of the Adaptive Domain Environment for Operating Systems (ADEOS) project [11]. ADEOS provides a hardware abstraction layer allowing a real-time kernel and a general-purpose kernel to co-exist. The role of this layer is to channel hardware interrupts to the operating system kernels at the next higher layer in the architecture. These operating systems could be real-time or non real-time. The ADEOS system does not implement a real-time operating system, but provides the mechanism by which interrupts can be passed to a real-time operating system at the highest possible priority. The same interrupts may be passed to a non real-time operating system once the real-time task has been completed. At the moment, the RTAI developers are working towards replacing their micro-kernel architecture with the ADEOS approach.

The latter solution to the scheduler latency problem was proposed by the Kansas University's Real-Time (KURT) Linux project in 1997 [13] and by TimeSys with the TymeSys Linux in 1998 [14]. This approach consists in providing a set of kernel patches in order to implement the Posix 1003.1d [26] real-time extensions (high resolution timers, preemptible kernel, improved task scheduler, etc.) directly within the structure of a standard Linux. Similar patches, i.e. the so-called preemption [27] and low-latency [28][29] patches, put together in a single patch and available as a standard kernel configuration option (CONFIG_PREEMPT) beginning from Linux release 2.5.4-pre6, are based on the idea of creating opportunities for the kernel scheduler to be run more often minimizing the time between the occurrence of an event and the running of the scheduler. In particular, the preemption patch modifies the spinlock macros and the interrupt return code so that, if it is safe to preempt the current process and a rescheduling request is pending, the scheduler is called. On the other hand, the low-latency patch introduces explicit preemption points in kernel code blocks that may execute for long stretches of time: the idea is to find places that iterate over large data structures and figure out how to safely introduce a call to the scheduler if the loop has gone over a certain threshold and a scheduling pass is needed. A standard kernel compiled enabling the CONFIG_PREEMPT patch presents a maximum latency and jitter of the order of a few tens of milliseconds also when particularly demanding activities, as discussed in Section 2.2.2., are executed [30]. These performances satisfy the requirements of typical soft real-time applications and, as discussed in the following, of many hard real-time embedded systems in the field of robotics and industrial automation. Currently, Ingo Molnar and Thomas Gleixner are working to a new real-time preemption patch (CONFIG_RT_PREEMPT option) [31] and to a generic clock event layer with high resolution support [32]. Thanks to this two new features Linux should gain actual hard real-time capabilities and be reliably used for developing even more demanding robotics and industrial automation applications.

2.2.2. Standard Linux for Real-Time: Limitations and Advantages

The advantages of using a pure Linux for application development over the dual Linux/real-time kernel method can be summarized by the fact that there is *no need of writing kernel code*. Indeed, because the threads of the real-time layer cannot directly access Linux services, for the same reasons that prohibit Linux processes from behaving deterministically, the developer has to design its application divided in two separate, although collaborating, worlds: "pure" real-time threads with no access to Linux services which are loaded as kernel modules, and non-real-time threads that can have full access to Linux system calls. Communications resources such as shared memory or FIFOs are needed to allow interchanges between these two worlds. Although it is certainly possible to separate real-time and non-real-time kernel instances, doing so increases application complexity, introducing dangers to the system integrity related to the running of user written software as kernel modules. Using the patched kernel solution it is possible to run real-time applications in user space, allowing their coding and start-up like a normal Unix process (which is much easier than inserting modules into the kernel) and avoiding any problem to overall system stability in the case of crashes due to programming bugs. In addition, using a pure GNU/Linux environment allows the user to use a unique set of tools for debugging and performance analysis, without any need of focusing on either the real-time parts or on the GNU/Linux-based application components at any given time. Anyway, as mentioned before, to obtain an actual real-time behavior by activating the CONFIG_PREEMPT option in the standard kernel the programmer has to avoid certain activities that can cause unacceptable system latencies triggering long non-preemptible paths that still remains in the kernel. A comprehensive list of these activities is reported in [30] and shortly discussed in the following just to show how they are usually not used in typical real-time embedded system programming applications:

- Memory stress: accessing large amounts of memory can force the kernel to invoke the page fault handler that contains long non-preemptible sections of code. If the memory size is sufficiently large (not too a demanding requirement nowadays) so that the overall embedded system (file system, operating system threads and user applications) could be completely loaded in memory, page faults can never happen;

- Caps-lock and console switch stress: the keyboard and console drivers contain long non-preemptible paths. Most embedded control systems are headless, i.e. they have no keyboard and monitor. In the case a user interaction with keyboard and monitor is required, a connection from a remote computer by means of a serial line or network and a tool as ssh (Secure SHell) can be used;

- I/O stress: when the kernel or drivers have to transfer large data chunks, they usually move them inside non-preemptible code sections. Hence, system calls that move large amounts of data can cause long latencies. This issue can be addressed by using either ad hoc written drivers or, when possible, direct memory-mapped or port-mapped I/O in user space bypassing kernel drivers;

- Procfs stress: the /proc file system is a pseudo file system used by Linux to exchange data between kernel and user applications. Concurrent accesses to the shared data structures must be protected by non-preemptible sections. If the size of the exchanged data is rather large, accessing to procfs can significantly increase the latency. Most embedded applications don't need the use of the /proc file system.

- Fork stress: creating processes can generate high latencies due to the fact that the new process is created inside a non-preemptible section and involves copying large amounts of data. In the application, all the necessary process creations can be executed at the start-up, before entering the main control loop. If a thread needs a delayed start, it can wait for a synchronization semaphore.

2.2.3. Embedding the System

One of the most important requirements of a typical real-time control system is the possibility of embedding it in an easy way in order to increase robustness, reliability and reducing costs. In the last years, the dramatic improvements in size and performance of solid-state memories, such as compact flash (CF) memory cards and RAM, have made less critical the construction of customized embedded GNU/Linux systems. BusyBox [33], a software package which provides a fairly complete POSIX environment, makes very easy to customize and build embedded GNU/Linux systems. By using this tool, a minimal GNU/Linux system can be quickly and simply built by adding the device nodes, configuration files, and the kernel the user is interested in, in suitable directories, i.e. /dev, /etc and /boot. The created file system with the desired characteristics (e.g. having remote network access by means of ssh and sftp, a web server running in background, etc.), containing the daemon and tool executables together with the related libraries, configuration files and device nodes, is then copied in a suitable single image file, which, once compressed, can be transferred in a bootable CF disk mounted on the CPU board and seen by the kernel as a bootable IDE disk. In particular, the CF disk can be divided in two partitions: the former, read-only, containing the compressed file image, the kernel and a few device nodes necessary for the booting process, the latter, read-write, containing the user application (put in this part so that it could be easily updated) and that can also be used to memorize log files. By using the LInux LOader (LILO) [34] utility it is possible to instruct a CPU board without hard disks to boot a GNU/Linux system directly from a CF disk. In order to avoid any latency due to slow accesses to hard or CF disks, during the start-up sequence the compressed file image is uncompressed and mounted in RAM, whose remaining part is utilized as system memory for running OS tools and applications. This particular type of solution, being the RAM much faster and less subject to failures than a hard disk, is remarkably interesting both from the reliability and the speed point of view.

On the basis of the above considerations in embedding and making GNU/Linux real-time, a platform was developed for supporting embedded real-time control systems. In particular, an approach relying on embedding standard GNU/Linux, as shown in the previous section, programming avoiding the above-mentioned activities causing system latencies, and using a suitable timer to generate interrupt signals, was implemented. In the following a detailed discussion of the reasons of the choice of particular hardware and software tools, together with the presentation of a set of C++ classes designed in order to simplify and standardize the development of control system applications, will be given.

2.2.4. Hardware

The selected hardware components are standard industrial PC-derived computers and relevant I/O boards. In particular, CPU boards supplied with both a PC/104+ and a PCI bus (i.e. supporting I/O cards for both bus types) are used. These hardware components, which have proved in the last decades to be a consolidated and mature standard, give the developer the opportunity of finding, at a low cost, a high number of I/O boards and CPUs with very high computing power. Moreover, within the chipset of PC-derived computers is usually present a real-time clock (RTC) that, suitably programmed, can be used as a good time basis for the control application. It is worthwhile noticing that by using PC-compatible industrial hardware it is also possible to perform a large part of the control system development and debugging on any common desktop or laptop computer running GNU/Linux (of course working at lower frequencies and without using I/O or utilizing a simulated one) before proceeding with tests on the actual system in lab or at field. Actually, thanks to the extreme GNU/Linux portability and flexibility, the same program developed on a desktop/laptop can be moved and executed on a target appliance without any modification.

2.2.5. Software

- **Real Time Clock (RTC) as a time basis for real-time** An indispensable feature for an embedded real-time control system is the presence of at least one high precision timer able to generate events used to timing the system. Indeed, since Linux native timing resolution is rather poor (usually 10 ms on the current versions), high frequency interrupt signals can be generated by using the RTC, a particular clock usually built into the chipset of all PCs and PC-derived computers. The main function of the RTC is to keep the date and time while the computer is turned off. However it can also be programmed to generate interrupt signals from a slow 2 Hz frequency to a relatively fast 8192 Hz one, in increments of powers of two. From the programmer point of view, the RTC is seen as a particular read only character device (/dev/rtc), on which a user thread that need to be synchronized can execute a blocking read. When the RTC generates the interrupt signal the read call unblocks returning the number of interrupts received since the last read. At high frequencies, or under high loads, the user thread must check this number to determine if there has been any interrupt "pileup". The interrupt frequency, starting and stopping of the RTC can be easily programmed into the RTC via various ioctl calls.

- **POSIX threads** Real-time data acquisition and control modules are implemented as POSIX threads (in the following called *pthreads* for short). GNU/Linux provides two different libraries to manage threads which give the user a set of primitives for creating, running, stopping and resuming pthreads, synchronizing their operations, and specifying their scheduling policies and priorities: LinuxThreads [35] and NPTL [36]. LinuxThreads was the first library implementing POSIX threads available for Linux. Unfortunately, LinuxThreads had a number of issues with true POSIX compliance, particularly in the areas of signal handling, scheduling, and interprocess synchronization primitives. Nowadays LinuxThreads has been superseded by NPTL, which implements a more complete and efficient implementation of POSIX threads for Linux. Anyway, using either LinuxThreads or NPTL, there are three possible types of scheduling policies for pthreads: SCHED_FIFO, SCHED_RR and SCHED_OTHER and the priorities range from 0 (lowest) to 99 (highest). Linux manages pthreads having SCHED_FIFO scheduling policy in a fully preemptive and priority-based way, guaranteeing that at any given instant the highest priority ready SCHED_FIFO pthread is the one that is executed by the CPU. SCHED_RR pthreads are treated by the scheduler in a similar way but with a round-robin algorithm, and are rarely used. Actually, normal Linux threads are simply pthreads having priority 0 (the lowest) and SCHED_OTHER scheduling policy, i.e. a Unix-like time-sharing one. Being the priority of SCHED_FIFO pthreads always greater than 0 they never can be preempted by a Linux thread. Thus it is possible to implement real-time applications delegating time-critical tasks that cannot be interrupted to pthreads having SCHED_FIFO scheduling policy (referred in the following as real-time pthreads), while $SCHED\_OTHER$ Linux threads can be active only when there are no real-time pthreads running or ready to run.

- **Custom real-time scheduler and thread, data and communication management**
In order to simplify and standardize the development of control system applications, a set of C++ classes encapsulating real-time and Linux threads, a custom scheduler devoted to manage their timing requests and message queues for inter-thread communications were implemented using LinuxThreads library primitives. The taxonomy of the thread classes in shown in Figure 1, where the abstract base class (ABC) Thread embodies the main characteristics of a generic thread and acts as a base class on which other classes can be built.

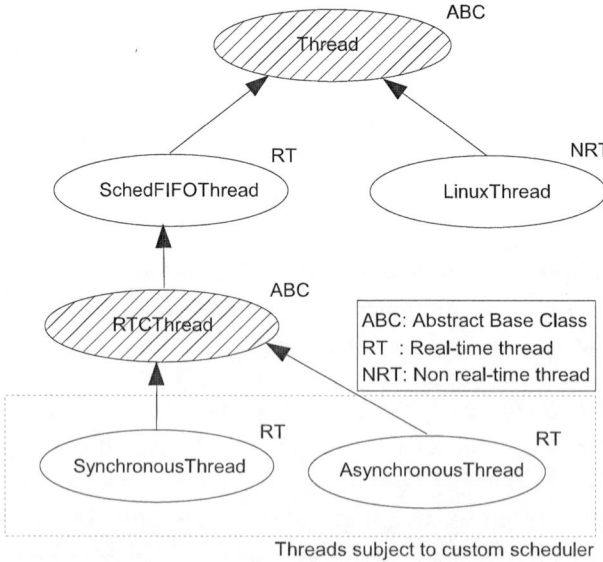

Figure 1. Thread classes taxonomy.

The Thread class derives SchedFIFOThread and LinuxThread classes, representing threads with scheduling policy and priority ($SCHED\_FIFO$, $1\ldots99$) and

($SCHED\_OTHER$, 0) respectively, i.e. real-time pthreads and normal Linux threads. It is worthwhile noticing that the class SchedFIFOThread implements a generic pthread that is usually waiting for a specific event. The main characteristics of a generic real-time pthread subject to the custom scheduler management are embodied by the abstract base class RTCThread, derived from SchedFIFOThread class. RTCThread class derives the SynchronousThread and AsynchronousThread classes representing respectively real-time pthreads that need to be periodically scheduled at a requested frequency and real-time pthreads that need to be awaken after a desired time interval. The custom scheduler is simply a real-time pthread having the highest available priority that is cyclically activated by an interrupt generated by the RTC at a prefixed frequency and whose aim is to manage requests from AsynchronousThreads and SynchronousThreads. A few other classes were designed and implemented to provide in an easy way inter-thread and network communications.

3. Application: Marine Robotics

3.1. Overview

3.1.1. Unmanned Marine Vehicles

Marine robotics has been traditionally characterised by the development of unmanned underwater vehicles, both tethered Remotely Operated Vehicles (ROVs) when video feedback and strict interactions with the operating environment are required and untethered autonomous underwater vehicles (AUVs) when the coverage of large areas is needed. For an overview of the principal ROVs and AUVs prototypes the reader can refer to [37].

More recently, attention focused on the design and development of unmanned surface vehicles (USVs) too in spite of the strong limitations on their operational exploitation due to legal and liability issues [38]. The main applications of USVs have been bathymetric surveys, environmental monitoring and sampling, coastal defense, and supporting of Autonomous Underwater Vehicle (AUV) operations. An example is given by the development of a family of USV at the MIT AUV Lab [39] since the early 1990's. ARTEMIS, a small scale replica of a fishing trawler, used for generating bathymetric maps by executing DGPS way-point-defined surveys supervised by a human operator through a radio modem, was followed by the catamarans for automated bathymetry ACES (Autonomous Coastal Exploration System) and its evolution AutoCat, based on the commercially available Float CatTM hulls. In the last years, basic trials of in-field autonomous operation of unmanned marine vehicles in accordance with conventions for safe and proper collision avoidance as prescribed by the Coast Guard Collision Regulations have been executed by using a couple of SCOUT (Surface Craft for Oceanographic and Undersea Testing) autonomous kayak test platforms developed by MIT [40]. Meanwhile the Lisbon IST-ISR Dynamical Systems and Ocean Robotics Laboratory is carrying on the development of a flotilla of autonomous marine vehicles such as Delfim, a small autonomous catamaran for testing the concept of an autonomous surface craft capable of working in close cooperation with an autonomous underwater vehicle [41], and Caravela [42], an autonomous robotic research vessel capable of performing a large number of oceanographic missions without support from a permanent, dedicated crew. In Europe again, the University of Rostock developed Measuring Dolphin for carrying measuring devices in the field of oceanography, water ecology and hydrology [43], while the University of Plymouth developed Springer, an autonomous catamaran designed to sense, monitor and track water pollution [44]. A detailed presentation of the Charlie USV, developed by CNR-ISSIA in Italy, will be given in the following.

As far as military applications are concerned, the main operational advance has been given by the QinetiQ Ltd shallow water influence mine sweeping system (SWIMS) [45], employed by the Royal Navy to support mine countermeasures operations in Iraq in 2003. SWIMS basically consisted in the development of a conversion kit to transform existing Combat Support Boats, already operated by the British Army, in remote controlled vessels. Other military USVs such as the testbed developed at SSC San Diego [46], based on the Bombardier SeaDoo Challenger 2000, and the Israeli Protector USV [47] focus research and development efforts in the integration of sensors for over-the-water obstacle detection and avoidance as, for instance, radar, electro-optic sensors and machine vision technologies.

3.1.2. Control Architectures: Recent Advances

During the Nineties also the field of marine robotics reflected the competition between behaviour-oriented layered control and hierarchical intelligent control [48][49] architectures. The result were very interesting control architectures able to support mission management and control systems too such as the control system developed by the Lisbon IST-ISR for its family of AUVs and USVs, including a Petri net based mission management system, originally designed for the MARIUS AUV [50], and the MOOS control architecture, which includes a mission control system named Helm, designed and exploited for the MIT unmanned marine vehicles [51].

In any case, a clear distinction between the so-called, using the classical intelligent control nomenclature, Execution, Coordination and Organisation levels, allowing the definition of well-established interfaces between the synchronous, continuous time navigation, guidance and control modules (Execution level), and the asynchronous, discrete event systems managing mission coordination and control, made possible the integration of architectural levels developed by different research teams. This was, for instance, the case of the mission control of the NPS Phoenix AUV [52] and CNR-IAN Romeo ROV [53], both running their own original execution level, performed by the IST-ISR Petri net based mission controller. In the new century, following the experience of the CNRS-LAAS in autonomous mobile robots [54], the Execution Controller, i.e. a layer devoted to guarantee the consistency of the execution level data in the presence of task conflicts and dependencies, bridging the gap between the world of control systems and the one of artificial intelligence, has been introduced by CNR-ISSIA in the control architecture of its UMVs, i.e. the Romeo ROV [55] and Charlie USV.

In particular, this research pointed out that basic relationships, i.e. mutual exclusion and dependency, between navigation, guidance and control (NGC) tasks are embedded by the topological structure of the execution level, i.e. by the graph representing I/O relationships between tasks and variables [56]. Thus, once defined a set of basic rules stating data consistency, *from bottom upward activation* of NGC tasks, and the hierarchical structure of the control architecture, it is possible to automatically generate a couple of Petri nets modelling

and controlling the execution of guidance and control tasks and motion estimation tasks respectively [57].
According to this scheme the Execution Level synchronously executes the navigation, guidance and control tasks according to their current state, evaluating and communicating to the Execution Controller the result of estimator initializations. In addition, the Execution level receives reference variable and algorithm parameter values from the user and performs memory mapped I/O, e.g. analog and digital I/O, and send commands and collects data from asynchronous instrument drivers. On the other hand, the Execution Controller manages the activation and deactivation of NGC tasks according to the basic rules defining the control architecture structure and guaranteeing data consistency.

3.2. Requirements

According to the research trends discussed above, the basic requirements for an embedded real-time system able to support the control architecture of unmanned marine vehicles have been defined.
From the point of view of real-time performances, the time constants typical of marine vehicles do not require high performances in terms of sampling rates: 10 Hz is usually sufficient as, for instance, in the case of the Jason II ROV, where a conventional Linux distribution has been adopted for supporting the implementation of the robot control system [6]. Thus, the requirement of a system able to support scheduling times of the order of 1 KHz has been introduced for keeping the system open to future improvements mainly in the fields of manipulation and image processing.
As far as the structure of the control architecture is concerned, the software platform has to support both real-time synchronous threads, e.g. execution level, and device drivers handling analog, digital, serial and Ethernet I/O, and non real-time modules executing potentially unbounded search algorithms, e.g. execution control level. Moreover, the need of integrating new synchronous modules for system monitoring and event handling, and asynchronous ones for path-planning, mission control and supervision has to be considered.

3.3. Testbed: Charlie USV

The proposed GNU/Linux based embedded real-time architecture has been originally integrated with the control system of the Charlie USV. This vehicle was originally designed and developed by CNR-ISSIA or sampling sea surface microlayer using a Harvey-like [58] cylinder and collecting data on the air-sea interface in Antarctica [59], and then exploited in preliminary studies for anti-intrusion operations in protected waters.
The vessel, with a catamaran-like shape, is 2.40 m long, 1.70 m wide and weighs about 300 Kg in air (see Figure 2).

3.3.1. Electro-mechanical Architecture

The Charlie USV is propelled by two DC thrusters whose revolution rate is controlled by a couple of servo-amplifiers, closing a hardware thruster revolution rate control loop with time constant negligible with respect to the system. With respect to the original model,

Figure 2. Charlie USV. The vessel is 2.50 m long, 1.70 m wide, and weighs about 300 Kg in air.

where the steering system was based on the differential revolution rate of the two propellers, the vehicle has been upgraded with a rudder-based steering system, constituted by two rigidly connected rudders, positioned behind the propellers, and actuated by a Faulhaber 3564K024B-K1155 brushless motor with a reduction ratio of 415:1. The vessel navigation package is constituted by a GPS Ashtech GG24C integrated with a KVH Azimuth Gyrotrac, able to compute the True North given the measured Magnetic North and the GPS-supplied geographic coordinates. Electrical power supply is provided by four 12 V @ 40 Ah lead batteries integrated with four 32 W triple junction flexible solar panels. A Breezecom wireless LAN link at 2 Mbps connects the vehicle to an operator station.

Its control system is implemented on an Advantech PCM-9574 SBC (Single Board Computer), equipped with an Intel Pentium III at 750 MHz CPU with 512 MB of RAM, 4 RS-232 serial ports, 1 CF card slot and Ethernet at 100 Mbps, and connected to three PC-104 modules (Advantech PCM-3724, Advantech PCM-3718HG and Diamond Ruby-MM-1612) performing digital I/O, analog input and analog output respectively.

3.3.2. Control Architecture and GNU/Linux Real-Time System Integration

The software architecture of the Charlie2005 control system, shown in Figure 3, basically consists of the *Execution Level*, running synchronous navigation, guidance and control tasks, the *Execution Controller*, which guarantees the consistency of task dependencies and constraints [56], and a set of drivers, implementing the interfaces with physical I/O devices. Communications with the operator station are performed through a stream socket as far as pilot commands are concerned and a datagram socket for telemetry. In addition a *dispatcher* for addressing user commands to the corresponding modules was introduced. [1]

---

[1]The *dispatcher* thread sends low level actuator commands to the *rudder driver*, reference variables and parameter values to the *execution level*, and task activation/deactivation commands to the *execution controller*.

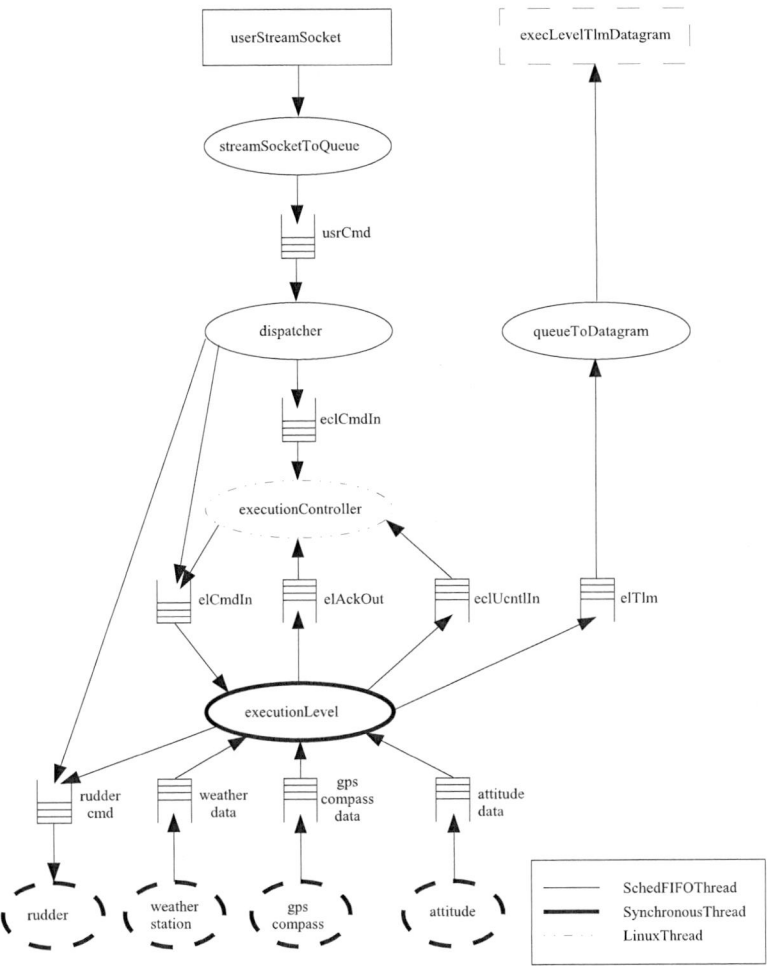

Figure 3. Charlie control system software architecture.

As discussed in Section 2.2.5. SchedFIFOThreads are suitable for implementing real-time threads blocking on asynchronous events. This is the case of the *dispatcher*, as well as the *streamSocketToQueue* and *queueToDatagram* threads, managing data transfers from command stream socket and to telemetry datagram socket respectively. These threads block on reading input sockets or queues. In a similar way, the drivers of the RS-232 serial connected devices, i.e. rudder controller, weather station, GPS and compass, and attitude sensors, are implemented as SchedFIFOThreads encapsulating the management of serial channels. Finally, the *executionLevel* thread is a SynchronousThread running at 8 Hz and the *executionController* thread, implementing a potentially unbounded Petri net search algorithm, is a LinuxThread. The custom scheduler runs at 1024 Hz. The control system modules can be integrated using the proposed RTC-based real-time infrastructure as shown in Figure 3, where a *dispatcher*[2] for addressing user commands to the corresponding module has been

---

[2]The *dispatcher* sends low level actuator commands to the *rudder driver*, reference variable and parameter value to the *execution level*, and task activation/deactivation commands to the *execution controller*.

introduced. SchedFifoThreads are suitable for implementing real-time threads blocking on asynchronous events. This is the case of the *dispatcher*, as well as the *streamSocket2Queue* and *queue2datagram* threads, managing data transfers from command stream socket and to telemetry datagram socket respectively. These tasks blocks on reading input socket or queue. In a similar way, the drivers of the RS-232 serial connected devices, i.e. rudder controller, weather station, GPS and compass, and attitude sensors, are implemented as SchedFifoThread incapsulating the management of serial interrupts. Finally, the Execution Level is a real-time synchronous thread (RTCThread) and the Execution Controller, implementing a potentially unbounded Petri net search algorithm, is a non real-time thread of LinuxThread type. The custom scheduler runs at 1024 Hz, while the execution level is scheduled at 8 Hz.

3.3.3. Experimental Results

Extensive laboratory tests, carried out also on more complex control systems for industrial automation [60], have demonstrated that the system guarantees an error in the scheduling time lower than $250\mu s$ managing Ethernet, RS-232 serial ports, and analog and digital I/O, thus, largely satisfying real-time requirements for low dynamics guidance and control of marine robots, and able, in perspective, to support higher dynamics systems such as electro-mechanical arms mounted on the vehicles. An example of the obtained results is given in Figure 4 left, showing the histogram of the error in the sampling time of the custom scheduler, scheduled at 1024 Hz, during a three hour and a half mission of the Charlie USV.

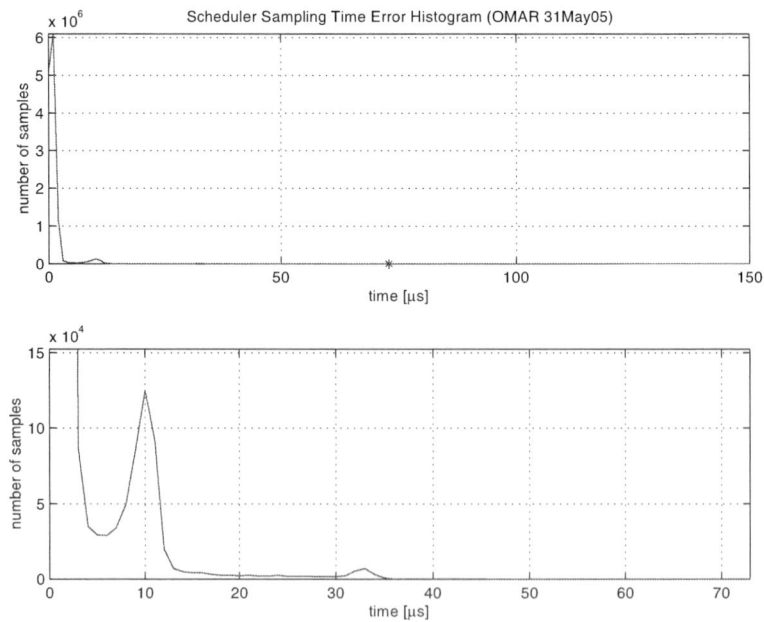

Figure 4. Real-time custom scheduler sampling time error histogram.

4. Conclusion

In this chapter an embedded real-time platform based on COTS hardware and free software has been presented showing its capability in supporting advanced intelligent control system applications as in the case of unmanned marine vehicles. In particular, research demonstrated that using GNU/Linux, the RTC timer and a few C++ classes encapsulating the behavior of usual control application threads it is possible to implement robotics real time applications with timing requirement up to the order of 1 ms in an easy way and with virtually no expense. Current developments, such as the new real-time preemption patch (CONFIG_RT_PREEMPT option) [31] provided by Ingo Molnar and Thomas Gleixner's generic clock event layer with high resolution support [32], indicate that GNU/Linux is more and more gaining hard real-time capabilities, getting an operating system for developing even more demanding robotics and industrial automation applications.

Acknowledgments

Due to its nature, underwater robotics research requires to be performed by a team of research scientists and engineers.
The research presented in this chapter is the result of the work carried out by the Autonomous robotic systems and control group of CNR-ISSIA in Genova, Italy. Thus, a special mention is for the other members of the group, Riccardo Bono, Giorgio Bruzzone and Edoardo Spirandelli, who gave fundamental contributions to the robot hardware and software development, maintenance and operation at sea.
The work presented in this chapter has been partially funded by Regione Liguria - Obiettivo 2 (2000-2006) Sottomisura 1.4B- L. 598/94 art. 11 "Interventi per la ricerca industriale e lo sviluppo precompetitivo" in the project "Embedded real-time platform for industrial automation and robotics" and by PRAI-FESR Regione Liguria prot.5 "Coastal and harbour underwater anti-intrusion system".

References

[1] http://www.gnu.org.

[2] http://www.kernel.org.

[3] http://www.windriver.com.

[4] http://www.lynuxworks.com.

[5] http://www.qnx.com.

[6] L. Whitcomb, J. Howland, D. Smallwood, D. Yoerger, T. Thiel, A new control system for the next generation of US and UK deep submergence oceanographic ROVs, in: *Proc. of the 1st IFAC Workshop Guidance and Control of Underwater Vehicles*, Newport, UK, 2003, pp. 137–142.

[7] N. Mc Guire, Embedded GNU/Linux - drawing the big-picture, in: *Proc. of International Conference on Industrial Technology*, Vol. 2, 2003, pp. 1237–1242.

[8] A. Lennon, Embedding Linux, *IEE Review* **47** (3) (2001) 33–37.

[9] http://www.fsmlabs.com.

[10] http://www.aero.polimi.it/~rtai.

[11] http://home.gna.org/adeos.

[12] http://www.xenomai.org.

[13] http://www.ittc.ku.edu/kurt.

[14] http://www.timesys.com.

[15] http://mca2.sourceforge.net.

[16] http://www.comedi.org.

[17] http://www.orocos.org.

[18] L. Wang, Factory automation systems: evolution and trends, in: *Proc. of AUTOTEST*, 2002, pp. 880–886.

[19] http://www.osaca.org.

[20] http://www.omac.org.

[21] http://www.sml.co.jp/OSEC.

[22] http://www.osadl.org.

[23] http://www.realtimelinuxfoundation.org.

[24] R. Fox, E. Kasten, K. Orji, C. Bolen, C. Maurice, J. Venema, Real-time results without real-time systems, *IEEE Trans. on Nuclear Science* **51** (3) (2004) 571–575.

[25] http://www.fsmlabs.com/rtlinuxfree.html.

[26] http://www.opengroup.org.

[27] http://www.kernel.org/pub/linux/kernel/people/rml.

[28] http://www.zipworld.com.au/~akpm/linux/schedlat.html.

[29] http://people.redhat.com/~mingo.

[30] L. Abeni, A. Goel, C. Krasic, J. Snow, J. Walpole, A measurement-based analysis of the real-time performance of Linux, in: *Proc. of Eight IEEE Real-Time and Embedded Technology and Applications Symposium*, 2002.

[31] http://people.redhat.com/mingo/realtime-preempt.

[32] http://www.tglx.de/projects/hrtimers/ols2006-hrtimers.pdf.

[33] http://www.busybox.net.

[34] http://lilo.go.dyndns.org.

[35] http://pauillac.inria.fr/~xleroy/linuxthreads.

[36] http://people.redhat.com/drepper/nptl-design.pdf.

[37] M. Caccia, Mobile Robots: New Research, Nova Science Publishers Inc., 2005, Ch. *Handling uncertainty in underwater robotics: the example of ROV bottom-following*, pp. 143–176.

[38] *Issues concerning the rules for the operation for autonomous marine vehicles* (AMV) - A consultation paper (7 August 2006), http://sig.sut.org.uk/urg_uris/URG_AMV_paper.pdf.

[39] J. Manley, A. Marsh, W. Cornforth, C. Wiseman, Evolution of the autonomous surface craft AutoCat, in: *Proc. of Oceans'00*, Vol. 1, 2000, pp. 403–408.

[40] M. Benjamin, J. Curcio, COLREGS-based navigation in Unmanned Marine Vehicles, in: *IEEE Proceedings of AUV-2004*, 2004.

[41] A. Pascoal, et al., Robotic ocean vehicles for marine science applications: the european asimov project, in: *Proc. of Oceans 2000*, 2000.

[42] DSORlab, http://dsor.isr.ist.utl.pt/Projects/Caravela/.

[43] J. Majohr, T. Buch, Advances in unmanned marine vehicles, *IEE Control Series*, 2006, Ch. Modelling, simulation and control of an autonomous surface marine vehicle for surveying applications Measuring Dolphin MESSIN, pp. 329–352.

[44] T. Xu, J. Chudley, R. Sutton, Soft computing design of a multi-sensor data fusion system for an unmanned surface vehicle navigation, in: *Proc. of 7th IFAC Conference on Manoeuvring and Control of Marine Craft,* 2006.

[45] S. Cornfield, J. Young, Advances in unmanned marine vehicles, *IEE Control Series,* 2006, Ch. Unmanned surface vehicles - game changing technology for naval operations, pp. 311–328.

[46] J. Ebken, M. Bruch, J. Lum, Applying UGV technologies to unmanned surface vessel's, in: *SPIE Proc. 5804*, Unmanned Ground Vehicle Technology VII, 2005.

[47] Protector - Unmanned Naval Patrol Vehicle, http://www.israeli-weapons.com/weapons/naval/protector/Protector.html.

[48] K. Valavanis, G. Saridis, Intelligent Robotic Systems: Theory, Design and Applications, Kluwer Academic Press, 1992.

[49] P. Antsaklis, K. Passino, Introduction to intelligent control systems with high degree of autonomy, *An introduction to intelligent and autonomous control,* Kluwer Academic Publishers, Boston, MA, 1993.

[50] P. Oliveira, A. Pascoal, V. Silva, C. Silvestre, The mission control system of MARIUS AUV: System design, implementation, and tests at sea, *Int. J. on Sys. Science - Special Issue on Underwater Robotics* **29** (10) (1998) 1065–1080.

[51] P. Newman, MOOS - a mission oriented operating suite, Technical report, Massachussetts Institute of Technology (2002).

[52] A. Healey, D. Marco, R. McGhee, P. Oliveira, A. Pascoal, V. Silva, C. Silvestre, Implementation of a CORAL/Petri net strategic level on the NPS Phoenix vehicle, in: *Proc. of 6th International Advanced Robotics program Workshop on Underwater Robotics*, Toulon, France, 1996.

[53] G. Bruzzone, R. Bono, M. Caccia, G. Veruggio, C. Ferreira, C. Silvestre, P. Oliveira, A. Pascoal, Internet mission control of the Romeo unmanned underwater vehicle using the CORAL mission controller, in: *Proc. of MTS/IEEE Oceans'99*, Vol. 3, Seattle, USA, 1999, pp. 1081–1087.

[54] R. Alami, R. Chatila, S. Fleury, M. Ghallab, F. Ingrand, An architecture for autonomy, *International Journal of Robotic Research* **17** (4) (1998) 315–337.

[55] M. Caccia, R. Bono, G. Bruzzone, G. Veruggio, Unmanned underwater vehicles for scientific applications and robotics research: the ROMEO project, *Marine Technology Society Journal* **24** (2) (2000) 3–17.

[56] M. Caccia, P. Coletta, G. Bruzzone, G. Veruggio, Execution control of robotic tasks: a Petri net-based approach, *Control Engineering Practice* **13** (8) (2005) 959–971.

[57] M. Caccia, G. Bruzzone, Execution control of robotic tasks for marine systems, in: *Proc. of IFAC World Congree 2005*, 2005.

[58] G. Harvey, Microlayer collection from the sea surface. a new method and initial results, *Limnol Oceanogr.* **11** (1966) 608–613.

[59] M. Caccia, R. Bono, G. Bruzzone, G. Bruzzone, E. Spirandelli, G. Veruggio, A. Stortini, G. Capodaglio, Sampling sea surface with SESAMO, *IEEE Robotics and Automation Magazine* **12** (3) (2005) 95–105.

[60] G. Bruzzone, M. Caccia, *Notes on standard GNU/Linux for embedded real-time robotics and manufacturing control systems,* Tech. Rep. SRAC-TR06/01, CNR-ISSIA Genova (2006).

In: Robotics Research Trends
Editor: Xing P. Guô, pp. 267-287
ISBN 1-60021-997-7
© 2008 Nova Science Publishers, Inc.

Chapter 8

A Mobile Haptic Interface for Bimanual Manipulations in Extended Remote/Virtual Environments

Angelika Peer[*], *Thomas Schauß*[†],
Ulrich Unterhinninghofen[‡], *and Martin Buss*[§]
Institute of Automatic Control Engineering
Technische Universität München

Abstract

The concept of a new mobile haptic interface for unconstrained bimanual manipulation is presented. This device, which has been developed at the High-Fidelity Telepresence and Teleaction Research Centre, Munich, Germany, allows locomotion and haptic interaction at the same time. In contrast to most existing haptic interfaces, it is therefore not restricted to desktop applications but also enables bimanual manipulation tasks with high interaction forces in extended remote or virtual environments. The design of this mobile haptic interface is based on a modular system consisting of two components: two haptic interfaces and a mobile platform. While the haptic interfaces only cover parts of the human arm workspace, the mobile platform extends these to arbitrarily large remote environments. A special design and control concept of the haptic interfaces makes it possible to decouple translational from rotational movements. This decoupling helps to significantly simplify the control algorithms which handle the interaction between the single components. The mobile platform which carries the two haptic interfaces must be positioned in such a way that the manipulability of both haptic interfaces is maximized. Different optimization strategies are presented and compared. The motion of the mobile platform must be synchronized with the control of the haptic interfaces in order to hide the platform motion from the operator. Finally, experimental results are presented.

---

[*]E-mail address: angelika.peer@tum.de
[†]E-mail address: thomas.schauss@mytum.de
[‡]E-mail address: ulrich.unterhinninghofen@tum.de
[§]E-mail address: mb@tum.de

1. Introduction

Haptic interfaces can be described as bidirectional human-system-interfaces. On the one hand they provide the operator with force/torque information from virtual or remote environments and on the other hand they are used to read the motion/force input of the operator. In recent years haptic devices have received a lot of attention: They found their way into applications such as medical training, rehabilitation, virtual prototyping, telesurgery, telemaintenance as well as micromanipulation.

However most existing haptic interfaces are limited in their degrees of freedom (d.o.f.), have only a small workspace or/and a low output capability (velocity, acceleration, and force/torque capability). Thus tasks which require 6 d.o.f. manipulations with high interaction forces (high output capability) in extended remote or virtual environments (big workspace) are not possible.

In order to overcome these limitations, a concept for a new mobile haptic interface has been developed. The mobile haptic interface consists of two different main components: two haptic interfaces and a mobile platform. While the haptic interfaces only cover parts of the human arm workspace, the mobile platform extends these to arbitrarily large remote environments.

Possible applications for such a system are e.g. the control of a mobile teleoperator in a telepresence scenario or the haptic exploration in extended virtual environments as e.g. a virtual museum, see Fig. 1.

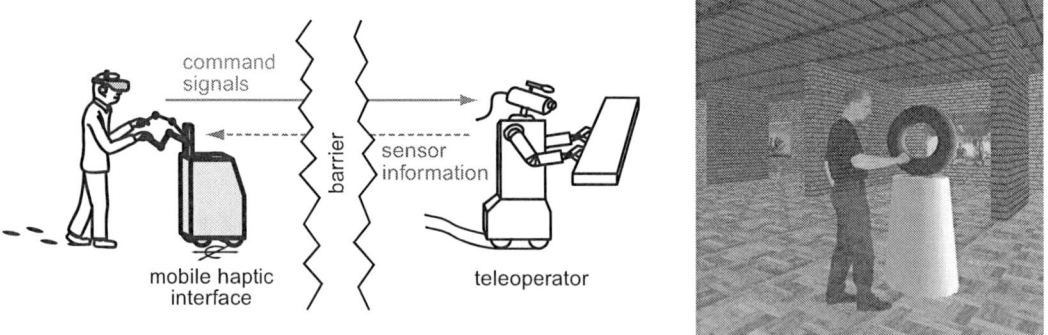

Figure 1. Application scenarios of a mobile haptic interface: teleoperation system (left), exploration of a virtual museum (right).

This paper is structured as follows: The following Section 2. starts with a review of the state of the art in the field of stationary and mobile haptic devices. This section will be followed by a design rationale and description of the new mobile haptic interface. Section 4. presents different control algorithms which have been applied to the new device and Section 5. reports some experimental validation results. Finally, Section 6. concludes the paper and shows directions of future research.

2. State of the Art

In the past years, the interest in the development of haptic devices for telerobotic applications has dramatically increased. A huge number of different kinds of haptic devices has been developed and partly commercialized. But almost all of these devices are stationary devices with quite a small workspace and a moderate force level. In order to increase the workspace of such devices, usually hand controlled input devices such as a joystick or mouse are used [8] or some indexing technique is applied. If control by the operator's hand is not possible, as in the case of bimanual manipulation, these devices can also be substituted by a special kind of foot pedal [2, 14].

Since the operator can not move around, none of these approaches provide a proprioceptive perception of locomotion. As shown in [3], such incomplete or false proprioceptive cues result in a deterioration of the natural orientation and navigation capabilities of a human operator.

More realistic locomotion interfaces such as treadmills and tracking systems for human operator locomotion can be found in the field of virtual reality applications. These systems allow the human operator to freely move around in the remote environment but do not provide any force feedback information. Thus, simultaneous manipulation and locomotion is not possible.

A known approach to circumvent this problem and allow simultaneous manipulation and locomotion is to use body grounded haptic interfaces such as exoskeletons. But as reported in [16], working with exoskeletons is very fatiguing since the range of human arm movements is restricted and/or long time operations are not possible because of the high weight of the system. In addition, mounting application specific end-effectors is extremely difficult.

A much more advanced locomotion interface has been proposed in [10, 11] and later adapted in [4]. They mounted a stationary haptic interface on a mobile platform. Since in this case the weight of the haptic interface is fully supported by the platform, the operator fatigue can be significantly reduced. But these systems allow only one-handed manipulation and their haptic interfaces are limited to either 3 or 4 d.o.f. In [12] the first bimanual mobile haptic interface for haptic grasping in large virtual environments was presented, but again haptic interfaces with only 3 d.o.f. were used.

The research presented here focuses on the development of a bimanual mobile haptic interface with the ability to manipulate in full 6 d.o.f. Since the design is based on two independently controlled modules - mobile platform and haptic interfaces - haptic interfaces with at least 6 d.o.f. are required. In addition a high output capability and bandwidth is necessary to represent stiff environments.

Stationary haptic interfaces that achieved a sufficient development status are mostly characterized by highly lightweight mechanical designs requiring no active force feedback control to provide a good backdrivability. E.g. the PHANToM family [9] belongs to that kind of systems. Only a few devices, e.g. the PHANToM Premium as well as the DELTA Haptic Device [5] show an improved but still moderate output capability. As the device workspace and therefore also the device size increase force sensing is necessary to compensate for the increased friction and inertia. The HapticMASTER [20] is an example for such a haptic device which provides 100 N continuous force, but is limited to 3 d.o.f. The

6 d.o.f. device Mirage F3D-35 Haptic System [15] satisfies the force requirements (peak forces of about 100 N), but is limited to a quite small workspace. The most advanced haptic interfaces are the Virtuose 6D40-40 with 30 N continuous force and a workspace of the human arm reach as well as the INCA 6D of Haption with 40 N continuous force and an almost unlimited operational workspace. While the former is too bulky to be mounted on a mobile platform, which is necessary for unconstrained bimanual manipulation, the latter seems to be only suitable for one handed operations.

In order to fill this gap a new bimanual admittance type haptic interface called ViSHARD7 has been developed, which is mountable on a mobile platform. Thus it is not restricted only to desktop applications but also enables bimanual manipulation tasks with high interaction forces in extended remote or virtual environments.

3. System Setup

The mobile haptic interface comprises two haptic interfaces with seven degrees of freedom ViSHARD7) each covering the most important working range of the human arm. These two haptic interfaces are mounted on a mobile platform which extends the usable workspace to arbitrarily large environments. Using this setup, the operator can freely move around in the virtual or remote environment and interact with it at the same time. An overview of the complete system is shown in Fig. 2. It represents a typical application scenario where the operator is immersed in a remote or virtual environment by wearing a head mounted display (HMD). This section provides a description of the single components, haptic interfaces and mobile platform, and specifies their connections.

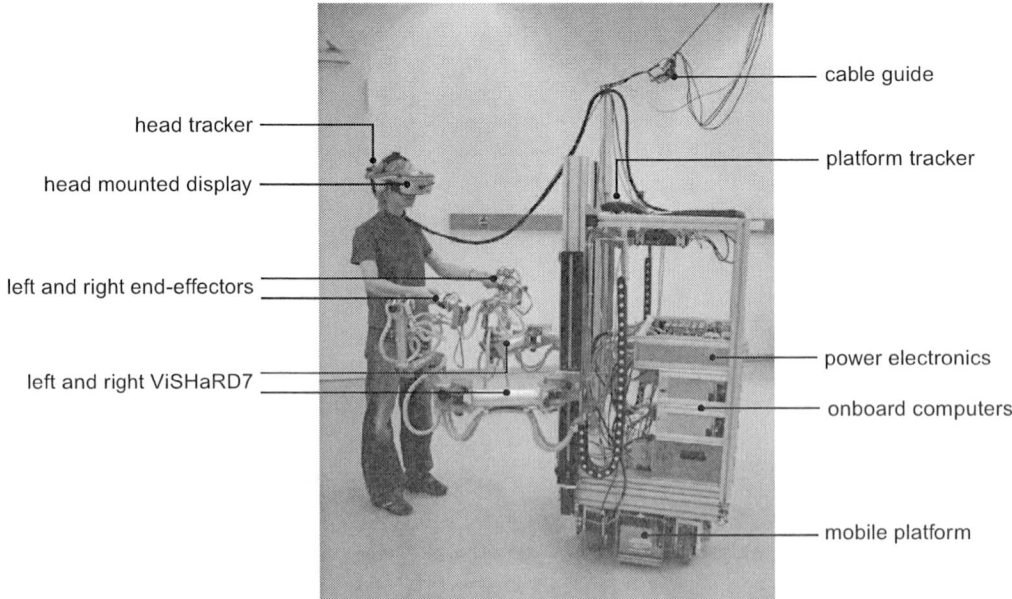

Figure 2. Hardware setup in a typical application scenario.

3.1. Haptic Interfaces

The kinematic structure of one of the haptic interfaces, called VISHARD7 (Virtual Scenario Haptic Rendering Device with 7 actuated d.o.f.), is illustrated in Fig. 3. It shows the reference configuration with all joint angles q_i defined to be zero. The corresponding link length design is summarised in Tab. 1. Fig. 4 shows a typical operational configuration.

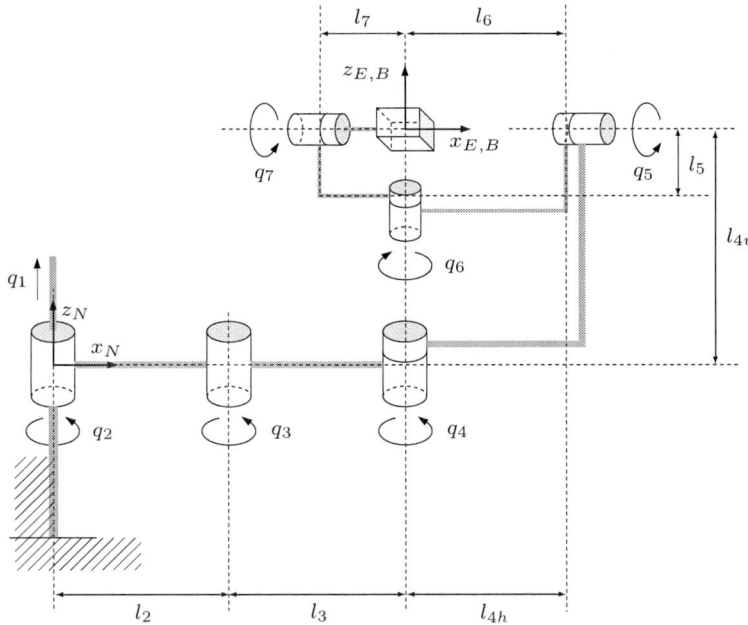

Figure 3. Kinematic model of VISHARD7.

The first joint has been designed as linear axis and enables vertical motions in z_N-direction. Joint 2 and 3 are arranged in a SCARA configuration and allow positioning in the x_N-y_N plane. As known in the literature, the maximum manipulability of such a two-link planar arm can be achieved for a construction with equal joint lengths. Thus, the link lenghts two and three have been set to $l_2 = l_3 = 0.35$ m.

The SCARA part is in a singular configuration when link 2 and 3 are collinear. Hence, configurations near the base have to be omitted. Joint 4 is used to prevent singular configurations in the wrist formed by joints 5, 6, and 7. Singularities in the wrist arise when the axes of joint 5 and 7 are collinear, which can be avoided by a rotation of joint 4. An adequate inverse kinematics algorithm guarantees singularity free operation.

VISHARD7 has been designed in such a way that joint 4, 5, 6, 7 intersect in a single point, where the angular d.o.f. are mechanically decoupled from the translational ones. As already mentioned in [19], such a mechanical decoupling of the angular from the translational d.o.f. has several advantages: The natural dynamics of the orientational d.o.f. is reduced and the torque capability of the rotational actuators can be chosen to match the capability of a human wrist so that no additional safety mechanisms are required. In the case of designing a mobile haptic interface consisting of two independently working components (haptic interface and mobile platform) such a construction can furthermore significantly simplify the algorithms which take care of the interaction between these two components.

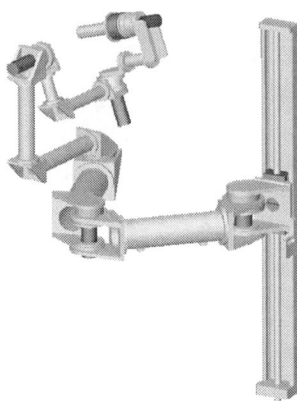

Figure 4. Typical operational configuration.

Table 1. Link length design of ViSHaRD7

Link i	Length
l_1	0.6 m
$l_2 = l_3$	0.35 m
$l_{4h} = l_6$	0.2155 m
l_{4v}	0.3411 m
l_5	0.082 m
l_7	0.0654 m

The link length design guarantees a reachable workspace of almost a half cylinder with radius and height of 0.7 m. Thereby, possible collisions with the arm itself and the platform are considered. In contrast to this reachable workspace, the specifications of the dextrous workspace of the device are given in Table 2.

The actuation torque of all rotational joints is provided by DC-motors coupled with harmonic drive gears offering zero backlash. For the linear axis, a *LM Guide Actuator* of THK has been chosen, which guarantees high rigidity and high accuracy. A brushless DC-motor, which carries the whole weight of all movable parts, is used to drive this linear axis. Since brushless DC-motors usually have better thermal properties than comparable DC-motors this results in a more compact design. An additional brake holds the haptic interface in a fixed position when no motor currents are provided. While all DC-motors of the rotational joints are supplied by Copley amplifiers configured in torque mode, the brushless DC-motor is driven by a 4QEC servo amplifier DES 70/10 of maxon motor with sinusoidal commutation and digital current control.

In order to permit force feedback control, the device is equipped with a six-axis JR3 force-torque sensor providing a bandwidth of 8 kHz at a comparatively low noise level. The

Table 2. Target specifications of ViSHARD7

Property	Value
transl. workspace	$h = 0.6$ m $d = 0.1$ m $r_1 = 0.2$ m $r_2 = 0.6$ m
rot. workspace*	pitch, roll: $\pm 360°$ yaw: $\pm 60°$
peak force	vertical: 533 N horizontal: 155 N
peak torque	pitch, yaw: 11 Nm roll: 4.8 Nm
trans. velocity	vertical: 0.895 m/s horizontal: 1.1 m/s
rot. velocity*	pitch, yaw: 4.3 rad/s roll: 8.9 rad/s
trans. acceleration	vertical: 9.2 m/s^2 horizontal: 13.5 m/s^2
rot. acceleration*	pitch, yaw: 183 rad/s^2 roll: 318 rad/s^2
maximum payload**	34 kg
mass of moving parts	≈ 13 kg

* numbers refer to a device controlled by inverse function, see Sec. 4.1.1.

** calculated for zero steady state human operator input force

joint angles of the rotational joints are measured by digital MR-encoders with a resolution of 4,096 counts per revolution, resulting in a high position resolution when multiplied with the gear ratio. The position of the linear axis is measured at the drive end by using a Scancon Encoder with a resolution of 30,000 counts per revolution (quadrature encoder). The combination of a slope of 10 mm/round of the linear axis and a maximum motor speed of 5,370 rpm allows translational velocities of up to 0.895 m/s. The maximum payload of the linear axis is 340 N and can be calculated considering the limit of the average torque of the motor, the slope of the linear axis, and the mass of all moving parts.

Matlab/Simulink Real-Time-Workshop is used to automatically generate code from Simulink models, which is then executed on a RTAI real time operating system. All models run with a sampling rate of 1 kHz. Data acquisition is performed by using Sensoray S626 PCI-I/O boards.

3.2. Mobile Platform

The mobile platform of the mobile haptic interface is used to position the haptic interfaces in such a way that their workspace is not exceeded. Therefore, the mobile platform must follow the motions of the human operator. Consequently, it must be able to move in any direction and turn around its own axis. An omnidirectional mobile platform is used to fulfill these requirements.

An omnidirectional, holonomic platform offers the highest degree of maneuverability. Most holonomic platforms are based on Omni-Wheels or Meccanum-Wheels. In principle these wheels consist of small rollers or cones mounted on the circumference of a larger wheel. This allows the wheels to be driven like a standard wheel but it also allows sidewards movements with low friction. With a proper wheel arrangement, true omnidirectional, holonomic behavior can be achieved. The disadvantage, however, is that the irregularities in the shape result in noise and vibrations during platform motions. These vibrations are very distracting in a device designed for haptic interaction.

An alternative design consists in using a platform with four conventional wheels which can be independently driven and steered. This also allows omnidirectional maneuvering with low vibrations but at the cost of non-holonomic constraints. In order to realize abrupt changes of the driving direction the wheels need to be turned in the new direction. This can lead to delays in the realization of a given desired trajectory. However, in the given application, small deviations from the desired trajectory can be compensated by the haptic interfaces.

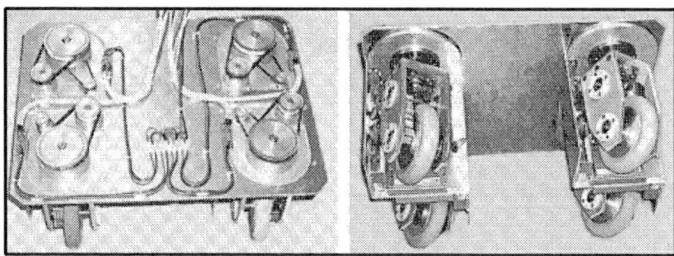

Figure 5. Hardware setup of mobile platform without payload

Following the above outlined rationale, an omnidirectional platform based on four conventional, independently driven and steered wheels is employed as base for the mobile haptic interface (see Fig. 5). The developed platform has a maximum payload of about 200 kg. Maximum speed is around 1.5 m/s, maximum driving power is 600 W. For further details on the mechanical design, the reader is referred to [6].

4. Control Design

The objective of the presented mobile haptic interface is to allow haptic interaction and locomotion at the same time. These two requirements can be attributed to the two different hardware components haptic interfaces ViSHaRD7 and omnidirectional mobile platform, respectively. As the dynamic properties of haptic interfaces and mobile platform differ significantly, their motions can be regarded as decoupled. Thus, the rendering of the desired environment admittance is solely performed by the haptic interfaces. In this section the control algorithms of the two different components - haptic interfaces and mobile platform - are presented and the coordination between them is discussed.

4.1. Control of ViSHaRD7

Realization of a human haptic interaction with a remote environment requires controlling of the motion-force relation between the operator and the haptic interfaces. This can be achieved by either controlling the interaction force of the device with the operator (impedance display mode) or the device motion (admittance display mode).

In order to provide effective compensation of the disturbances due to friction and the natural device dynamics, an admittance control strategy has been implemented for ViSHaRD7. In contrast to impedance control, which is frequently used for light and highly backdrivable devices, admittance control is particularly well suited for robots with high dynamics and non-linearities. The high gain inner control loop closed on motion allows for an effective elimination of the nonlinear device dynamics. Interested readers may have a look at [17] for a more detailed analysis of haptic control schemes.

In admittance control the minimum target mass and inertia of the haptic display is bounded by stability: When the human operator touches the device and free space motion should be rendered, the device needs to accelerate very quickly. This again implies very high control gains, which cause potential stability problems during free space motion. Thus in free space motion a minimum mass and inertia necessary for stability of the master control have to be implemented. While the minimum mass is realized in form of a double integrator

$$\boldsymbol{f}_N = \boldsymbol{M}_T \ddot{\boldsymbol{x}}_N, \qquad (1)$$

the implementation of the minimum rotational inertia M_R is based on the well known Euler's dynamical equations of rotation, see [13] for more details. This master dynamics relates the interaction force $\boldsymbol{F}_{\text{ext}}$ to the reference end-effector velocity $\dot{\boldsymbol{x}}_r$. An algorithm for inverse kinematics resolution calculates the reference joint velocities $\dot{\boldsymbol{q}}_r$. The joint angles $\boldsymbol{q}_r$ finally are the reference input to a conventional position control law, e.g. independent joint controllers or a computed torque scheme.

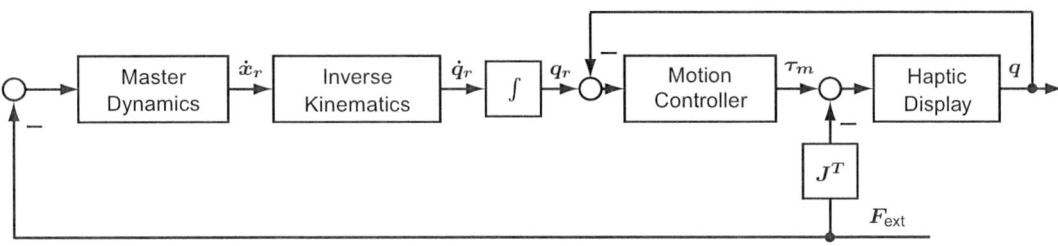

Figure 6. Admittance control scheme.

4.1.1. Inverse Kinematics

The inverse kinematics, the mapping of the end-effector to the joint motion, can be either realized on the position

$$q = f(x) \tag{2}$$

or on the velocity level.

$$\dot{q} = f(\dot{x}), \tag{3}$$

whereby $q, \dot{q} \in \mathbf{R}^n$ are the joint angle and velocity and $x, \dot{x} \in \mathbf{R}^m$ the end-effector position and velocity. Since for VISHARD7 $n > m$ the manipulator is redundant with respect to the end-effector task. This redundancy allows changing the internal configuration without changing the position and orientation of the end-effector. This implies that no unique solution for the inverse kinematics problem given by (2) and (3) can be derived.

To solve this problem for the haptic input device VISHARD7 a simple inverse function for the whole haptic interface has been defined, which decouples translational from rotational movements and thus simplifies the interaction with the mobile platform.

A simple inverse function is defined when using the following mapping from joint angles to Cartesian positions:

$$q_1 = \left(\frac{2\pi}{0.01}\right) z, \tag{4}$$

$$q_2 = \arctan 2(y, x) + \cos^{-1}\left(\frac{x^2 + y^2}{2l\sqrt{x^2 + y^2}}\right), \tag{5}$$

$$q_3 = \cos^{-1}\left(1 - \frac{x^2 + y^2}{2l^2}\right) + \pi, \tag{6}$$

where (x, y, z) is the end-effector position with respect to the haptic interface base coordinate system S_N, q_i are the joint angles of the i-th joint, and l is the link length of link 2 and 3. By setting joint angle 4 to $q_4 = q_{4,0} - \sum_{i=2}^{3} q_i$, a decoupling of translational and rotational motions can be achieved. It should be noted that this special inverse function implies a singular configuration at the point $x = y = 0$, which has to be omitted.

For the rotational part an inverse kinematics solution operating at the angular velocity level has been applied. In a first step the time derivative of the end-effector orientation (given by means of yzx-Euler-angles $[\alpha, \beta, \gamma]$) can be calculated from the angular velocity of the end-effector ω_B:

$$\begin{pmatrix} \dot{\alpha} \\ \dot{\beta} \\ \dot{\gamma} \end{pmatrix} = \begin{pmatrix} 0 & \frac{\cos\gamma}{\cos\beta} & -\frac{\sin\gamma}{\cos\beta} \\ 0 & \sin\gamma & \cos\gamma \\ 1 & -\frac{\sin\beta\cos\gamma}{\cos\beta} & \frac{\sin\beta\sin\gamma}{\cos\beta} \end{pmatrix} \omega_B. \tag{7}$$

Choosing the Euler-angles in such a way that they correspond to the joint angles q_5, q_6 and q_7 the inverse function for the rotational part is given with:

$$q_5 = \alpha, \tag{8}$$

$$q_6 = -\beta + \pi/2, \tag{9}$$

$$q_7 = \gamma. \tag{10}$$

This inverse kinematics solution has a singular configuration for $\beta = k\pi/2$ with $k \in \mathbf{N}$, which however can be easily avoided by introducing a joint limitation for q_6.

4.2. Control of the Mobile Platform

The mobile platform is operated in velocity control mode. As the operator motions cannot be reliably predicted, a control scheme which does not rely on a preplanned path is used.

The kinematic structure of the mobile platform is depicted in Fig. 7a. The controller of the mobile platform needs to ensure the coordinated motion of all wheels, i.e. all wheel normals must be either parallel or intersect in a single point. The respective wheel configurations are called admissible wheel configurations (AWC). All AWCs can be represented on the surface of a unit sphere and can be described by using two spherical angles: the azimuth angle η represents the direction of the translational motion and the altitude ζ is a measure for the relative amount of rotational motion. In Fig. 7b the unit sphere model is illustrated. All configurations on the equator ($\zeta = 0$) correspond to pure translational motion while configurations at one of the poles ($\zeta = \pm\pi/2$) represent pure rotational motion. The absolute motion speed is described by a generalized velocity ω.

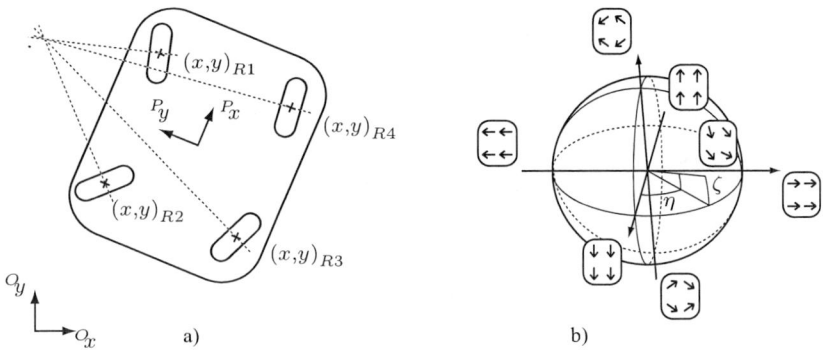

Figure 7. (a) Kinematic model of the mobile platform (b) Unit sphere representation of wheel configurations.

A desired velocity in Cartesian coordinates $\dot{\boldsymbol{x}} = (\dot{x}, \dot{y}, \dot{\psi})$ is first translated into the triple (η, ζ, ω). This triple is then used to calculate the desired steering angles φ_i and wheel velocities v_i of the four wheels $i \in \{1 \ldots 4\}$ by solving

$$v_i \begin{pmatrix} \cos \varphi_i \\ \sin \varphi_i \end{pmatrix} = \omega \begin{pmatrix} \cos \zeta \cos \eta - \kappa_G y_{Ri} \sin \zeta \\ \cos \zeta \sin \eta + \kappa_G x_{Ri} \sin \zeta \end{pmatrix}, \qquad (11)$$

where (x_{Ri}, y_{Ri}) are the coordinates of wheel i with respect to the platform center point. For translational motions, $|\omega|$ is equal to the translational velocity $|\dot{\boldsymbol{x}}|$; for rotational motions, $|\kappa_G \omega|$ is equal to the angular velocity $|\dot{\psi}|$. Finally, the desired steering angles and wheel velocities are implemented by the local controllers in each wheel-module. A more in-depth description of the mobile platform control can be found in [11].

4.3. Position Optimization of the Mobile Platform

The core idea of the presented mobile haptic interfaces consists in using the mobile platform to position the two haptic interfaces at the optimal position in the environment. The position is optimal when it provides the operator with maximum manipulability at all times. Therefore, a mathematical description has to be derived which calculates the optimal platform position. This optimal position is calculated in the platform coordinate system, i.e. it represents a relative position which can be easily transformed to the desired platform velocity by a linear P-controller. The desired velocity is the control input to the platform controller described in the previous section.

In order to simplify the optimization problem, only the planar degrees of freedom are considered. This is possible because the mobile platform can only perform planar motions, i.e. translations in x– and y–direction and rotations around the z–axis. Furthermore, the planar degrees of freedom are also decoupled in the kinematics of the haptic interfaces (cf. Fig. 3). Consequently, only the joint angles q_2 and q_3 of both arms are needed to compute the optimal relative platform position.

4.3.1. Manipulability Measure

When maximizing the manipulability of the haptic interfaces, different types of manipulability and different manipulability measures can be considered. Most importantly, one can distinguish between force manipulability and velocity manipulability. In the former case, the configuration dependent ability to exert forces is measured, whereas in the latter case the ability to generate velocity is described. In a device with serial kinematics such as ViSHaRD7, the force manipulability cannot degenerate. In contrast, the velocity manipulability degenerates close to singular configurations (see [21]). Therefore, the optimization strategy is designed to maximize the velocity manipulability of the haptic interfaces. To this end, a manipulability measure based on a singular value decomposition of the Jacobian is used.

The manipulability of the haptic interfaces is bounded by the maximum joint velocities. The resulting maximum velocities in Cartesian space are computed using the Jacobian $\boldsymbol{J}(q_2, q_3)$ of the SCARA part of ViSHaRD7. The smallest singular value $\sigma_m(q_2, q_3)$ of $\boldsymbol{J}(q_2, q_3)$ is commonly used as a measure of manipulability. It describes how fast the end-effector can move in an arbitrary direction without allowing the joint velocities (q_2, q_3) to leave a unit circle. Therefore, maximizing the smallest singular value $\sigma_m(q_2, q_3)$ also maximizes the allowable Cartesian velocity of the end-effector.

The Jacobian is defined by:

$$\boldsymbol{J} = \begin{pmatrix} \frac{\delta x}{\delta q_2} & \frac{\delta x}{\delta q_3} \\ \frac{\delta y}{\delta q_2} & \frac{\delta y}{\delta q_3} \end{pmatrix} = \begin{pmatrix} -l_2 \sin(q_2) - l_3 \sin(q_2 + q_3) & -l_3 \sin(q_2 + q_3) \\ l_2 \cos(q_2) + l_3 \cos(q_2 + q_3) & l_3 \cos(q_2 + q_3) \end{pmatrix}. \quad (12)$$

In order to calculate the maximum Cartesian velocities with respect to the given maximum joint velocities $\dot{q}_{2,max}, \dot{q}_{3,max}$, a scaled Jacobian is used:

$$\boldsymbol{R} = \text{diag}(\dot{q}_{2,max}, \dot{q}_{3,max}), \quad (13)$$
$$\tilde{\dot{q}} = \boldsymbol{R}^{-1}\dot{q}, \text{ where } q = (q_2, q_3)^T, \quad (14)$$
$$\boldsymbol{J}\dot{q} = (\boldsymbol{JR})\tilde{\dot{q}} = \tilde{\boldsymbol{J}}\tilde{\dot{q}}, \quad (15)$$
$$\tilde{\boldsymbol{J}} = \boldsymbol{JR} = \boldsymbol{J}\begin{pmatrix} \dot{q}_{2,max} & 0 \\ 0 & \dot{q}_{3,max} \end{pmatrix}. \quad (16)$$

The smallest singular value $\tilde{\sigma}_m(q_2, q_3)$ of the scaled Jacobian $\tilde{\boldsymbol{J}}(q_2, q_3)$ is the maximum speed with which the manipulator can move in an arbitrary horizontal direction while the joint velocities stay within the given constraints.

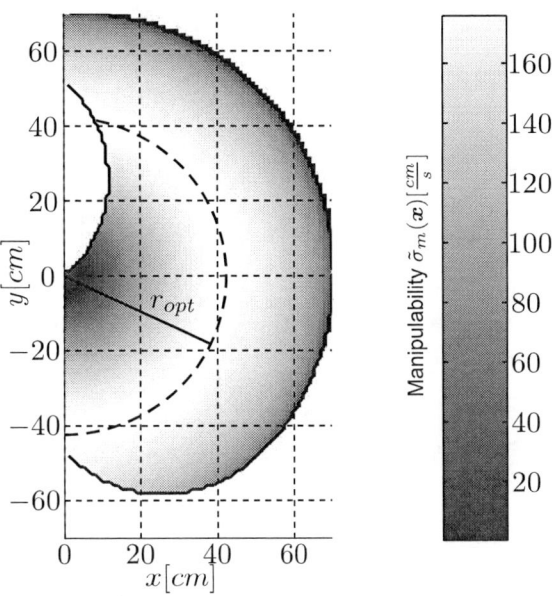

Figure 8. Manipulability and circle with maximum manipulability in the base coordinate system of ViSHaRD7. On the dashed circle the manipulability $\tilde{\sigma}_m(x)$ is maximized.

Fig. 8 shows the planar velocity manipulability $\tilde{\sigma}_m(x)$ of one ViSHaRD7 for all reachable end-effector positions. It is affected by angle q_3, only. Thus, the manipulability is constant on circles around joint 2 and the maximum manipulability is given on a circle with $r_{opt} = 42$ cm.

4.3.2. Maximizing Manipulability

As shown in the previous section, the manipulability is optimal when the end-effector position is located on a circle with radius r_{opt}. This criterion yields a solution for the optimal angle q_3 of both haptic interfaces. q_{r2} and q_{l2} should be chosen in such a way that their minimum distance to the joint limits is maximized. This is achieved when both joint angles, q_{r2} and q_{l2}, are equal. The resulting configuration is symmetric and the corresponding platform position can be obtained by simple geometric calulations: the platform must be aligned parallel to the connecting line from x_L to x_R and its center point must lie on the perpendicular bisector of the connecting line (see Fig. 9).

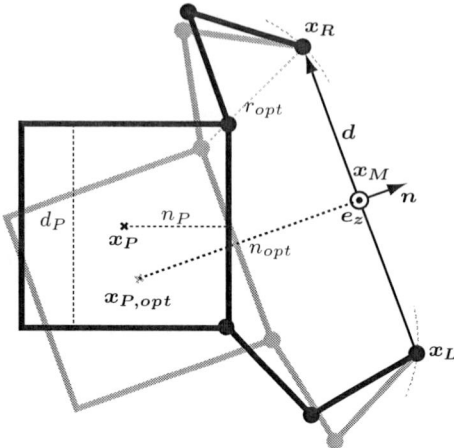

Figure 9. Geometric solution for optimal platform positioning.

The end-effector positions x_L and x_R are used to compute the connecting vector d and midpoint x_M:

$$d = x_R - x_L, \tag{17}$$

$$x_M = \frac{x_R + x_L}{2}. \tag{18}$$

The vector from the optimal platform position $x_{P,opt}$ to the midpoint x_M is obtained by calculating its direction which is perpendicular to d pointing away from the platform and the optimal distance n_{opt}:

$$n = \frac{d}{\|d\|} \times e_z \tag{19}$$

$$n_{opt} = \sqrt{r_{opt}^2 - \left(\frac{\|d\| - d_P}{2}\right)^2} + n_P \tag{20}$$

Finally, the optimal platform position $x_{P,opt}$ is calculated:

$$x_{P,opt} = x_M - n_{opt} \cdot n \tag{21}$$

As the platform can only perform planar motions, the z-component of $x_{P,opt}$ is ignored. The optimal platform orientation $\psi_{P,opt}$ is identical to the direction of the normal vector n:

$$\psi_{P,opt} = \angle n \tag{22}$$

4.3.3. Including Human Arm Workspace in Optimization Strategy

The approach presented in the previous section always converges to a configuration of the haptic interfaces which provides the operator with the maximum velocity manipulability.

This solution works well for slow motions of the operator where the dynamics of the platform can be neglected. In this case, the end-effector positions will always be close to the points of optimal manipulability. For fast operator motions, however, the mobile platform cannot reposition the haptic interfaces fast enough to avoid a significant degradation of the manipulability. For even faster motions, the end-effectors can reach the boundaries of the admissible workspace.

It is, therefore, desireable to take the different dynamics of human motions into consideration. Analogously to the motions of the mobile haptic interface, human motions can be decomposed into fast motions of limited range which are performed by using the arms only and slower, but unlimited motions performed by using the legs. According to this idea, the mobile platform should always be positioned in such a way that the current workspace of the human arms is mostly covered by the workspace of the two haptic interfaces. This optimization goal requires maximizing the overlap between the workspaces of the human arms and the haptic interfaces. However, this optimization problem cannot be solved in real-time due to its high complexity. Furthermore, the human arms can generate velocities far higher than the maximum velocities of the haptic interface in almost any configuration. But then, only moderate velocities are used in typical application scenarios of the mobile haptic interface.

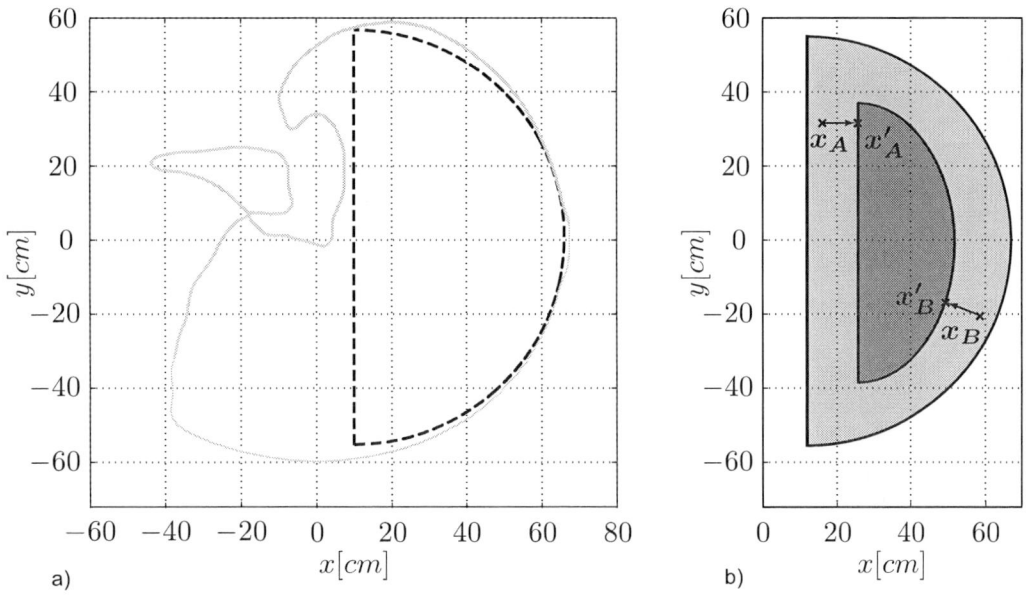

Figure 10. a) Workspace of human right arm in shoulder height (solid line) and approximation by a semicircle (dashed line) b) Mapping from actual to effective end-effector positions. The point of origin is coincident with the shoulder of the human arm.

Therefore, a simplified approach to take the human workspace into account is presented. Fig. 10a shows the computed workspace of the human arm based on a physiological model (see [7]). The relevant workspace can be well approximated by a semicircle.

If the operator holds the end-effectors close to the workspace boundaries of his/her own arm, the respective haptic interface does not need to provide the full motion range in the direction of this boundary. To this end, the position optimization described in Sec. 4.3.2. is no longer applied to the actual end-effector positions x_L and x_R, but instead to shifted end-effector positions x'_L and x'_R. The method for obtaining these effective end-effector positions is illustrated in Fig. 10b: end-effector positions in the outer region (light-gray) are moved to the border of the inner region (dark-gray). In the inner region, no change is applied.

The advantages of this approach can be seen in figure 11: In condition a), the operator holds the end-effectors close to his/her body. Consequently, his/her arms can only perform small position changes away from the mobile platform, but large position changes in direction of the mobile platform. To account for this asymmetry, the mobile platform is positioned farther away from the end-effectors (cf. $x_A \rightarrow x'_A$ in Fig.10b). Condition c) shows the opposite condition, where the operator arms are fully extended and the haptic interface is positioned closer to the end-effectors (cf. $x_B \rightarrow x'_B$ in Fig.10b). Condition b) represents the nominal case without position mapping.

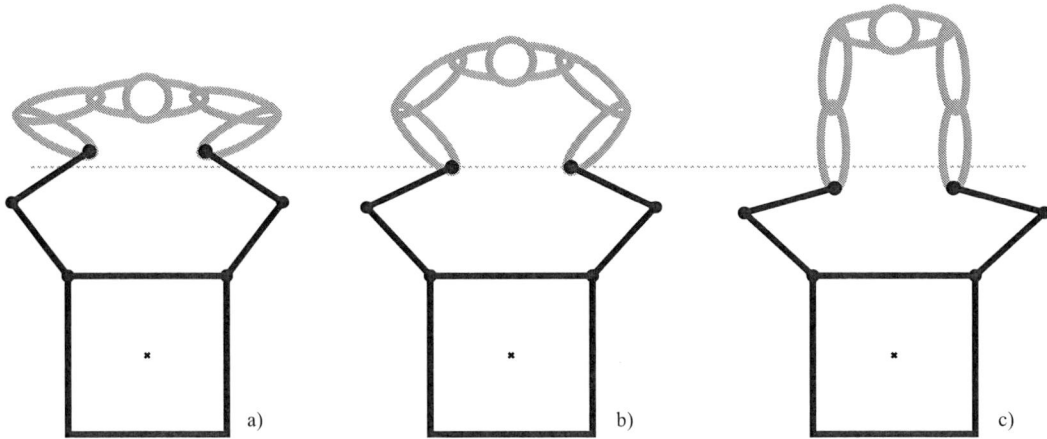

Figure 11. Mobile platform positioning for different arm postures: a) bent arms, b) neutral position, c) extended arms.

Including the human arm workspace in the position optimization strategy offers advantages when the operator often makes fast motions using the full workspace of his/her arms. However, if the operator performs abrupt motions using his legs, the performance can in some cases be deteriorated because the haptic interfaces are used closer to the workspace limits. Furthermore, it should be noted that including the human arm workspace requires the position of the operator to be tracked. In most application scenarios, the additional effort can be neglected because the position of the operator is already needed to correctly position the teleoperator or avatar.

4.4. Overall Control Structure

The complete control structure of the mobile haptic interface is illustrated in Fig. 12. The interaction forces with the operator are measured for both hands by force/torque-sensors. The controllers of the two haptic interfaces implement the desired master dynamics using an admittance control structure. The resulting end-effector positions are transformed to the platform coordinate system and used to calculate the optimal platform position. The actual end-effector positions x_R, x_L are transformed to account for limitations of the workspace of the human operator as described in the above subsection. From these effective end-effector positions x'_R, x'_L the optimal platform position $x_{P,opt}$ is determined through simple geometric calculations. The desired optimal platform position is transformed to a desired velocity by employing a linear P-controller, and some non-linear conditioning elements like a small dead-band zone and a velocity and acceleration limiter. Finally, the desired velocity $\dot{x}_{P,opt}$ is input to the platform velocity controller.

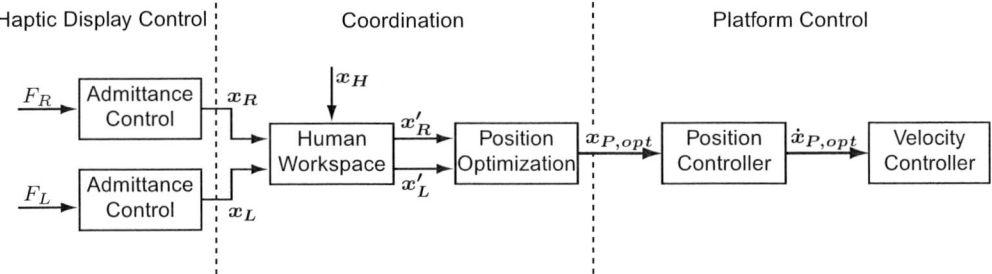

Figure 12. Overall control structure of the mobile haptic interface.

5. Experimental Validation

In this section experiments are described which have been conducted in order to validate the designed system. The mobile haptic interface is employed in tasks like exploring large environments, performing local manipulation, carrying objects over large distances, etc.

The evaluation is conducted using a virtual reality (VR). The evaluation task is to move an object along a predefined path which is presented in the VR. The trajectory is chosen such that it includes straight walking, arcs with different curvature, and a zigzag course resulting in abrupt changes of the walking direction. This experiment is well suited for assessing the maneuverability of the mobile haptic interface.

5.1. Setup

The experiments require an extensive virtual environment in which the operator moves using the mobile haptic interface. Therefore, they are conducted in a laboratory environment of approx. 5 m × 5 m which is fully covered by an acoustic tracking system allowing to record motions of operator and mobile haptic interface. The resolution of the tracking system is about 1 mm. The operator wears a head mounted display (HMD) which gives visual feedback according to the current operator position and gaze direction. The scene is displayed with a resolution of 1280×1024 pixel at a framerate of 30 Hz. The visual rendering of the virtual environment is performed on the onboard computers of the mobile haptic interface. This reduces the cabling effort.

The virtual room only contains some structured walls which the operator needs to orient himself/herself and a structured floor with the desired trajectory (see Fig. 13).

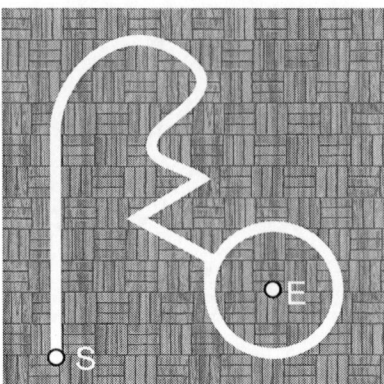

Figure 13. Floor of virtual room with commanded trajectory.

The test persons are instructed to move along the path starting from point S as fast and accurately as possible. During the movement on the circle, the object must be orientated to the end point E, resulting in a sidewards movement of the test person. The experiment is carried out with persons without any prior experience in the use of haptic interfaces as well as with persons who are acquainted with the use of haptic devices.

5.2. Results

During the experiment the travelled path of the operator is recorded in VR coordinates. Two exemplary trajectories taken from an unexperienced and an experienced user are shown in Fig. 14. Obviously, there is a significant training effect in using the mobile haptic interface.

The total travelled distance is 21.2 m for the unexperienced subject and 16.0 m for the experienced subject. The difference is caused by the smaller path deviations and the smaller circle radius for the experienced user. The task execution times for walking along the path are 74 s and 81 s, respectively. The observed results are in accordance with the general experience that operators who are used to employing haptic interfaces tend to operate more slowly but also more precisely. The resulting average speed is between 0.2 m/s and 0.29 m/s which is equivalent to speeds up to a quarter of normal walking speed. As ratios of execution times between telepresent and direct task execution are typically above 10:1, this ratio of 4:1 is comparatively low.

6. Conclusion

A mobile haptic interface allowing simultaneous haptic interaction and natural locomotion in arbitrarily large remote or virtual environments has been presented. The device combines the ability to operate bimanually in full six degrees-of-freedom, high interaction forces and torques, and the useability in extensive scenarios. While devices exist which also provide a solution to one of the mentioned criteria, no other system has been previously presented in the literature which integrates all these aspects into a single device.

For this mobile haptic interface a modular design concept consisting of two different main components - two haptic interfaces and a mobile platform - has been proposed. While

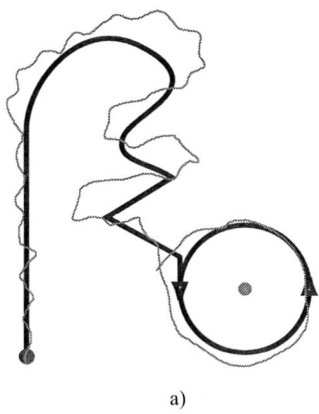 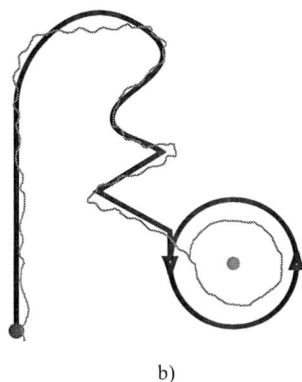

a)   b)

Figure 14. Travelled trajectory: a) unexperienced subject, b) experienced subject.

the haptic interfaces only cover parts of the human arm workspace, the mobile platform extends these to arbitrarily large remote environments. The presented design and control concepts of the haptic interfaces enable a decoupling of translational from rotational movements, which significantly simplifies the control algorithms which handle the interaction between haptic interfaces and mobile platform. In order to ensure a good manipulability of the haptic interfaces the mobile platform has to be positioned appropriately. Two different optimization strategies to position the mobile platform have been investigated: While the first approach, based on the maximization of the manipulability of the two haptic interfaces, is suitable only for slow motions of the human operator, the second approach also includes the human workspace in the optimization strategy and thus can also be used for fast operator motions. The experimental results clearly show the good maneuverability of the developed system.

Future research challenges are a more detailed evaluation of the mobile haptic interface and the integration into a teleoperation scenario.

Acknowledgments

This work is supported in part by the German Research Foundation (DFG) within the collaborative research center SFB453 "High-Fidelity Telepresence and Teleaction. Special thanks go to J. Gradl, H. Kubick, T. Lowitz, and T. Stoeber for their excellent work during the robot construction phase.

References

[1] ACROE and Laboratoire ICA. Ergos. (2006). http://www-acroe.imag.fr/ergos-technologies/.

[2] Caldwell, D.G., Wardle, A., Kocak, O., & Goodwin, M. (1996). Telepresence Feedback and Input Systems for a Twin Armed Mobile Robot. *IEEE Robotics & Automation Magazine*, **3**:29–38.

[3] Darken, R.P. (1999). Spatial orientation and wayfinding in large-scale virtual spaces ii. *Presence*, **8**(6):3–6.

[4] Formaglio, A., Giannitrapani, A., Franzini, M., Prattichizzo, D., & Barbagli, F. (2005). Performance of mobile haptic interfaces. In *44th IEEE European Control Conference on Decision and Control (CDC-ECC '05)*:8343–8348.

[5] Grange, S., Conti, F., Rouiller, P., Helmer, P., & Baur, C. (2001). Overview of the Delta Haptic Device. In *Eurohaptics*.

[6] Hanebeck, U.D., & Saldic, N. (1999). A modular wheel system for mobile robot applications. In *Proceedings of the IEEE/RSJ International Conference on Intelligent Robots and Systems (IROS)*:17–23.

[7] Klopcar, N., & Lenarcic, J. (2005). *Kinematic Model for Determination of Human Arm Reachable Workspace*. Jozef Stefan Institut, Ljubljana, Slovenien.

[8] Lee, D., Martinez-Palafox, O., & Spong, M.W. (2006). Bilateral Teleoperation of a Wheeled Mobile Robot over Delayed Communication Network. In *Proceedings of the 2006 IEEE International Conference on Robotics and Automation*, Orlando, Florida.

[9] Massie, T., & Salisbury, J. (1994). The PHANTOM haptic interface: A device for probing virtual objects. In *Proceedings ASME Winter Annual Meeting: Dynamic Systems and Control Division*, **55**:295–301.

[10] Nitzsche, N., Hanebeck, U., & Schmidt, G. (2003). Design issues of mobile haptic interfaces. *Journal of Robotic Systems*, **20**(9):549–556.

[11] Nitzsche, N., & Schmidt, G. (2004). A Mobile Haptic Interface Mastering a Mobile Teleoperator. In *Proceedings of the IEEE/RSJ International Conference on Intelligent Robots and Systems IROS*, Sendai, Japan.

[12] de Pascale, M., Formaglio, A., & Prattichizzo, D. (2006). A mobile platform for haptic grasping in large environments. *Virtual Reality Journal*, **10**:11–23.

[13] Peer, A., Komoguchi, Y., & Buss, M. Towards a Mobile Haptic Interface for Bimanual Manipulations. (2007). In *Proceedings of the IEEE/RSJ International Conference on Intelligent Robots and Systems*, submitted.

[14] Peer, A., Unterhinninghofen, U., & Buss, M. (2006). Tele-assembly in Wide Remote Environments. In *2nd International Workshop on Human-Centered Robotic Systems*, Munich, Germany.

[15] Quanser Inc. MirageF3D35. (2006). http://www.quanser.com/english/downloads/products/Specialty/Mirage_PIS_060806.pdf.

[16] Schiele, A., & Visentin, G. (2003). The esa human arm exoskeleton for space robotics. In *7th International Symposium on Artificial Intelligence, Robotics and Automation in Space*, Nara, Japan.

[17] Ueberle, M., & Buss, M. (2004). Control of kinesthetic haptic interfaces. In *Proc. IEEE/RSJ International Conference on Intelligent Robots and Systems, Workshop on Touch and Haptics*.

[18] Ueberle, M., Mock, N., & Buss, M. (2007). Design, control, and evaluation of a hyper-redundant haptic interface. In M. Ferre, M. Buss, R. Aracil, C. Melchiorri, & C. Balaguer (Eds.), *Advances in Telerobotics: Human System Interfaces, Control, and Applications*. Springer, STAR series.

[19] Ueberle, M., Mock, N., Peer, A., Michas, C., & Buss, M. (2004). Design and Control Concepts of a Hyper Redundant Haptic Interface for Interaction with Virtual Environments. In *Proceedings of the IEEE/RSJ International Conference on Intelligent Robots and Systems IROS, Workshop on Touch and Haptics*, Sendai, Japan.

[20] Van der Linde, R.Q., Lammertse, P., Frederiksen, E., & Ruiter, B. (2002). The HapticMaster, a new high-performance haptic interface. In *Eurohaptics '02*:1–5.

[21] Wen, J.T.-Y., & Wilfinger, L.S. (1999). Kinematic manipulability of general constrained rigid multibodysystems. *IEEE Transactions on Robotics and Automation*, **15**(3):558 – 567.

Chapter 9

COOPERATIVE MULTIPLE ROBOTS COLLISION AVOIDANCE PROBLEM BASED ON BERNSTEIN-BÉZIER PATH TRACKING

Igor Škrjanc and Gregor Klančar[*]
Faculty of Electrical Engineering, University of Ljubljana,
Tržaška 25, SI-1000 Ljubljana, Slovenia

Abstract

In this chapter a new cooperative collision-avoidance method for multiple nonholonomic robots with constraints and known start and goal velocities based on Bernstein-Bézier curves is presented. In the presented examples the velocities and accelerations of the mobile robots are constrained and the start and the goal velocity are defined for each robot. This means that the proposed method can be used as subroutine in a huge path-planning problem in the real time, in a way to split the whole path in smaller partial paths. The reference path of each robot from the start pose to the goal pose, is obtained by minimizing the penalty function, which takes into account the sum of all the paths subjected to the distances between the robots, which should be bigger than the minimal distance defined as the safety distance, and subjected to the velocities and accelerations which should be lower than the maximal allowed for each robot. When the reference paths are defined the model predictive trajectory tracking is used to define the control. A prediction model derived from linearized tracking-error dynamics is used to predict future system behavior. A control law is derived from a quadratic cost function consisting of the system tracking error and the control effort. The results of the simulation, real experiments and some future work ideas are discussed.

Keywords: Mobile Robots, Collision Avoidance, Path Planning, Bernstein- Bézier Curves, Predictive Control

1. Introduction

Collision avoidance is one of the main issues in applications for a wide variety of tasks in industry, human-supported activities, and elsewhere. Often, the required tasks cannot be

---

[*]E-mail addresses: igor.skrjanc@fe.uni-lj.si, gregor.klancar@fe.uni-lj.si, Phone: +386 1 4768764, Fax: +386 1 4264631

carried out by a single robot, and in such a case multiple robots are used cooperatively. The use of multiple robots may lead to a collision if they are not properly navigated. Collision-avoidance techniques tend to be based on speed adaptation, route deviation by one vehicle only, route deviation by both vehicles, or a combined speed and route adjustment. When searching for the best solution that will prevent a collision many different criteria are considered: time delay, total travel distance or time, planned arrival time, etc. Our optimality criterion will be the minimal total travel distance of all mobile robots involved in the task, subject to a minimal safety distance between all the robots and subject to velocity and acceleration constraints of each mobile robot.

In the literature many different techniques for collision avoidance have been proposed. The first approaches proposed avoidance, when a collision between robots is predicted, by stopping the robots for a fixed period or by changing their directions. The combination of these techniques is proposed in [1] and [22]. The behavior-based motion planning of multiple mobile robots in a narrow passage is presented in [21]. Intelligent learning techniques were incorporated into neural and fuzzy control for mobile-robot navigation to avoid a collision as proposed in [9], [13]. Also, some adaptive navigation techniques for mobile robots navigation appeared, as proposed in [5].

In our case we are dealing with cooperative collision avoidance where all the robots are changing their paths cooperatively to achieve the goal. The control of multiple mobile robots to avoid collisions in a two-dimensional free-space environment is mainly separated into two tasks, the path planning for each individual robot to reach its goal pose as fast as possible and the trajectory tracking control to follow the optimal path.

The second part of the task is to design the control that will ensure the perfect trajectory tracking of the real mobile robots. Several controllers were proposed for mobile robots with nonholonomic constraints. An extensive review of nonholonomic control problems can be found in [12]. In trajectory-tracking control a reference trajectory is usually obtained by using a reference robot; therefore, all kinematics constraints are implicitly considered by a reference trajectory. From the reference trajectory a feed-forward system of inputs combined with a feedback control law are mostly used [20], [17], [2]. Lyapunov stable time-varying state-tracking control laws were pioneered by [8] and [19]. The stabilization to the reference trajectory requires a nonzero motion condition. Many variations and improvements to this state-tracking controller followed in subsequent research [4], [18]. A tracking controller obtained with input-output linearization is used in [20], a saturation feedback controller is proposed in [7] and a dynamic feedback linearization technique is used in [17].

In the field of mobile robotics predictive approaches to path-tracking seem to be very promising because the reference trajectory is known beforehand. However, the solution of the control problem is normally obtained by a minimization of some cost function. In [16] a generalized predictive control is chosen to control the mobile robot, minimizing the quadratic cost function. A generalized predictive controller using the Smith predictor to cope with an estimated system time-delay is presented in [15]. In [14] a model predictive control based on a linear, time-varying description of the system is used. The multi-layer neural network predictive-controller scheme to a path-tracking problem is proposed in [6].

The proposed reference-tracking control is based on a prediction, where the main idea of the control law is to minimize the difference between a future trajectory following the errors of the robot and the reference path. The proposed control law is analytically derived;

therefore, it is computationally effective and can be easily used in fast real-time implementations. The main advantage of the proposed predictive controllers is an explicitly obtained analytical control law that enables fast real-time implementations.

The chapter is organized as follows. In Section 2. the problem is stated. The concept of path planning is shown in Section 3.. The idea of optimal collision avoidance for multiple mobile robots based on Bézier curves is discussed in Section 4.. In Section 5. the proposed model predictive controller is derived. The experimental results of the obtained collision-avoidance control are presented in Section 6. and the conclusion is given in Section 7..

2. Statement of the Problem

The collision-avoidance control problem of multiple nonholonomic mobile robots is proposed in a two-dimensional free-space environment. The simulations are performed for a small two-wheel differentially driven mobile robot of dimension $7.5 \times 7.5 \times 7.5$ cm.

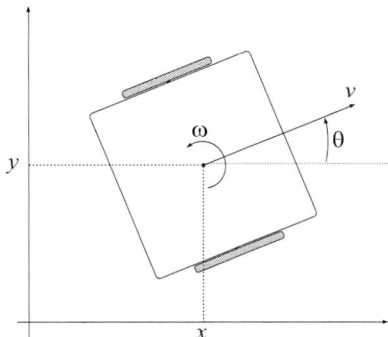

Figure 1. The generalized coordinates of the mobile robot.

The architecture of our robots has a nonintegrable constraint in the form $\dot{x} \sin \theta - \dot{y} \cos \theta = 0$ resulting from the assumption that the robot cannot slip in a lateral direction where $q(t) = [x(t)\ y(t)\ \theta(t)]^T$ are the generalized coordinates, as defined in Figure 1. The kinematics model of the mobile robot is

$$\dot{q}(t) = \begin{bmatrix} \cos \theta(t) & 0 \\ \sin \theta(t) & 0 \\ 0 & 1 \end{bmatrix} \begin{bmatrix} v(t) \\ \omega(t) \end{bmatrix} \quad (1)$$

where $v(t)$ and $\omega(t)$ are the tangential and angular velocities of the platform. During low-level control the robot's velocities are bounded within the maximal allowed velocities, which prevents the robot from slipping.

The danger of a collision between multiple robots is avoided by determining the strategy of the robots' navigation, where we define the reference path to fulfil certain criteria. The reference path of each robot from the start pose to the goal pose is obtained by minimizing the penalty function, which takes into account the sum of all the absolute maximal times subjected to the distances between the robots, which should be larger than the defined safety distance and maximal allowed velocities of each mobile robot.

3. Path Planning Based on Bernstein-Bézier Curves

Given a set of control points $P_0, P_1, \ldots, P_b$, the corresponding Bernstein-Bézier curve (or Bézier curve) is given by

$$\mathbf{r}(\lambda) = \sum_{i=0}^{b} B_{i,b}(\lambda)\mathbf{p}_i$$

where $B_{i,b}(\lambda)$ is a Bernstein polynomial, λ is a normalized time variable ($\lambda = t/T_{max}$, $0 \leq \lambda \leq 1$) and $\mathbf{p}_i$, $0 = 1, \ldots, b$ stands for the local vectors of the control point P_i ($\mathbf{p}_i = P_{i_x}\mathbf{e}_x + P_{i_y}\mathbf{e}_y$, where $P_i = (P_{i_x}, P_{i_y})$ is the control point with coordinates P_{i_x} and P_{i_y}, and $\mathbf{e}_x$ and $\mathbf{e}_y$ are the corresponding base unity vectors). The absolute maximal time T_{max} is the time needed to pass the path between the start control point and the goal control point. The Bernstein-Bézier polynomials, which are the base functions in the Bézier-curve expansion, are given as follows:

$$B_{i,b}(\lambda) = \binom{b}{i} \lambda^i (1-\lambda)^{b-i}, \ i = 0, 1, \ldots, b$$

which have the following properties: $0 \leq B_{i,b}(\lambda) \leq 1$, $0 \leq (\lambda) \leq 1$ and $\sum_{i=0}^{b} B_{i,b} = 1$.

The Bézier curve always passes through the first and last control point and lies within the convex hull of the control points. The curve is tangent to the vector of the difference $\mathbf{p}_1 - \mathbf{p}_0$ at the start point and to the vector of the difference $\mathbf{p}_b - \mathbf{p}_{b-1}$ at the goal point. A desirable property of these curves is that the curve can be translated and rotated by performing these operations on the control points. The undesirable properties of Bézier curves are their numerical instability for large numbers of control points, and the fact that moving a single control point changes the global shape of the curve. The former is sometimes avoided by smoothly patching together low-order Bézier curves.

The properties of Bézier curves are used in path planning for nonholonomic mobile robots. In particular, the fact of the tangentiality at the start and at the goal points and the fact that moving a single control point changes the global shape of the curve. Let us assume the starting pose of the mobile robot is defined in the generalized coordinates as $\mathbf{q}_0 = [x_0, y_0, \theta_0]^T$ and the velocity in the start pose as v_0. The goal pose is defined as $\mathbf{q}_b = [x_b, y_b, \theta_b]^T$ with the velocity in the goal pose as v_b, where b stands for the order of the Bézier curve. This means that the robot starts in position $P_0(x_0, y_0)$ with orientation θ_0 and velocity v_0 and has the goal defined with position $P_b(x_b, y_b)$, the orientation θ_b and the velocity v_b.

Let us define five control points P_0, P_1, P_2, P_3 and P_4 which uniformly define the fourth order Bézier curve of 4zh order. The control points $P_1(x_1, y_1)$ and $P_3(x_3, y_3)$ are defined to fulfill the velocity and orientation requirements in the path. The need for flexibility of the global shape and the fact that moving a single control point changes the global shape of the curve imply the introduction of control point denoted as $P_2(x_2, y_2)$. By changing the position of point P_2 the global shape of the curve changes. This means that having in mind the flexibility of the global shape of the curve and the start and the goal pose of the mobile robot, the path can be planned by four fixed points and one variable point. The Bézier curve is now defined as a sequence of points P_0, P_1, P_2, P_3 and P_4 in Figure 2, where D stands for the distance between the start and the goal control point. The Bernstein polynomials

of the fourth order ($B_{i,b}$, $i = 0, \ldots, b$, $b = 4$), and the control points define the curve as follows:
$$\mathbf{r}(\lambda) = B_{0,4}\mathbf{p}_0 + B_{1,4}\mathbf{p}_1 + B_{2,4}\mathbf{p}_2 + B_{3,4}\mathbf{p}_3 + B_{4,4}\mathbf{p}_4 \tag{2}$$

or

$$\mathbf{r}(\lambda) = (1-\lambda)^4 [x_0\ y_0]^T + 4\lambda(1-\lambda)^3 [x_1\ y_1]^T + 6\lambda^2(1-\lambda)^2 [x_2\ y_2]^T + \\ +4\lambda^3(1-\lambda)[x_3\ y_3]^T + \lambda^4 [x_4\ y_4]^T \tag{3}$$

The control point P_2 will be defined using optimization, and the control points P_1 and P_3 are defined from the boundary velocity conditions.

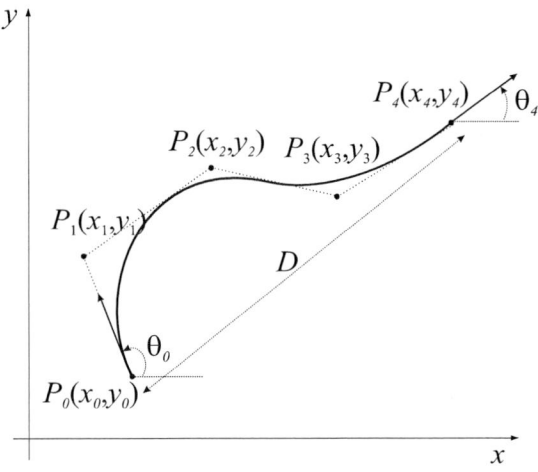

Figure 2. The Bézier curve

Let us therefore defined the velocity as the derivation of the path vector $\mathbf{r}(\lambda)$ according to normalized time λ as follows:

$$\mathbf{v}(\lambda) = \frac{d\mathbf{r}(\lambda)}{d\lambda} = \sum_{i=0}^{b-1} b(\mathbf{p}_{i+1} - \mathbf{p}_i) B_{b-1,i} \tag{4}$$

in the normalized time λ. In the case of the 4th order ($b = 4$) curve, the velocity becomes:

$$\mathbf{v}(\lambda) = 4(\mathbf{p}_1 - \mathbf{p}_0)B_{3,0} + 4(\mathbf{p}_2 - \mathbf{p}_1)B_{3,1} + 4(\mathbf{p}_3 - \mathbf{p}_2)B_{3,2} + 4(\mathbf{p}_4 - \mathbf{p}_3)B_{3,3} \tag{5}$$

The velocity vectors in the start position ($\lambda = 0$) and in the goal position ($\lambda = 1$) than become:

$$\begin{aligned}\mathbf{v}(0) &= 4\mathbf{p}_1 - 4\mathbf{p}_0 \\ \mathbf{v}(1) &= 4\mathbf{p}_4 - 4\mathbf{p}_3\end{aligned} \tag{6}$$

This means that the vectors to the control points $\mathbf{p}_1$ and $\mathbf{p}_3$ are defined as follows:

$$\mathbf{p}_1 = \mathbf{p}_0 + \frac{1}{4}\mathbf{v}(0)$$

$$\mathbf{p}_3 = \mathbf{p}_4 - \frac{1}{4}\mathbf{v}(1) \tag{7}$$

According to the orientation of the robot in the start and goal positions θ_s and θ_g and given start and required tangential velocities of the robot v_0 and v_4, the velocity vector can be written in x and y components as follows:

$$\begin{aligned}\mathbf{v}(0) &= [v_x(0)\ v_y(0)]^T = [v(0)\cos\theta_0\ v(0)\sin\theta_0]^T \\ \mathbf{v}(1) &= [v_x(1)\ v_y(1)]^T = [v(1)\cos\theta_4\ v(1)\sin\theta_4]^T\end{aligned} \tag{8}$$

Using Eqs. 7 and 8, the control points P_1 and P_3 are uniformly defined. The only unknown control point remains P_2 which will be defined by optimization to obtain the optimal path which will be collision safe.

4. Optimal Collision Avoidance Based on Bernstein-Bézier Curves

In this subsection a detailed presentation of cooperative multiple robots collision avoidance based on Bézier curves will be given by taking into account the velocity constraints of the mobile robots. Let as assume the number of robots equals n. The i-th robot is denoted as R_i and has the start position defined as $P_{0i}\ (x_{0i}, y_{0i})$ and the goal position defined as $P_{4i}\ (x_{4i}, y_{4i})$. The normalized time variable of i-th robot is denoted as $\lambda_i = t/T_{max_i}$, where T_{max_i} stands for the absolute maximal time of the i-th robot. The reference path will be denoted with the Bézier curve $\mathbf{r}_i(\lambda_i) = [x_i(\lambda_i), y_i(\lambda_i)]^T$. In Figure 3 a collision avoidance for $n = 2$ is presented for reasons of simplicity.

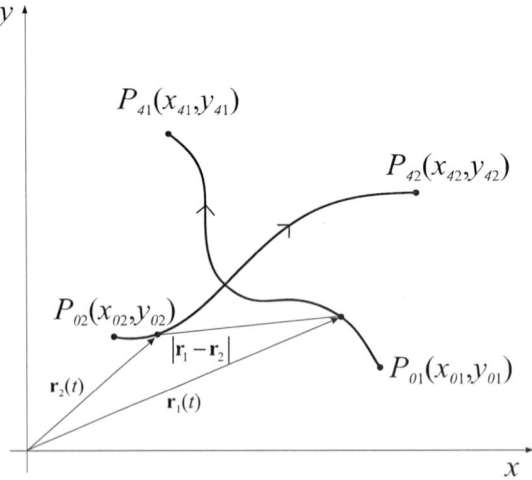

Figure 3. Collision avoidance based on Bernstein-Bézier

The safety margin to avoid a collision between two robots is, in this case, defined as the minimal necessary distance between these two robots. The distance between the robot R_i and R_j is $r_{ij}(t) = |\ \mathbf{r}_i(t) - \mathbf{r}_j(t)\ |$, $i = 1, \ldots, n$, $j = 1, \ldots, n$, $i \neq j$. Defining the minimal necessary safety distance as d_s, the following condition for collision avoidance is obtained $r_{ij} \geq d_s$, $0 \leq \lambda \leq 1$, i,j. Fulfilling this criteria means that the robots will never meet in the same region defined by a circle with radius d_s, which is called a non-overlapping criterion. At the same time we would like to minimize the sum of traveled paths s_i of all robots. The length of the path at the normalized time $s_i(\lambda_i)$ is defined as $s_i(\lambda_i) = \int_0^{\lambda_i} v_i(\lambda_i) d\lambda_i$, where $v_i(\lambda_i)$ stands for the tangential velocity of the ith robot in the normalized variable λ_i

$$v_i(\lambda_i) = |\dot{\mathbf{r}}(\lambda_i)| = \left(\dot{x}_i^2(\lambda_i) + \dot{y}_i^2(\lambda_i)\right)^{\frac{1}{2}}$$

where $\dot{x}_i(\lambda_i)$ stands for $\frac{dx_i(\lambda_i)}{d\lambda_i}$ and $\dot{y}_i(\lambda_i)$ for $\frac{dy_i(\lambda_i)}{d\lambda_i}$.

To define the feasible reference path that will be collision safe and will satisfy the maximal velocity v_{max_i} an the the maximal acceleration a_{max_i} of the mobile robot, the real time should be introduced. The relation between the tangential velocity and acceleration in normalized time framework and the real tangential velocity and acceleration is the following

$$v_i(t) = \frac{1}{T_{max_i}} v_i(\lambda_i) \quad , \quad a_i(t) = \frac{1}{T_{max_i}^2} a_i(\lambda_i)$$

The length of the path of the robot R_i from the start control point to the goal point is now calculated as:

$$s_i = \int_0^1 \left((\dot{x}_i^2(\lambda_i)) + \dot{y}_i^2(\lambda_i))\right)^{\frac{1}{2}} d\lambda_i$$

Assuming that the start P_{0i}, the goal P_{4i} and P_{1i} and P_{3i} control points are known, the global shape and length of each path can be optimized by changing the flexible control point P_{2i}. The collision-avoidance problem is now defined as an optimization problem as follows:

$$\begin{aligned}
& minimize \ \sum_{i=1}^n s_i \\
& subject\ to \\
& d_s - r_{ij}(t) \leq 0, \ \forall i,j, \ i \neq j, \ 0 \leq t \leq \max_i(T_{max_i}) \\
& v_i(t) - v_{max_i} \leq 0, \ \forall i, \ 0 \leq t \leq \max_i(T_{max_i}) \\
& a_i(t) - a_{max_i} \leq 0, \ \forall i, \ 0 \leq t \leq \max_i(T_{max_i})
\end{aligned} \qquad (9)$$

The minimization problem is called an *inequality optimization* problem. Methods using penalty functions transform a constrained problem into an unconstrained problem. The constraints are placed into the objective function via penalty parameter in a way to penalize any violation of the constraints. In our case the following penalty function should be used to have the unconstrained optimization problem

$$\begin{aligned}
& minimize \quad F = \sum_i s_i + \\
& + c_1 \sum_{ij} \max_{ij}\left(0, 1/r_{ij}(t) - 1/d_s\right) + \\
& + c_2 \sum_i \max_i\left(0, v_i(t) - v_{max_i}\right) + \\
& + c_3 \sum_i \max_i\left(0, a_i(t) - a_{max_i}\right), \\
& i,j, \ i \neq j, \ 0 \leq t \leq \max_i(T_{max_i})
\end{aligned} \qquad (10)$$

$subject\ to$

$\mathbf{P}_2, \mathbf{T}_{max}$

where c_1, c_2 and c_3 stand for a large scalar to penalize the violation of constraints and the solution of the minimization problem $\min_{\mathbf{P}_2} F$ is a set of n control points $\mathbf{P}_2 = \{P_{21}, \ldots, P_{2n}\}$ and $\mathbf{T}_{max}$ a set of n maximal times $\mathbf{T}_{max} = \{T_{max_1}, \ldots, T_{max_n}\}$. Each optimal control point P_{2i}, $i = 1, \ldots, n$ uniformly defines one optimal path, which ensures collision avoidance in the sense of a safety distance and will be used as a reference trajectory of the ith robot and will be denoted as $\mathbf{r}_i(\lambda)$. The optimal solution is also subjected to the time, because also the velocities an accelerations of the robots are taken into account in the penalty function 10.

5. Path Tracking

The previously obtained optimal collision-avoidance path for the ith robot is defined as $\mathbf{r}_{ri}(t) = [x_{ri}(t), y_{ri}(t)]^T$, $i = 1, \ldots, n$. In this section the development of a predictive path-tracking controller [11] will be presented. The path-tracking control is realized as a sum of the feed-forward and feed-back controls. The feed-forward control for the ith robot is calculated from a feasible reference path $\mathbf{r}_{ri}(t) = [x_{ri}(t), y_{ri}(t)]^T$, which enables us to reach a desired pose. The feed-forward control inputs $v_{ri}(t)$ and $\omega_{ri}(t)$ are derived using a kinematic model (1). The tangential velocity $v_{ri}(t)$ is calculated as follows

$$v_{ri}(t) = \left(\dot{x}_{ri}^2(t) + \dot{y}_{ri}^2(t) \right)^{\frac{1}{2}} \tag{11}$$

The tangent angle of each point on the path is

$$\omega_{ri}(t) = \frac{\dot{x}_{ri}(t)\ddot{y}_{ri}(t) - \dot{y}_{ri}(t)\ddot{x}_{ri}(t)}{\dot{x}_{ri}^2(t) + \dot{y}_{ri}^2(t)} = v_{ri}(t)\kappa(t) \tag{12}$$

where $\kappa(t)$ is the path curvature. The necessary condition in the path-design procedure is a twice-differentiable path and a nonzero tangential velocity $v_{ri}(t) \neq 0$.

If for some time t the tangential velocity is $v_{ri}(t)=0$, the robot rotates at a fixed point with the angular velocity $\omega_{ri}(t)$ calculated from an explicitly given $\theta_{ri}(t)$.

The feedback control law is derived from a linear time-varying system obtained by an approximate linearization around the trajectory. The obtained linearization is shown to be controllable as long as the trajectory does not come to a stop, which implies that the system can be asymptotically stabilized by smooth time-varying linear or nonlinear feedback ([19]). The tracking error $\mathbf{e}(t) = [e_1(t)\ e_2(t)\ e_3(t)]^T$ of a mobile robot expressed in the frame of the real robot reads

$$\mathbf{e} = \begin{bmatrix} \cos\theta & \sin\theta & 0 \\ -\sin\theta & \cos\theta & 0 \\ 0 & 0 & 1 \end{bmatrix} (\mathbf{q}_{ri} - \mathbf{q}). \tag{13}$$

In Fig. 4 the reference robot ideally follows the reference path, but the real robot has some error when following the reference trajectory. Therefore, the control algorithm should be designed to force the robot to follow the reference path precisely.

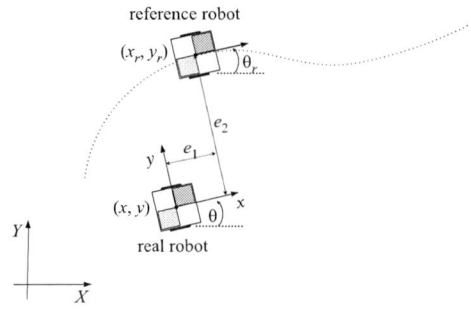

Figure 4. Robot following error transformation.

Considering the robot kinematics (1) and derivating relations (13) the following kinematics model is obtained

$$\dot{\mathbf{e}} = \begin{bmatrix} \cos e_3 & 0 \\ \sin e_3 & 0 \\ 0 & 1 \end{bmatrix} \begin{bmatrix} v_{ri} \\ \omega_{ri} \end{bmatrix} + \begin{bmatrix} -1 & e_2 \\ 0 & -e_1 \\ 0 & -1 \end{bmatrix} \mathbf{u} \quad (14)$$

where $\mathbf{u} = [v\ \omega]^T$ is the velocity input vector and v_{ri} and ω_{ri} are already defined in (11) and (12). The robot input vector $\mathbf{u}$ is further defined as the sum of the feed-forward and feedback control actions ($\mathbf{u} = \mathbf{u}_F + \mathbf{u}_B$) where the feed-forward input vector, $\mathbf{u}_F$, is obtained by a nonlinear transformation of the reference inputs $\mathbf{u}_F = [v_{ri} \cos e_3\ \omega_{ri}]^T$ and the feedback input vector, is $\mathbf{u}_B = [u_{B_1}\ u_{B_2}]^T$, which is the output of the controller defined in section 5.1..

Using the relation $\mathbf{u} = \mathbf{u}_F + \mathbf{u}_B$ and rewriting (14) results in the following tracking-error model

$$\dot{\mathbf{e}} = \begin{bmatrix} 0 & \omega & 0 \\ -\omega & 0 & 0 \\ 0 & 0 & 0 \end{bmatrix} \mathbf{e} + \begin{bmatrix} 0 \\ \sin e_3 \\ 0 \end{bmatrix} v_{ri} + \begin{bmatrix} -1 & 0 \\ 0 & 0 \\ 0 & -1 \end{bmatrix} \mathbf{u}_B . \quad (15)$$

Furthermore, by linearizing the error dynamics (15) around the reference trajectory ($e_1 = e_2 = e_3 = 0$, $u_{B_1} = u_{B_2} = 0$) the following linear model results

$$\dot{\mathbf{e}} = \begin{bmatrix} 0 & \omega_{ri} & 0 \\ -\omega_{ri} & 0 & v_{ri} \\ 0 & 0 & 0 \end{bmatrix} \mathbf{e} + \begin{bmatrix} -1 & 0 \\ 0 & 0 \\ 0 & -1 \end{bmatrix} \mathbf{u}_B \quad (16)$$

which in the state-space form is $\dot{\mathbf{e}} = \mathbf{A}_c \mathbf{e} + \mathbf{B}_c u_B$. According to Brockett's condition [3] a smooth stabilization of the system (1) or its linearization is only possible with time-varying feedback. In the following the obtained linear model is used in the derived predictive control law.

5.1. Model Predictive Control Based on a Robot Tracking-Error Model

To design the controller for trajectory tracking the system (16) will be written in discrete-time form as

$$\mathbf{e}(k+1) = \mathbf{A}\mathbf{e}(k) + \mathbf{B}u_B(k)$$

where $\mathbf{A} \in \mathbb{R}^n \times \mathbb{R}^n$, n is the number of state variables and $\mathbf{B} \in \mathbb{R}^n \times \mathbb{R}^m$, m is the number of input variables. The discrete matrix $\mathbf{A}$ and $\mathbf{B}$ can obtained as follows

$$\mathbf{A} = \mathbf{I} + \mathbf{A}_c T_s, \quad \mathbf{B} = \mathbf{B}_c T_s \tag{17}$$

which is a good approximation during a short sampling time T_s.

The idea of the moving-horizon control concept is to find the control-variable values that minimize the receding-horizon quadratic cost function (in a certain interval denoted with h) based on the predicted robot-following error:

$$J(u_B, k) = \sum_{i=1}^{h} \epsilon^T(k,i) \mathbf{Q} \epsilon(k,i) + \mathbf{u}_B^T(k,i) \mathbf{R} \mathbf{u}_B(k,i) \tag{18}$$

where $\epsilon(k,i) = \mathbf{e}_{ri}(k+i) - \mathbf{e}(k+i|k)$ and $\mathbf{e}_{ri}(k+i)$ and $\mathbf{e}(k+i|k)$ stands for the reference robot following-trajectory and the robot-following error, respectively, and $\mathbf{Q}$ and $\mathbf{R}$ stand for the weighting matrices where $\mathbf{Q} \in \mathbb{R}^n \times \mathbb{R}^n$ and $\mathbf{R} \in \mathbb{R}^m \times \mathbb{R}^m$, with $\mathbf{Q} \geq 0$ and $\mathbf{R} \geq 0$.

5.1.1. Output Prediction in the Discrete-Time Framework

In the moving time frame the model output prediction at the time instant h can be written as:

$$\mathbf{e}(k+h|k) = \Pi_{j=1}^{h-1} \mathbf{A}(k+j|k) \mathbf{e}(k) + \sum_{i=1}^{h} \left(\Pi_{j=i}^{h-1} \mathbf{A}(k+j|k) \right) \mathbf{B}(k+i-1|k) \mathbf{u}_B(k+i- \\ + \mathbf{B}(k+h-1|k) \mathbf{u}_B(k+h-1) \,. \tag{19}$$

Defining the robot-tracking prediction-error vector

$$\mathbf{E}^*(k) = \begin{bmatrix} e(k+1|k)^T & e(k+2|k)^T & \ldots & e(k+h|k)^T \end{bmatrix}^T$$

where $\mathbf{E}^* \in \mathbb{R}^{n \cdot h}$ for the whole interval of observation (h) and the control vector

$$\mathbf{U}_B(k) = \begin{bmatrix} \mathbf{u}_B^T(k) & \mathbf{u}_B^T(k+1) & \ldots & \mathbf{u}_B^T(k+h-1) \end{bmatrix}^T$$

and

$$\mathbf{\Lambda}(k,i) = \Pi_{j=i}^{h-1} \mathbf{A}(k+j|k)$$

the robot-tracking prediction-error vector is written in the form

$$\mathbf{E}^*(k) = \mathbf{F}(k) \mathbf{e}(k) + \mathbf{G}(k) \mathbf{U}_B(k) \tag{20}$$

where

$$\mathbf{F}(k) = [\mathbf{A}(k|k) \ \mathbf{A}(k+1|k)\mathbf{A}(k|k) \ \ldots \ \mathbf{\Lambda}(k,0)]^T , \tag{21}$$

and
$$\mathbf{G}(k) = \begin{bmatrix} g_{11} & 0 & \cdots & 0 \\ g_{21} & g_{22} & \cdots & \vdots \\ \vdots & \vdots & \ddots & \vdots \\ g_{n1} & g_{n2} & \cdots & g_{nh} \end{bmatrix} \quad (22)$$

$$g_{11} = \mathbf{B}(k|k), \; g_{21} = \mathbf{A}(k+1|k)\mathbf{B}(k|k)$$
$$g_{22} = \mathbf{B}(k+1|k), \; g_{n1} = \mathbf{\Lambda}(k,1)\mathbf{B}(k|k)$$
$$g_{n2} = \mathbf{\Lambda}(k,2)\mathbf{B}(k+1|k), \; g_{nh} = \mathbf{B}(k+h-1|k)$$

and $\mathbf{F}(k) \in \mathbb{R}^{n \cdot h} \times \mathbb{R}^n$, $\mathbf{G}(k) \in \mathbb{R}^{n \cdot h} \times \mathbb{R}^{m \cdot h}$.

The objective of the control law is to drive the predicted robot trajectory as close as possible to the future reference trajectory, i.e., to track the reference trajectory. This implies that the future reference signal needs to be known. Let us define the reference error-tracking trajectory in state-space as

$$\mathbf{e}_{ri}(k+i) = \mathbf{A}_{ri}^i \mathbf{e}(k) \quad (23)$$

for $i = 1, \ldots, h$. This means that the future control error should decrease according to dynamics defined with the reference model matrix $\mathbf{A}_{ri}$. Defining the robot reference-tracking error vector

$$\mathbf{E}_{ri}^*(k) = \begin{bmatrix} \mathbf{e}_{ri}(k+1)^T & \mathbf{e}_{ri}(k+2)^T & \ldots & \mathbf{e}_{ri}(k+h)^T \end{bmatrix}^T$$

where $\mathbf{E}_{ri}^* \in \mathbb{R}^{n \cdot h}$ for the whole interval of observation (h) the following is obtained

$$\mathbf{E}_{ri}^*(k) = \mathbf{F}_{ri}\mathbf{e}(k), \; \mathbf{F}_{ri} = \begin{bmatrix} \mathbf{A}_{ri} & \mathbf{A}_{ri}^2 & \ldots & \mathbf{A}_{ri}^h \end{bmatrix}^T \quad (24)$$

and $\mathbf{F}_{ri} \in \mathbb{R}^{n \cdot h} \times \mathbb{R}^n$

5.1.2. Control Law

The idea of MPC is to minimize the difference between the predicted robot-trajectory error and the reference robot-trajectory error in a certain predicted interval.

The cost function is, according to the above notation, now written as

$$J(U_B) = (\mathbf{E}_{ri}^* - \mathbf{E}^*)^T \overline{\mathbf{Q}} (\mathbf{E}_{ri}^* - \mathbf{E}^*) + \mathbf{U}_B^T \overline{\mathbf{R}} \mathbf{U}_B . \quad (25)$$

The control law is obtained by the minimization ($\frac{\partial J}{\partial U_B} = 0$) of the cost function and becomes

$$\mathbf{U}_B(k) = \left(\mathbf{G}^T \overline{\mathbf{Q}} \mathbf{G} + \overline{\mathbf{R}}\right)^{-1} \mathbf{G}^T \overline{\mathbf{Q}} \left(\mathbf{F}_{ri} - \mathbf{F}\right) \mathbf{e}(k) \quad (26)$$

where

$$\overline{\mathbf{Q}} = \begin{bmatrix} \mathbf{Q} & 0 & \cdots & 0 \\ 0 & \mathbf{Q} & \cdots & 0 \\ \vdots & \vdots & \ddots & \vdots \\ 0 & 0 & \cdots & \mathbf{Q} \end{bmatrix}, \; \overline{\mathbf{R}} = \begin{bmatrix} \mathbf{R} & 0 & \cdots & 0 \\ 0 & \mathbf{R} & \cdots & 0 \\ \vdots & \vdots & \ddots & \vdots \\ 0 & 0 & \cdots & \mathbf{R} \end{bmatrix} . \quad (27)$$

This means that $\overline{\mathbf{Q}} \in \mathbb{R}^{n \cdot h} \times \mathbb{R}^{n \cdot h}$ and $\overline{\mathbf{R}} \in \mathbb{R}^{m \cdot h} \times \mathbb{R}^{m \cdot h}$. Let us define the first m rows of the matrix $\left(\mathbf{G}^T \overline{\mathbf{Q}} \mathbf{G} + \overline{\mathbf{R}}\right)^{-1} \mathbf{G}^T \overline{\mathbf{Q}} \left(\mathbf{F}_{ri} - \mathbf{F}\right) \in \mathbb{R}^{m \cdot h} \times \mathbb{R}^{n}$ as $\mathbf{K}_{mpc}$. Now the feedback control law of the model predictive control is given by

$$\mathbf{u}_B(k) = \mathbf{K}_{mpc} \cdot \mathbf{e}(k) \tag{28}$$

with $\mathbf{K}_{mpc} \in \mathbb{R}^m \times \mathbb{R}^n$.

6. Experimental Results

In this section the path planning results of the optimal cooperative collision avoidance strategy between two and three mobile robots are shown and experimental results obtained on real platform using model predictive trajectory tracking control are given. The study was made to elaborate the possible use in the case of a real mobile-robot platform. In the real platform we are faced with the limitation of control velocities and accelerations. The additional details about real set-up and movies of the experiments are available at our website [10]. The study was done for two and three mobile robots.

6.1. Case Study for Two Mobile Robots

The maximal allowed tangential velocity and maximal allowed acceleration of the first mobile robot are $v_{max_1} = 0.3 m/s$ and $a_{max_1} = 0.4 m/s^2$. The maximal allowed tangential velocity and maximal allowed acceleration of the second mobile robot are defined as $v_{max_2} = 0.25 m/s$ and $a_{max_2} = 0.4 m/s^2$.

The starting pose of the first mobile robot R_1 in generalized coordinates is defined as $\mathbf{q}_{01} = \left[0.2, 1, -\frac{\pi}{4}\right]^T$ and the goal pose as $\mathbf{q}_{41} = \left[1, 0.5, -\frac{3\pi}{4}\right]^T$. The boundary velocities of the first mobile robot are the start tangential velocity $v_1(0) = 0.10 m/s$ and the goal tangential velocity $v_1(T_{max1}) = 0.10 m/s$. The second robot R_2 starts in $\mathbf{q}_{02} = \left[1, 0.2, -\frac{3\pi}{4}\right]^T$ and has the goal pose $\mathbf{q}_{42} = \left[0.6, 1, \frac{-3\pi}{4}\right]^T$. The boundary velocities of the second mobile robot are the start tangential velocity $v_2(0) = 0.10 m/s$ and the goal tangential velocity $v_2(T_{max2}) = 0.10 m/s$. The x and y coordinates are defined in meters. The safety distance is defined as $d_s = 0.40m$.

The optimal set $\mathbf{P}_2$ can be found by using one of the unconstrained optimization methods, but the initial conditions are very important. The optimization should be started with initial parameters which ensure a feasible solution. We are optimizing the total sum of all paths which are subjected to the certain conditions according to the safety distances and velocities of the robots. The velocity condition implies the implementation of the maximal time for each robot into the optimization routine. This implies that the initial set $\mathbf{P}_2$ will be defined as

$$\mathbf{P}_2 = \{(x_{21}, y_{21}), (x_{22}, y_{22})\}$$

where x_{2i} and y_{2i} are defined as follows:

$$x_{2i} = \frac{x_{0i} + x_{4i}}{2}, \quad y_{2i} = \frac{y_{0i} + y_{4i}}{2}, \quad i = 1, 2$$

The initial maximal times are defined as $T_{max_1} = 10s$ and $T_{max_2} = 20s$ to fulfill the maximal velocities constraints. The penalty parameters are $c_1 = 2$ and $c_2 = 2$.

The obtained results of the optimization routine are the following $P_{21}(1.4552, 1.0113)$, $P_{22}(0.6138, 0.5883)$ and $T_{max_1} = 6.4374$ and $T_{max_2} = 6.4375$. The minimal value of penalty function F is 2.2511.

The calculated trajectories of both two robots that are cooperatively avoiding the collision are shown in Fig. 5. The robot shapes in Fig. 5 are draw over the planned trajectory each $0.5s$.

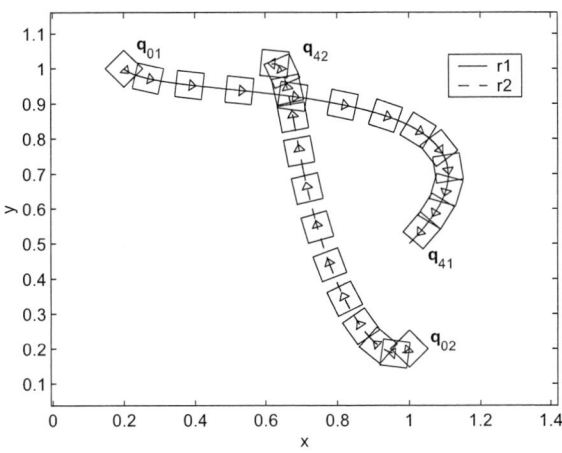

Figure 5. The paths of collision avoiding robots R_1 and R_2.

In Fig. 6 the distances between the mobile robots are shown. It is also shown that all the distances r_{12} satisfy the safety-distance condition. They are always bigger than prescribed safety distance d_s.

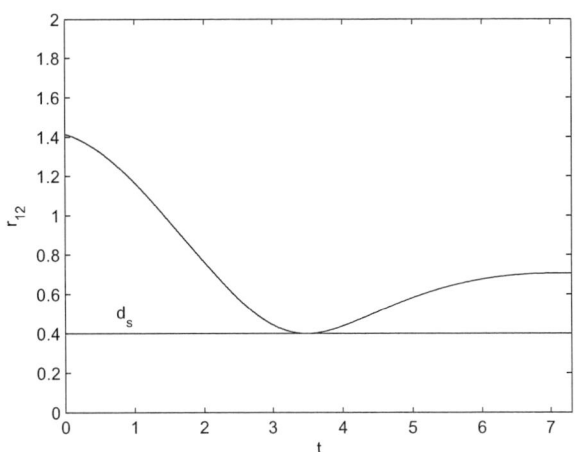

Figure 6. The distance r_{12} between robots R_1 and R_2.

The real tangential velocities profiles of avoiding robots R_1 and R_2 in time variable are given in Fig. 7. It is shown that the velocities profiles of both robots fulfill the boundary velocities requirements and also fulfill the allowed maximal velocities conditions.

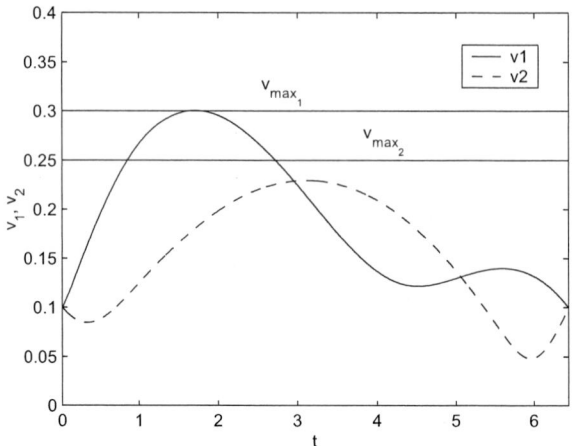

Figure 7. The real velocities of avoiding robots R_1 and R_2.

The acceleration profiles of robots R_1 and R_2 in time variable are given in Figure 8.

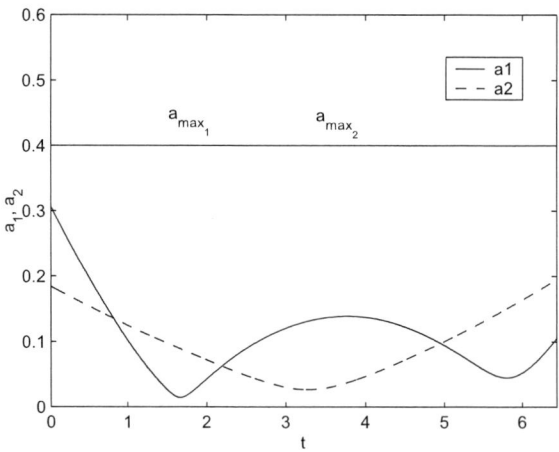

Figure 8. The real accelerations of avoiding robots R_1 and R_2.

All accelerations fulfill the allowed maximal accelerations conditions.

In Figure 9 the results of the experiment performed on small-sized real robots platform (size of each robot is $7.5cm \times 7.5cm \times 7.5cm\times$) is shown.

It can be observed that robots initial postures $\mathbf{q}_{01}$ and $\mathbf{q}_{02}$ have some initial error regarding to the planned trajectories. These initial errors were introduced intentionally to demonstrate the operation of the designed predictive controller. The predictive controller successfully drives the robot to follow the reference trajectories despite the noise in position (standard deviation of $2mm$) and orientation measurements (standard deviation of $0.1rad$).

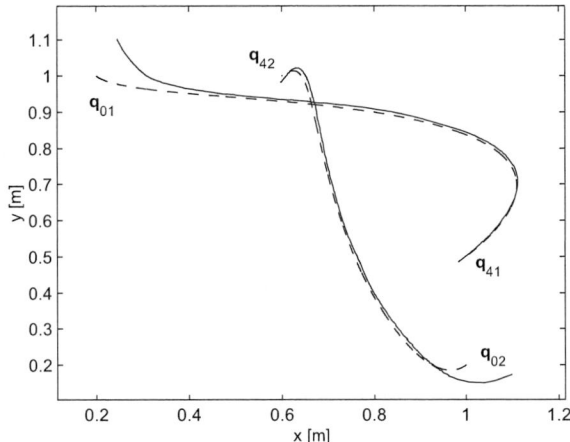

Figure 9. The control of collision avoiding robots R_1 and R_2 (solid line) on the reference trajectories (dashed line); real experiment.

6.2. Case Study for Three Mobile Robots

The maximal allowed tangential velocity of the mobile robots are $v_{max_i} = 0.8m/s$ and maximal allowed accelerations are $a_{max_i} = 0.5m/s^2$ where $i = 1, 2, 3$.

The starting pose of the first mobile robot R_1 in generalized coordinates is defined as $\mathbf{q}_{01} = \begin{bmatrix} 0.2, 1.4, -\frac{\pi}{4} \end{bmatrix}^T$ and the goal pose as $\mathbf{q}_{41} = \begin{bmatrix} 1.4, 0.2, -\frac{\pi}{4} \end{bmatrix}^T$. The boundary velocities of the first mobile robot are the start tangential velocity $v_1(0) = 0.40m/s$ and the goal tangential velocity $v_1(T_{max1}) = 0.4m/s$. The second robot R_2 starts in $\mathbf{q}_{02} = \begin{bmatrix} 1.4, 0.2, \frac{3\pi}{4} \end{bmatrix}^T$ and has the goal pose $\mathbf{q}_{42} = \begin{bmatrix} 0.2, 1.4, \frac{3\pi}{4} \end{bmatrix}^T$. The boundary velocities of the second mobile robot are the start tangential velocity $v_2(0) = 0.4m/s$ and the goal tangential velocity $v_2(T_{max2}) = 0.5m/s$. The third robot R_3 starts in $\mathbf{q}_{03} = \begin{bmatrix} 0.2, 0.2, \frac{\pi}{4} \end{bmatrix}^T$ and has the goal pose $\mathbf{q}_{43} = \begin{bmatrix} 1.4, 1.4, \frac{\pi}{4} \end{bmatrix}^T$. The boundary velocities of the third mobile robot are the start tangential velocity $v_3(0) = 0.4m/s$ and the goal tangential velocity $v_3(T_{max3}) = 0.4m/s$. The x and y coordinates are defined in meters. The safety distance is defined as $d_s = 0.35m$.

The optimal set $\mathbf{P}_2$ can be found by using one of the unconstrained optimization methods, but the initial conditions are very important. The optimization should be started with initial parameters which ensure a feasible solution. We are optimizing the total sum of all paths which are subjected to the certain conditions according to the safety distances, velocities and accelerations of the robots. The velocity condition implies the implementation of the maximal time for each robot into the optimization routine. This implies that the initial set $\mathbf{P}_2$ will be defined as

$$\mathbf{P}_2 = \{(x_{21}, y_{21}), (x_{22}, y_{22}), (x_{23}, y_{23})\}$$

where x_{2i} and y_{2i} are defined as follows:

$$x_{2i} = \frac{x_{0i} + x_{4i}}{2}, \quad y_{2i} = \frac{y_{0i} + y_{4i}}{2}, \quad i = 1, 2, 3$$

The initial maximal times are defined as $T_{max_i} = 5s$ ($i = 1, 2, 3$) to fulfill the maximal velocities constraints. The penalty function 10 parameters of are $c_i = 100$ ($i = 1, 2, 3$). The obtained results of the optimization routine are the following $P_{21}(1.61, 0.70)$, $P_{22}(0.97, 0.02)$, $P_{23}(0.63, 1.10)$ and $T_{max_1} = 4.5974$, $T_{max_2} = 4.5973$ and $T_{max_3} = 4.5973$. The minimal value of penalty function F is 5.2728.

The simulated positions of all three robots robots (R_1, R_3 and R_3) that are cooperatively avoiding the collision are shown in Figure 10. Robot R_1 start from upper left corner and finishes in lower right corner, robot R_2 starts from lower right and finishes in upper left corner and robot robot R_3 starts from lower left and finishes in upper right corner. The paths of these robots obviously cross. The robots (square shapes on the trajectories in Figure 10) are drawn each tenth sample time to illustrate their progress during the experiment. It can be observed that robots adjust their velocity profiles as well as trajectories in order to fulfill the design constraints (d_s, v_{max_i} and a_{max_i}).

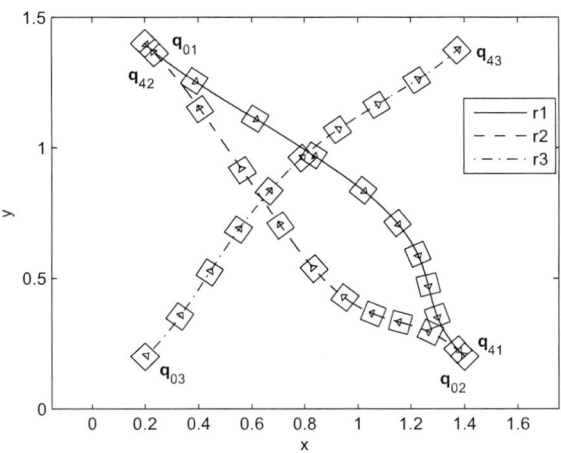

Figure 10. The paths of collision avoiding robots R_1, R_2 and R_3.

In Figure 11 the distances between the mobile robots are shown. It is also shown that all the distances (r_{12}, r_{13}, r_{23}) satisfy the safety-distance condition. They are always bigger than prescribed safety distance d_s.

The real tangential velocities profiles of avoiding robots R_1, R_2 and R_3 are given in Figure 12. It is shown that the velocities profiles of all three robots fulfill the boundary velocities requirements and also fulfill the allowed maximal velocities conditions.

The acceleration profiles of robots R_1, R_2 and R_3 are given in Figure 13. All accelerations fulfill the allowed maximal accelerations conditions.

In Figure 14 the results of the experiment performed on small-sized real robots platform (size of each robot is $7.5cm \times 7.5cm \times 7.5cm\times$) is shown.

It can be observed that robots initial postures $\mathbf{q}_{01}$, $\mathbf{q}_{02}$ and $\mathbf{q}_{03}$ have some initial error (they are not on the planned trajectory) the predictive controller successfully drives the robot to follow the reference trajectories despite the noise in position and orientation measurements (for details about noise rates see subsection 6.1.).

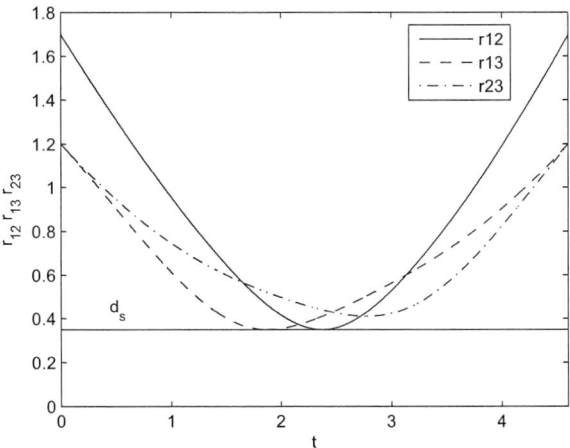

Figure 11. The distance r_{12}, r_{13}, r_{23} between robots R_1, R_2 and R_3.

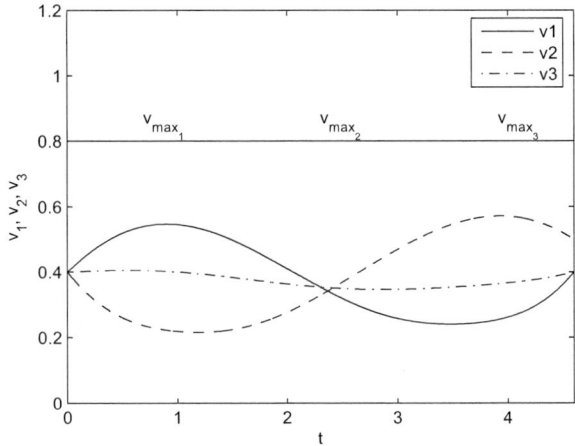

Figure 12. The real velocities of avoiding robots R_1, R_2 and R_3.

7. Conclusion

The optimal cooperative collision-avoidance approach based on Bézier curves allows us to include different criteria in the penalty functions. In our case the reference path of each robot from the start pose to the goal pose is obtained by minimizing the penalty function, which takes into account the sum of all traveled paths of the robots subjected to the distances between the robots, which should be bigger than the minimal distance defined as the safety distance, the maximal velocities of the robots and maximal allowed accelerations of the robots.

The model predictive trajectory tracking control is used to control the robots on the obtained reference paths. The predictive control law minimizes the quadratic cost function consisting of tracking errors and control effort. The solution to the control is analytically

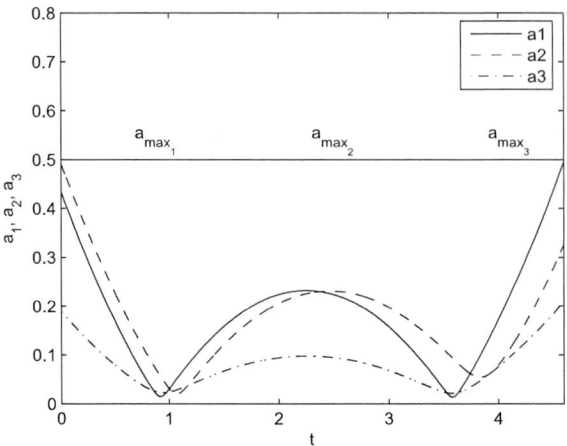

Figure 13. The real accelerations of avoiding robots R_1, R_2 and R_3.

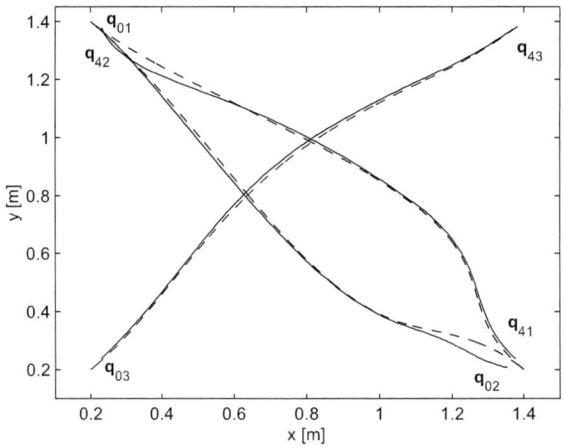

Figure 14. The control of collision avoiding robots R_1, R_2 (solid line) and R_3 on the reference trajectories (dashed line); real experiment.

derived, which enables fast real-time implementations. Future improvements will focus on an increased robustness of the presented algorithm to larger tracking errors, mainly resulting from the wrong initial robot posture. The concept of landing curves, which guaranties an exponential convergence to the reference trajectory, will be included.

The proposed cooperative collision-avoidance method for multiple nonholonomic robots based on Bézier curves and predictive reference tracking shows great potential and in the future will be implemented on a real large-scale mobile-robot Pioneer 3-AT platform.

References

[1] Arkin, R.C. Cooperation without communication: multiagent schema-based robot navigation, *Journal of Robotic Systems*. 1992, Vol. 9, No. 3, 351-364.

[2] Balluchi, A.; Bicchi, A.; Balestrino, A.; Casalino, G. Path Tracking Control for Dubin's Cars, in *Proceedings of the 1996 IEEE International Conference on Robotics and Automation*; Minneapolis, Minnesota, 1996; pp. 3123-3128.

[3] Brockett, R.W. Asymptotic stability and feedback stabilization, in *Differential Geometric Control Theory*, Brockett, R.W.; Millman, R.S.; Sussmann, H. J.; Eds.; Boston, MA: Birkhuser, 1983; pp. 181-191.

[4] Pourboghrat, F.; Karlsson, M.P. Adaptive control of dynamic mobile robots with nonholonomic constraints, *Computers and electrical Engineering*. 2002, Vol. 28, No. 4, 241-253.

[5] Fujimori, A.; Nikiforuk, P.N.; Gupta, M.M. Adaptive navigation of mobile robots with obsticle avoidance, *IEEE Transactions on Robotics and Automation*. 1999, Vol. 13, No. 4, 596-602.

[6] Gu, D.; Hu, H. Neural predictive control for a car-like mobile robot, *Robotics and Autonomous Systems*. 2002, Vol. 39, No. 2, 73-86.

[7] Lee, T.C.; Song, K.T.; Lee, C.H.; Teng, C.C. Tracking control of unicycle-modeled mobile robots using a saturation feedback controller, *IEEE Transactions on Control Systems Technology*. 2001, Vol. 9, No. 2, 305-318.

[8] Kanayama, Y.; Kimura, Y.; Miyazaki, F.; Noguchi, T. A stable tracking control method for an autonomous mobile robot, in *Proceedings of the 1990 IEEE International Conference on Robotics and Automation*; Cincinnati, OH, 1990; Vol. 1, pp. 384-389.

[9] Kim, C.G.; Triverdi, M.M. A neuro-fuzzy controller for mobile robot navigation and multirobot convoying, *IEEE Transaction on Systems, Man, and Cybernetics - PartB*. 1998, Vol. 28, No. 6, 829-840.

[10] Klančar, G. Optimal collision avoidance experiments, available at: http://msc.fe.uni-lj.si/PublicWWW/Klancar/ColisionAvoidance.html.

[11] Klančar, G.; Škrjanc, I. Tracking-error model-based predictive control for mobile robots in real time, *Robotics and Autonomous Systems*. 2007, http://dx.doi.org/10.1016/j.robot.2007.01.002, Article in press.

[12] Kolmanovsky, I.; McClamroch N.H. Developments in Nonholonomic Control Problems, *IEEE Control Systems*. 1995, Vol. 15, No. 6, 20-36.

[13] Kubota, N.; Morioka, T.; Kojima, F.; Fukuda, T. Adaptive behavior of mobile robot based on sensory network, *JSME Trans*. 1999, Vol. 65, 1006-1012.

[14] Kühne, F.; Gomes da Silva Jr., J. M.; Lages, W.F. Model Predictive Control of a Mobile Robot Using Linearization, *Mechatronics and Robotics 2004*; Aachen, Germany, 2004.

[15] Normey-Rico, J.E.; Gomez-Ortega J.; Camacho, E.F.; A Smith-predictor-based generalised predictive controller for mobile robot path-tracking, *Control Engineering Practice*. 1999, Vol. 7, No. 6, 729-740.

[16] Ollero, A.; Amidi, O. Predictive Path Tracking of Mobile robots. Application to the CMU Navlab, in *Proceedings of 5th International Conference on Advanced Robotics, Robots in Unstructured Environments (ICAR '91)*; Pisa, Italy, 1991; Vol. 2, 10811086.

[17] Oriolo, G.; Luca, A.; Vandittelli, M. WMR Control Via Dynamic Feedback Linearization: Design, Implementation, and Experimental Validation, *IEEE Transactions on Control Systems Technology*. 2002, Vol. 10, No. 6, 835-852.

[18] Raimondi, F.M.; Melluso, M. A new robust fuzzy dynamics controller for autonomous vehicles with nonholonomic constraints, *Robotics and Autonomous Systems*. 2005, Vol. 52, No. 2-3, 115-131.

[19] Samson, C. Time-varying feedback stabilization of car like wheeled mobile robot, *International Journal of Robotics Research*. 1993, Vol. 12, No. 1, 55-64.

[20] Sarkar, N.; Yun, X.; Kumar, V. Control of mechanical systems with rolling constraints: Application to dynamic control of mobile robot, *The International Journal of Robotic Research*. 1994, Vol. 13, No. 1, 55-69.

[21] Shan, L.; Hasegawa, T. Space reasoning from action observation for motion planning of multiple robots: mutal collision avoidance in a narrow passage, *J Robot Soc Japan*. 1996, Vol. 14, 1003-1009.

[22] Sugihara, K.; Suzuki, I. Distributed algorithms for formation of geometric patterns with many mobile robots, *Journal of Robotic Systems*. 1996, Vol. 13, No. 13, 127-139.

In: Robotics Research Trends
Editor: Xing P. Guô, pp. 309-331

ISBN: 1-60021-997-7
© 2008 Nova Science Publishers, Inc.

Chapter 10

ABLE: A STANDING STYLE TRANSFER SYSTEM FOR DISABLED LOWER LIMBS

Yoshikazu Mori
Dept. of Systems Engineering, Ibaraki Univ.,
Nakanarusawa-cho, Hitachi-shi, Ibaraki, Japan

Abstract

We have developed a standing style transfer system "ABLE" for a person with disabled legs. It realizes travel in a standing position even on uneven ground, standing up motion from a chair, and ascending stairs. ABLE consists of three modules: a pair of telescopic crutches, a powered lower extremity orthosis, and a pair of mobile platforms. We show the conceptual design of ABLE and the motion of each module. Cooperative operations using three modules are discussed through simulations. The standing up motion from a chair and ascending stairs, however, had a problem with adaptability to the environment and safety because it had executed the movement that relied on telescopic crutches. To solve these problems, we propose a new motion technique and compare it with the previous method. Some experimental results are also shown in this paper.

Index Terms—Robots, Handicapped aids, Electric control equipment, Wear

I. Introduction

Persons with disabled lower limbs are increasing globally. In Japan, their number was about 600,000 in 2001 [1]. Most of them use wheelchairs in daily life. Wheelchairs are now utilized widely as "second legs" because they are inexpensive and have simple mechanisms. Electric wheelchairs have come into wider use recently because of improvement of their controllability and running time. Notwithstanding, they engender several problems. Wheelchairs require much space during travel and facility use. It is difficult to ascend stairs. Therefore, a separate infrastructure for wheelchair users is inevitable. Moreover, medical failures must be addressed such as hematogenous disorder of legs caused by maintaining a sitting position, excretion failure, and arthropathy. Mental stress also occurs because of the low position of the eyes. However, most of these problems can be solved by the use of some equipment that transfers an ambulatory-disabled person with a standing position.

Some exoskeletonic power-assisted devices have been developed [2]–[5], such as the "HAL" device [4], [5]. This system is intended for persons with leg muscle atrophy; the system is controlled using the surface potential of the leg. Therefore, it is difficult for people to use the system if they have disabled lower limbs without that surface potential. This system also has an automatic control mode, but the user cannot operate it freely in that mode. Independence Technology LLC developed the "IBOT", an electric wheelchair that gets up and balances using only two wheels, such as an inverted pendulum. It also realizes a motion of ascending stairs. However, its movement requires much space. Moreover, medical problems caused by maintaining a sitting position remain unsolved.

Previous studies have examined telescopic crutches and auxiliary tools for binding the legs. At the National Rehabilitation Center for Persons with Disabilities, ambulation equipment has been developed which combines a pair of constant-length crutches with an auxiliary leg tool whereby the base of the leg performs a vertical motion [6], [7]. However, this equipment is intended for rehabilitation and its only available motion is to go straight.

This paper presents a conceptual design of standing style transfer system "ABLE" for those with disabled lower limbs. ABLE is mainly for the persons who have spinal cord injuries and cannot move hip joints and under: the level of spinal cord injury is L_1. ABLE consists of three modules: a pair of telescopic crutches, a powered lower extremity orthosis, and a pair of mobile platforms, and it realizes three basic operations that are indispensable for our daily life: traveling in a standing position, even on uneven ground; standing up motion from a chair; and ascending stairs.

In our previous studies, the motions using power of telescopic crutches: a standing up motion from a chair and ascending stairs, relied on the crutches [8]. However, this use of crutches is likely to lead to slipping and falling. Therefore, a new motion sequence is proposed [9]. The flexibility of the arms is made available by detaching the crutches from the armpits. This method is compared with the previous one in terms of stability and torque.

Section II presents the conceptual design of ABLE and the design of each module. Section III describes the cooperative operations using three modules. In practical use, safety and adaptability to various environments should be considered. We propose a new motion technique in Section IV. Experimental results are presented in Section V. Concluding remarks follow in Section VI.

II. Conceptual Design

What kinds of tools are required for a person with disabled legs to travel stably with a standing position? We propose a novel standing style transfer system ABLE that consists of three modules shown in Fig. 1: a pair of telescopic crutches, a powered lower extremity orthosis, and a pair of mobile platforms.

Telescopic crutches are useful to maintain body stability without taking an unnatural posture. It is possible to freely change the contact points to the ground according to different situations. Thereby, it eases the movement through narrow spaces, whereas stability is emphasized in a wide space. The powered lower extremity orthosis has an actuator on each joint. Mobile platforms enable the user to travel. Their size is nearly identical to that of shoes.

ABLE: A Standing Style Transfer System for Disabled Lower Limbs

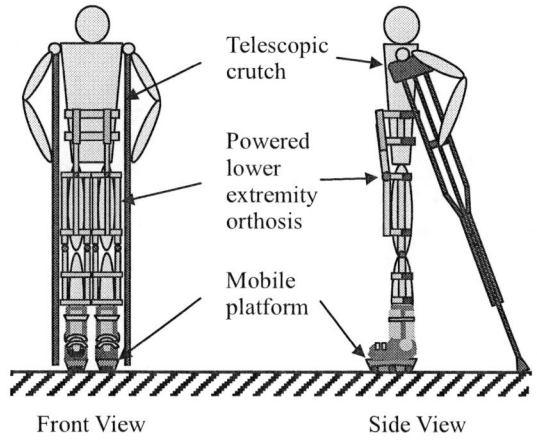

Figure 1. Standing style transfer system for lower limbs disabled.

Combining and coordinating these modules allows the following operations that are basic for a daily life: traveling in a standing position, even on uneven ground; a standing up motion from a chair; and ascending stairs. Next we show each module of ABLE.

A. Telescopic Crutch

The length of the telescopic crutch is adjustable by up/down switches attached to the grip through a computer. The switches serve not only to limit adjustment of the crutch length; it is also changeable according to the desired task. The main roles of this module are to maintain body balance and to supply power. Fig. 2 shows the telescopic crutch with an inclinometer. Table 1 shows telescopic crutch specifications.

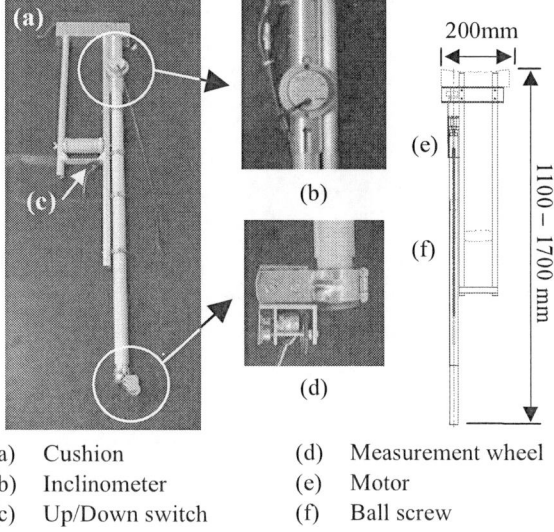

(a) Cushion
(b) Inclinometer
(c) Up/Down switch
(d) Measurement wheel
(e) Motor
(f) Ball screw

Figure 2. Telescopic crutch.

Table 1. Specifications of a telescopic crutch.

Weight	2.9 kg
Length	1100-1700 mm
Motor output	70 W
Maximum force	75.3 kgf
Maximum speed	621 mm/s
Lead of the ball screw	6 mm

B. Mobile Platform

Mobile platforms carry the user not only on flat floors, but also on uneven ground, or even outdoors, because they use crawlers for the transfer mechanism. These platforms also enable the user to turn, because they have a rotation board mechanism. There are two kinds of prospective mechanisms to turn the platform: one is the power wheeled steering mechanism and the other is a rotation board mechanism. We compared these mechanisms through experiments. Although the former was simple, it required much torque when loading a heavy weight and it was difficult to turn the platform accurately, for the distance of the two crawler belts was narrow.

Table 2. Specifications of a mobile platform.

Weight	4.1 kg
Size	$265 \times 87 \times 153$ mm
Maximum speed	5.6 km/h
Driving output	70 W
Steering output	20 W
Maximum angular velocity of rotation board	143°/s
Operational range of rotation board	$-45° \leq \theta \leq 45°$

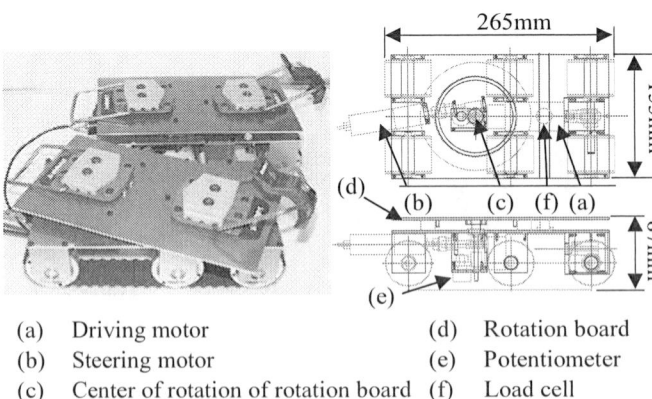

(a) Driving motor (d) Rotation board
(b) Steering motor (e) Potentiometer
(c) Center of rotation of rotation board (f) Load cell

Figure 3. Mobile platform.

Fig. 3 shows the mobile platforms. Only one motor drives the two crawlers through the worm gear. One load cell is installed between the base part and the rotation board. Table 2 shows the mobile platform specifications. The platform is fixed to the ski boot with a binding.

C. Powered Lower Extremity Orthosis

The powered lower extremity orthosis is for actively fixing, bending, and stretching each leg joint. This module has actuators at the hip and knee joints. As a user puts on a pair of ski boots, ankle joints are able to bend passively.

Fig. 4 shows the powered lower extremity orthosis. Table 3 shows the specifications. Because worm gears are used in each gearbox, each joint is not back drivable. Therefore, even if the power source is cutoff, the previous state remains. Comparing this mechanism with back drivable mechanisms, each joint consumes little energy if the user doesn't change his position.

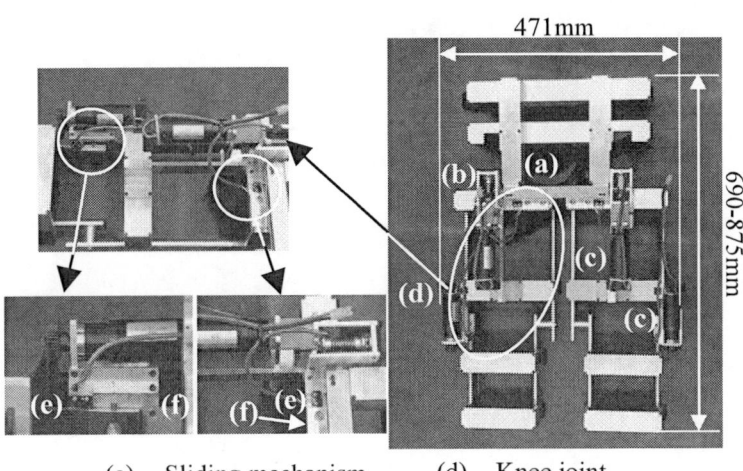

(a) Sliding mechanism (d) Knee joint
(b) Hip joint (e) Touch switches
(c) Motors (f) Stoppers

Figure 4. Powered lower extremity orthosis.

Table 3. Specifications of a powered lower extremity orthosis.

Weight	10.0 kg
Size	$(690-875) \times 471 \times 246$ mm
Motor output	70 W
Maximum angular velocity of each joint	90°/s
Range of a knee joint	$0° \leq \theta \leq 135°$
Range of a hip joint	$0° \leq \theta \leq 145°$

There is a stopper and a touch switch on each joint. They work together to keep the movable scope of the human joint.

The gearbox of the hip joint is in front of the leg, in order not to disturb the motion of swinging the crutches. However, this location causes a problem: the difference of the rotation centers between robot and human. If no additional mechanism is added, the hip joint will not rotate. To solve this problem, we added a sliding mechanism on the chest. If the joint is straight, the sliding mechanism is extended, and if it bends, the mechanism is shrunk, as shown in Fig. 5.

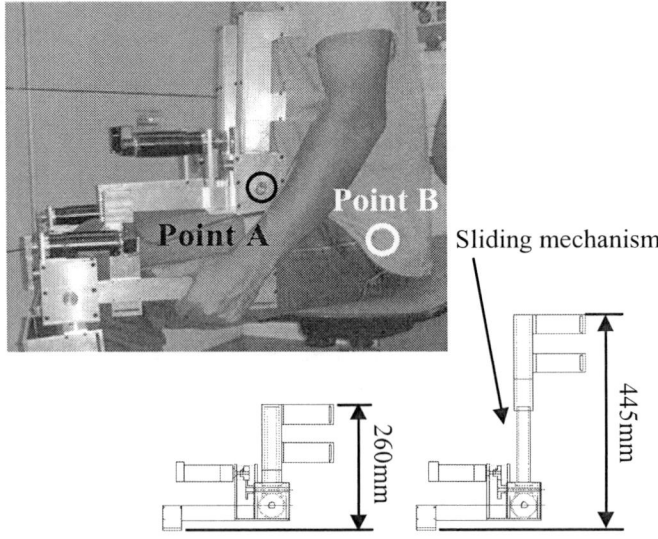

Figure 5. Difference of the rotation centers between robot (Point A) and human (Point B).

III. Cooperative Operation

This section presents three cooperative operations using the three modules described in the previous section. By combining and modifying these operations, the user is able to perform the functions in his daily life without special infrastructures.

Fig. 6 shows the model of a human with ABLE. The nomenclature is as follows:

Σ_f : Coordinate system relative to the floor.
Σ_a : Coordinate system relative to the ankle joint.
Σ_k : Coordinate system relative to the knee joint.
Σ_h : Coordinate system relative to the hip joint.
Σ_b : Coordinate system relative to the body.
Σ_s : Coordinate system relative to the shoulder joint.
Σ_c : Coordinate system relative to the crutch.
Σ_{G*1} : Coordinate system relative to the center of mass of the link *1.

$^{*1}\boldsymbol{p}_{*2} = [^{*1}x_{*2} \ ^{*1}y_{*2} \ ^{*1}z_{*2}]^T$: Vector from the origin in coordinate system Σ_{*1} to that in coordinate system Σ_{*2}.

$^{*1}L_{*2}$: Constant distance from the origin in coordinate system Σ_{*1} to that in coordinate system Σ_{*2}.

$^{*1}\theta_{*2}$: Angle from the x-axis in coordinate system Σ_{*1} to that in coordinate system Σ_{*2}.

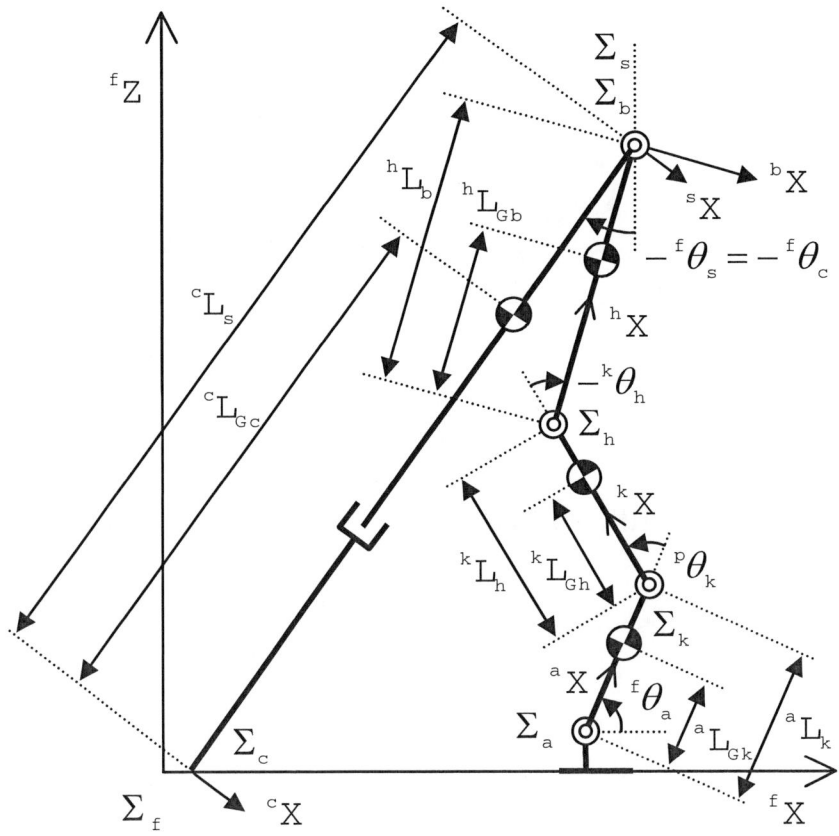

Figure 6. Model of a human with ABLE.

By applying the Newton-Euler equations on the condition of a quasi-static model, the forces and torques are derived as follows:

$$f_{fz} = \frac{\begin{bmatrix} M_{khbc}(md_h - nd_t) + \\ (^aL_{Gk}M_k + ^aL_kM_{hbc})\cos^f\theta_a + (^kL_{Gh}M_h + ^kL_hM_{bc})\cos^f\theta_k \\ + (^hL_{Gb}M_b + ^hL_bM_c)\cos^f\theta_h + (^cL_{Gc} - ^cL_s)M_c\cos^f\theta_s \end{bmatrix} g}{\begin{bmatrix} (md_h - nd_t) + \\ (^aL_k\cos^f\theta_a + ^kL_h\cos^f\theta_k + ^hL_b\cos^f\theta_h - ^cL_s\cos^f\theta_s) \end{bmatrix}} \quad (1)$$

$$F_c = \frac{f_{fz}}{\sin^f \theta_s} \qquad (2)$$

$$T_s = {}^c L_{Gc} M_c g \cos^f \theta_s + {}^c L_s (f_{fz} - M_c g) \cos^f \theta_s \qquad (3)$$

$$T_h = {}^h L_{Gb} M_b g \cos^f \theta_h - {}^h L_b (f_{fz} - M_c g) \cos^f \theta_h + T_s \qquad (4)$$

$$T_k = {}^k L_{Gh} M_h g \cos^f \theta_k - {}^k L_h (f_{fz} - M_{bc} g) \cos^f \theta_k + T_h \qquad (5)$$

$$T_a = {}^a L_{Gk} M_k g \cos^f \theta_a - {}^a L_k (f_{fz} - M_{hbc} g) \cos^f \theta_a + T_k, \qquad (6)$$

where M_{*1} is the mass of the link $*1$ and $M_{*1*2} = M_{*1} + M_{*2}$. M_k includes the mass of the mobile platform. d_h and d_t are the distances from the origin in coordinate system Σ_a to the heel and the toe of the mobile platform respectively. F_c is the floor reaction force, and f_{fz} is the perpendicular component of F_c. T_{*1} is the torque acting at the joint $*1$. g is the acceleration of gravity. m and n are the ratio of the load at the point of heel and toe respectively, and these are set at $m = n = 0.5$.

A. Traveling in a Standing Position

This action has two kinds of modes: straight motion and rotating motion. In both motions, at least one side of the crutches is on the ground in each mode to prevent falling over.

1) Straight Motion: During straight motion, all angles of the powered lower extremity orthosis are fixed and the mobile platform produces a propelling force to move the user. In this case, the crutches repeat telescopic motions so that the user may touch the ground with the crutches immediately when losing balance. It is desirable that the motions are automated. Automation of telescopic motions is realized by the installation of inclinometers to the telescopic crutch.

The optimum length of the crutch ${}^c L_s$ is derived from the joint angles of each joint, and the tilted angle of the crutch as follows:

$${}^f z_s = {}^f z_a + {}^a L_k \sin^f \theta_a + {}^k L_h \sin^f \theta_k + {}^h L_b \sin^f \theta_h \qquad (7)$$

$${}^c L_s = \frac{|{}^f z_s|}{\cos^f \theta_c}. \qquad (8)$$

The simulation of traveling straight on a flat surface in constant velocity is shown in Figure 7.

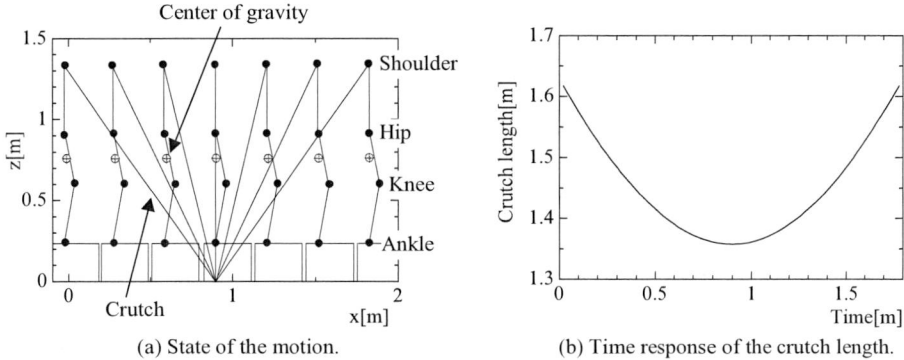

(a) State of the motion. (b) Time response of the crutch length.

Figure 7. Simulation of traveling in a standing position.

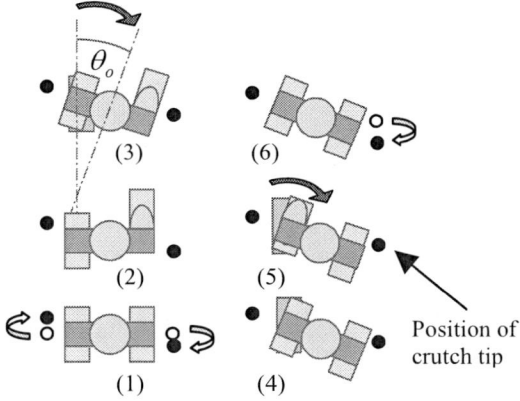

Figure 8. Sequence when rotating in a standing position.

2) Rotating Motion: When rotating, the powered lower extremity orthosis lifts the legs, the rotation board of the mobile platform rotates the base of the platform, and the crutches support the body, as shown in Fig. 8. In this figure, panels (1) to (6) show the process of rotating from the top view. The following numbers of each item correspond to the panel numbers.

(1) The process starts from the state of support of both legs. Each contact point of both crutches on the ground is changed as shown in this panel.
(2) The right leg is lifted from the ground by the powered lower extremity orthosis. The right knee is described in this panel as it is lifted.
(3) The rotation board of the left side is made to rotate for optional angles θ_0 by using the up/down switches attached to the grip of the crutch (see Fig. 2).
(4) The right leg is grounded again.
(5) The left leg is lifted and the rotation board of the left side is made to rotate for θ_0 against the ground. This operation is controlled automatically using a potentiometer and a load cell (see Fig. 3).
(6) The left leg is grounded, and then the contact point of the right crutch on the ground is changed.

B. Standing up and Sitting down

The telescopic crutches play an important role in standing up and sitting down motions, as shown in Fig. 9. They were placed under the armpits, and supported the body weight. The powered lower extremity orthosis controlled the angle of each joint in synchronization with the crutches. The mobile platforms maintained their positions.

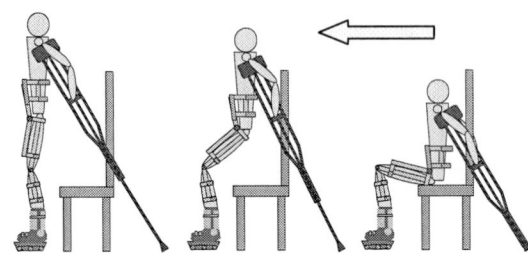

Figure 9. Standing up motion from a chair.

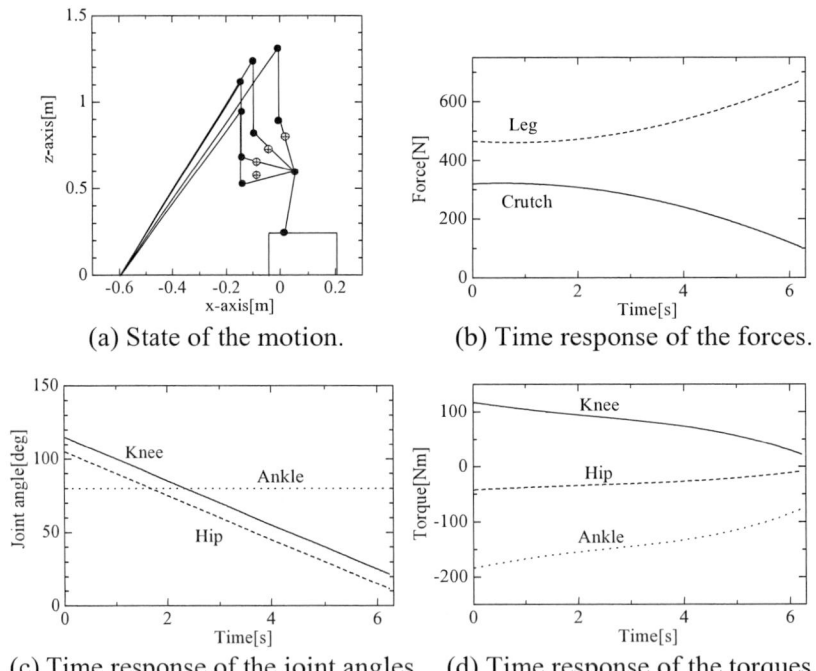

(a) State of the motion.   (b) Time response of the forces.

(c) Time response of the joint angles.   (d) Time response of the torques.

Figure 10. Simulation of standing up motion from a chair.

Fig. 10 shows a simulation of standing up from a seated position on the condition of a quasi-static model. The height of the chair is 40cm, and the distance from the ankle to the tip of the crutch is 60cm. Fig. 10(a) shows the state; Fig. 10(b) shows the time response of the floor reaction force for one side of the body; Figs. 10(c) and 10(d) show the time response of each joint angle and each joint torque respectively.

The user can sit down in the exact reverse motion of the standing up motion.

C. Ascending and Descending Stairs

Ascending and descending stairs motions are realized by applying the standing up motion from a chair. Basically, the mobile platforms do not move without approaching the stairs. A measurement wheel attached at the tip of the crutch measures the parameters of stairs (see Fig. 2).

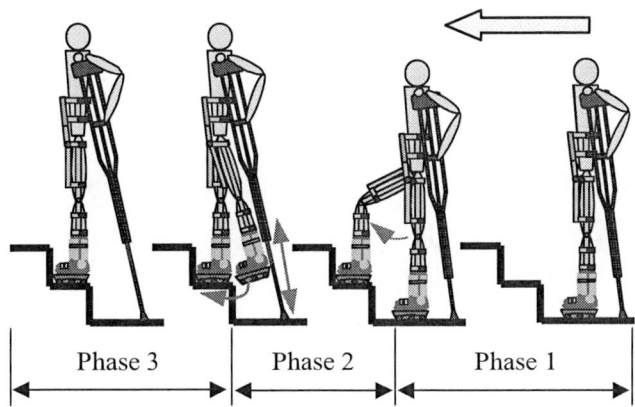

Figure 11. Ascending stairs motion.

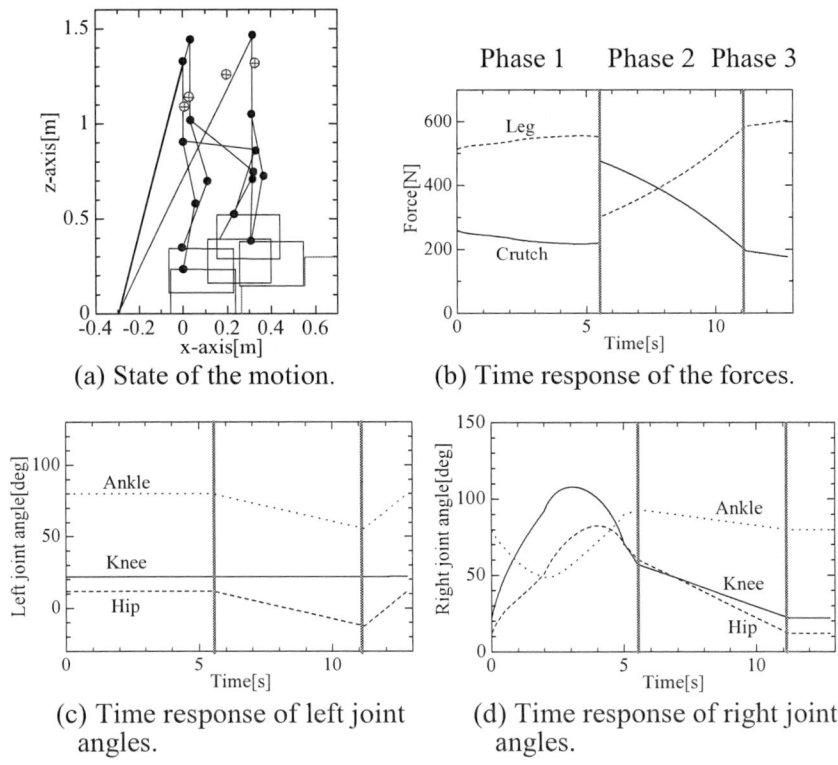

(a) State of the motion.

(b) Time response of the forces.

(c) Time response of left joint angles.

(d) Time response of right joint angles.

Figure 12. Continued on next page.

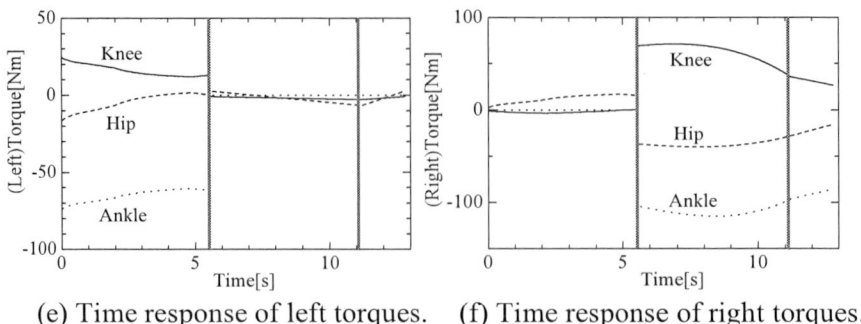

(e) Time response of left torques.   (f) Time response of right torques.

Figure 12. Simulation of ascending stairs motion.

Phases 1 to 3 of Fig. 11 show the sequence of ascending stairs. Fig. 11 shows a simulation of ascending stairs where the height of the stair $H=15$ cm and the depth $D=30$ cm: Fig. 12(a) shows the state; Fig. 12(b) shows the time response of the floor reaction force for one side of the body. Figs. 12(c) and 12(d) show the time response of the joint angles; Figs. 12(e) and 12(f) show the time response of the leg torque.

The descending stairs motion is harder than the ascending stairs motion because of a fear of falling. However, the descending stairs motion isn't so scary if the user descends stairs in the back direction. In this case, it is true that the operation is troublesome because of the difficulty of watching one's footing, but the sequence itself is the reversed sequence of the ascending stairs motion. We have tried the descending stairs motion in this sequence without the powered lower extremity orthosis, and have obtained the conviction of its feasibility and safety.

IV. New Motion Technique for Practical Use

A. Standing Up and Sitting Down

The use of the crutches in the previous method is likely to lead to slipping and falling because the crutches were located behind the body. In addition, the location of the crutches makes it possible to sit on a long bench, e.g. in a train.

We propose a new motion sequence as shown in Fig. 13.

(1) Each crutch is detached from the armpits and is in front of the body. The bust is bent over.
(2) The user is standing up, keeping the bust at same angle throughout the motion. The center of gravity is between feet.
(3) Each crutch is put under the armpits when the arms are stretched.
(4) The crutches are extended synchronizing with the standing motion of the powered lower extremity orthosis.

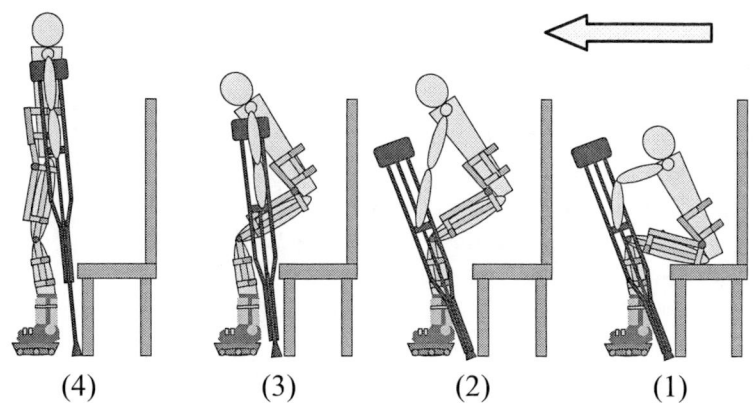

Figure 13. New standing up motion from a chair.

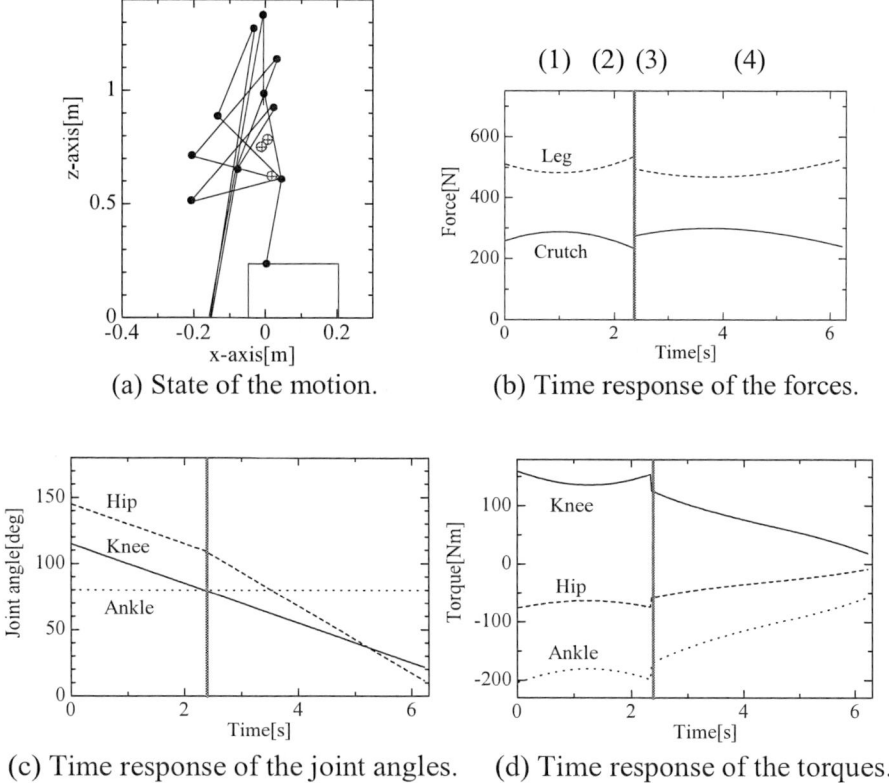

(a) State of the motion.

(b) Time response of the forces.

(c) Time response of the joint angles.

(d) Time response of the torques.

Figure 14. Simulation of new standing up motion from a chair.

The crutches are used not chiefly but rather as assistance. The telescopic motion of the crutches is not performed until Stage (3). The arms absorb the errors caused by the difference of location of the crutches. Therefore this method is adaptable to various environments.

Fig. 14 shows a simulation of the new standing up motion from a seated position on the condition of a quasi-static model. The height of the chair is the same as in Fig. 10, and the distance from the ankle to the crutch is 15cm. In comparison with Fig. 10, each value of the joint angle and joint torque are almost equal.

The new technique is superior to the previous one in terms of potential danger from a slip of the crutches. Fig. 10 (a) and Fig. 14 (a) show the difference of the position of the center of mass. In the case of the previous method, the user cannot maintain his posture in the period from 0 s to 4.1 s without crutches. On the other hand, in the case of the new technique, the user does not fall during the motion because the position of the center of mass is on the mobile platforms. Furthermore, in the case of the new technique, the crutch resists slipping because the maximum tilted angles of the crutch of the previous and new techniques are ${}^f\theta_c = -23.9°$ at time $t = 6.2$ s and ${}^f\theta_c = -10.3°$ at time $t = 2.4$ s respectively.

B. Ascending and Descending Stairs

In ascending stairs, similar to the case of seating, the telescopic crutches are behind the body. Therefore, it is dangerous because a user may tip over with the slip of the crutches in Phase 2 in Fig. 11.

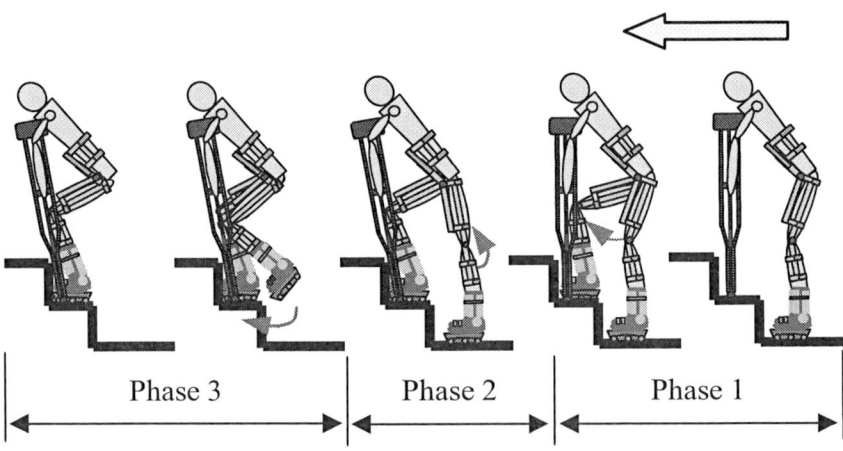

Figure 15. New ascending stairs motion.

Fig. 15 shows a new sequence for ascending stairs. The crutches are in front of the body and they are used as assistance. In Phase 2, the center of gravity of the body is shifted from the left foot to the right foot as the knee joint is stretched. The simulation results using this new sequence are shown in Fig. 16. In comparison with the previous method, the safety against slipping is enhanced in the new method. It is not necessary to control the crutches in the new method because both lengths of the crutches are constant throughout the motion sequence. In the new motion technique, large torque is required for the powered lower extremity orthosis. The crutches, however, consume no energy if the length of the crutch is in the minimum limit. So the overall difference in power consumption is not so great.

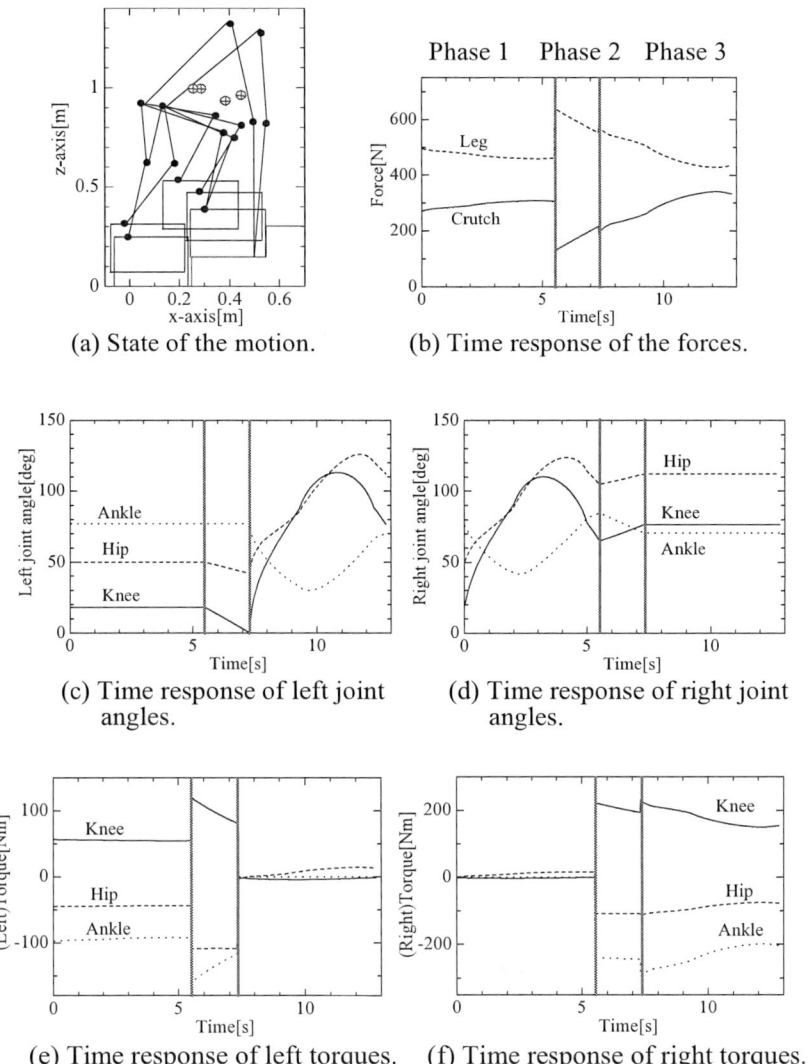

Figure 16. Simulation of new ascending stairs motion.

Similarly, in the case of the standing up motion, the new technique is superior to the previous one due to safety against the slipping of the crutches. Fig. 12 (a) and Fig. 16 (a) show the difference of the position of the center of mass. In the case of the previous method, the user cannot maintain his posture in the period from 0 s to 9.3 s without crutches. On the other hand, the user does not fall during the motion because the position of the center of mass is on the mobile platforms in the case of the new technique. The maximum tilted angles of the crutch of the previous and new techniques are $^{f}\theta_{c} = -21.7°$ in the period from 11.2 s to 12.8 s and $^{f}\theta_{c} = -3.0°$ in the period from 7.3 s to 12.8 s respectively. So the crutch resists slipping in the case of the new technique.

V. Experiment

We have tried five motions to verify the ability of ABLE. The subject was a man with no leg motion impairment whose height was 166cm and weight was 60kg.

Each module was controlled based on PD control theory. In Subsections A, B and C we used V55 CPU board (CPU 16MHz, Japan System Design CO., Ltd.). The sampling time was 30ms. In Subsections D and E we used ART-Linux as an operating system in view of the stability [10]. The PC was Libretto ff (CPU 233MHz, MMX Pentium). The interface board was a Ritech Interface Board (IF-0145-1, ZUCO Co., Ltd.). The command was transmitted via wireless LAN. The sampling time was 10ms.

A. Traveling in a Standing Position

When a pair of the mobile platforms traveled in a standing position at a constant velocity; a pair of the telescopic crutches was used for maintaining balance. Length of each crutch was automatically adjusted using an inclinometer. However, the output values of the inclinometer were disturbed at the moment when the crutch touched on the ground. Data were filtered by a low-pass filter that had the first Butterworth characteristics. Results are shown in Fig. 17.

Figs 17(a) and 17(b) show the relationship between the position of the body and the crutch length and angle respectively. Fig. 17(c) shows correspondence of the crutch angle and length, and it shows the tracking performance of the crutch. Crutches were controlled based on PD control theory. There were some errors, however, it did not matter practically. The time response of the telescopic motion of the crutch left room for improvement.

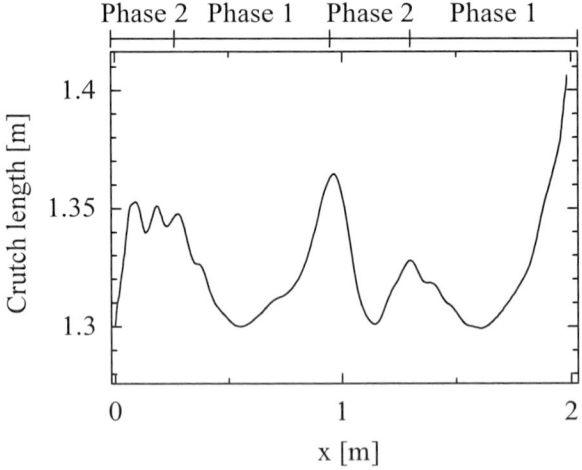

(a) Time response of the crutch length.

Figure 17. Continued on next page.

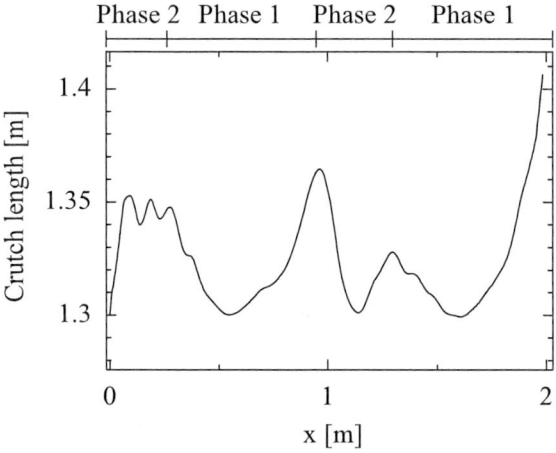

(a) Time response of the crutch length.

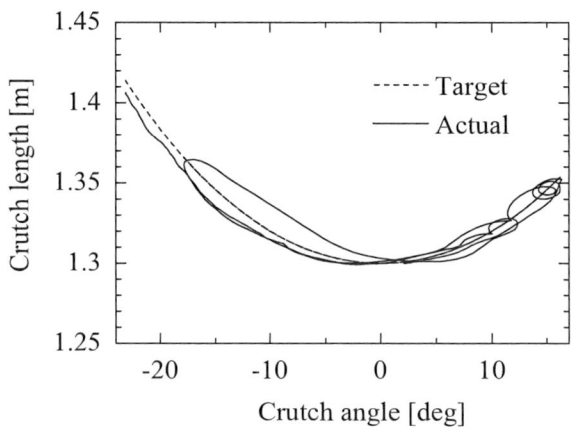

(c) Relationship between the crutch angle and the crutch length.

Figure 17. Experimental results when traveling in a standing position.

There is a danger to unavoidable accidents. The possibility of falling over is directly related to acceleration. Therefore, it will be necessary to control the mobile platforms so that the acceleration of them may not change abruptly. Bending the legs to some extent will be also effective to prevent falling over.

B. Rotating in a Standing Position

The sequence of the rotating motion is shown in Fig. 8. It is difficult to go straight if substantial rotational error remains after the rotating action. A returning operation to the neutral rotational angle would preferably be controlled automatically, not manually, at (5) of Fig. 8. The load cell installed between the base part and the rotation board senses the load weight. It informs the timing when the rotation board should be rotated. Figure 18 shows time responses of the load and the angle of the rotation board of the left side. "The boundary value" is a reference load that determines whether the leg is on or off the ground. The value was determined through experiments. The rotation board was controlled based on PD control theory. The final error of the rotation board was about 0.19 deg. In the experiment, the subject was quite stable from the beginning to the end.

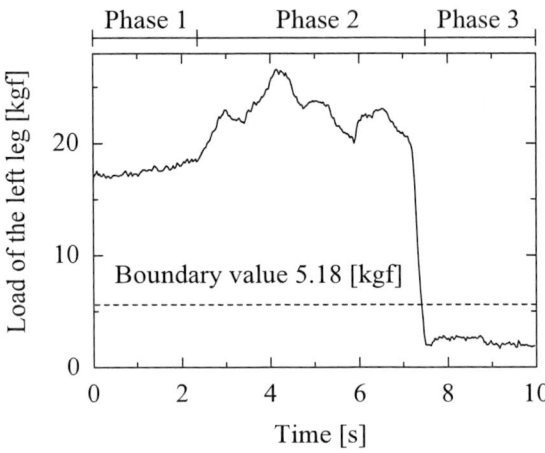

(a) Time response of the load of the left leg.

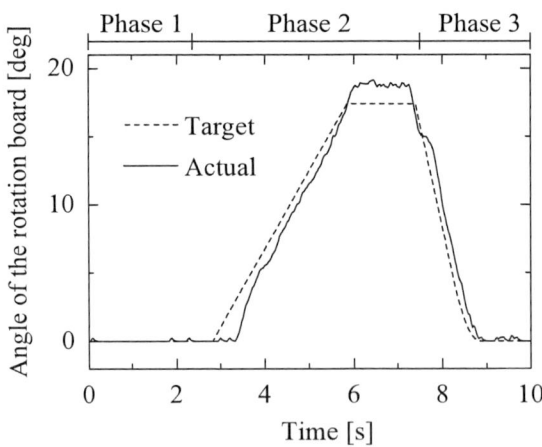

(b) Time response of the angle of the rotation board

Figure 18. Experimental results when rotating in a standing position.

C. Standing up from a Chair

This experiment, which elucidates a standing up motion from a chair, corresponds to the simulation shown in Fig. 9. The subject stood up from a chair using only a pair of crutches. The result is shown in Fig. 19.

A load plate that comprised three load cells measured the load to the left foot; the load to the left crutch was measured by one load cell installed on the crutch tip. An exoskeletonic measurement instrument measured waist, knee, and left leg ankle angles. In Fig. 19, the load is shown to undergo transition from the crutch to the foot smoothly during standing. One cause of the difference between Fig. 19 and Fig. 10 (b) is the position of the center of gravity. In the simulation, the center of gravity position is fixed on the center of the foot regardless of the body state. Another cause is the load against the chair, which is not considered in the simulation. Therefore, the sum of the ground reaction forces is not sufficiently large at the beginning in Fig. 19. This simulation is available to determine the optimal positions and the acceptable error ranges of the crutches and the legs.

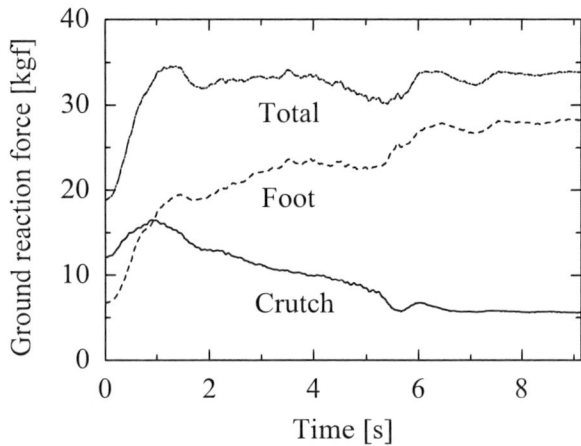

Figure 19. Experimental results when standing up motion from a chair.

D. New Standing up from a Chair

The snap shots of an experiment of standing up motion from a chair in the new technique are shown in Fig. 20. In Stages (1) and (3), the up/down switch attached to the right crutch was pressed to send the start signal to the PC. The experimental results of the time response of the angles and electric currents of each joint are shown in Fig. 21 and Fig. 22 respectively. Note that the time lag from 5s to 19s in Figs. 21 and 22 was caused by the motion to place the crutches under the armpits (see Fig. 13). In Fig. 22 the electric currents correspond with the torque of each joint. Although there are some tracking errors in Fig. 21, the subject succeed in standing up. The user required large arm force in the beginning of Stage (1), but no such force was needed afterwards. The user felt little fear of falling with the slip of the crutches.

Figure 20. Snap shots of an experiment of new standing up motion from a chair.

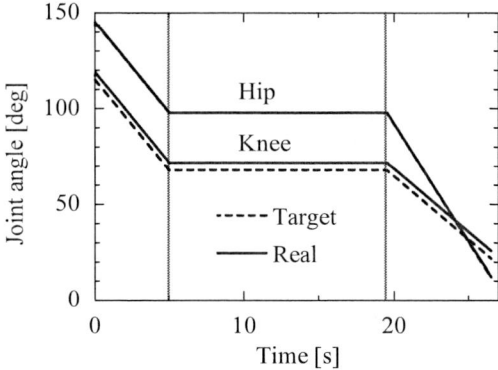

Figure 21. Experimental results of new standing up motion from a chair (Joint angles).

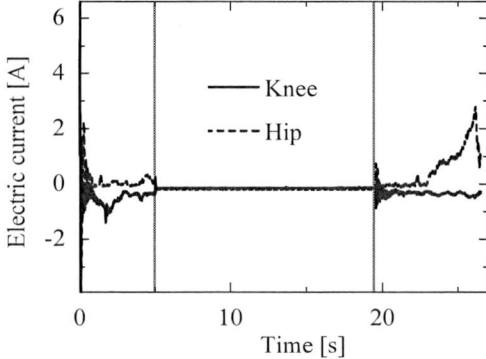

Figure 22. Experimental results of new standing up motion from a chair (Electric currents).

Figure 23. Passing through a ticket gate.

E. Passing through an Automated Ticket Gate

Fig. 23 shows the image of the subject passing through an automated ticket gate that was not designed for wheelchair users. After traveling straight for about 2m, the subject stopped just before the gate, took out a ticket, and passed through the gate in the usual way. The target velocity was 0.15m/s. The subject could proceed stably using the crutches.

VI. Conclusions

This study proposed a standing style transfer system "ABLE" that helps users with disabled lower limbs to enjoy mobility without special infrastructure. ABLE consists of a pair of telescopic crutches, a pair of mobile platforms, and a powered lower extremity orthosis. We showed the conceptual design of ABLE and the motion of each module.

We proposed a new technique for a standing up motion from a chair, and ascending stairs. The new sequence was inferior to the previous one in the point of necessary torque of a powered lower extremity orthosis. It was, however, superior in the points of safety and a less controlled module.

The experiment showed the results of various motions: traveling and rotating in a standing position, and standing up from a chair. We also showed the outdoor experiment of passing through a ticket gate.

In future work, we plan to repeat improvement of this system for practical use and to try ABLE in various actual environments.

Acknowledgments

This research was partially supported by the Ministry of Education, Culture, Sports, Science and Technology, Grant-in-Aid for Young Scientists (B), 2003, 15760176.

The author would like to thank Kazuhiro Takayama, Takeshi Zengo, Ruijyu Tsukamoto, Kyotaro Tomoda and Jun Okada who have been his lab members in Tokyo Metropolitan University and have contributed great efforts and lots of help.

References

[1] Ministry of Health, Labor and Welfare Official Web Site, "Physically handicapped child and person field study result," http://www.mhlw.go.jp/houdou/2002/08/h0808-2b.html, (in Japanese), 2002.
[2] Yobotics, "RoboWalker," http://www.yobotics.com/robowalker.
[3] T. Nakamura, K. Saito, Z. Wang and K. Kosuge, "Realizing Model-based Wearable Antigravity Muscles Support with Dynamics Terms," *Proc. of IEEE IROS*, 2005, pp.3443–3448.
[4] S. Lee and Y. Sankai, "Power Assist Control for Walking Aid with HAL-3 Based on EMG and Impedance Adjustment around Knee Joint," *Proc. of IEEE IROS*, 2002, pp.1499–1504.

[5] H. Kawamoto and Y. Sankai, "Power assist control for leg with hal-3 based on virtual torque and impedance adjustment," *Proc. of IEEE SMC*, TP1B3, 2002.
[6] K. Ikeda, T. Iwatsuki, and S. Kajita, "Basic Study on an Ambulatory Apparatus with Weight Bearing Control," *Journal of Mechanical Engineering Laboratory*, Vol. 52, No. 4, 1998, pp.1–8.
[7] Betto, H. Yano, S. Kaneko, H. Torii, S. Fujitani, "Development of ambulatory apparatus equipped with function of weight bearing control system (WBC Orthoses Series)," *Japanese Society of Prosthetics and Orthotics*, (in Japanese), 1998, pp.41–48.
[8] Y. Mori, K. Takayama, T. Zengo and T. Nakamura,"Development of Straight Style Transfer Equipment for Lower Limbs Disabled: Verification of Basic Motion," *Journal of Robotics and Mechatronics*, Vol.16, No.5, 2004, pp.456-463.
[9] Y. Mori, J. Okada and K. Takayama, "Development of a Standing Style Transfer System "ABLE" for Lower Limbs Disabled," IEEE/ASME Trans. on Mechatronics, Vol.11, No.4, 2006, pp.372-380.
[10] Woo-Keun Yoon,"PA-10 Control on Linux," *Proc. of 2003 JSME Conference on Robotics and Mechatronics*, (in Japanese), 2003, 2P1-3F-D7(1)-(2).

In: Robotics Research Trends
Editor: Xing P. Guô, pp. 333-355

ISBN: 1-60021-997-7
© 2008 Nova Science Publishers, Inc.

Chapter 11

STOCHASTIC ANALYSIS OF A REPAIRABLE STANDBY ROBOT-SAFETY SYSTEM

Shen Cheng and B.S. Dhillon
Department of Mechanical Engineering,
University of Ottawa, Ottawa, Ontario K1N 5N6, Canada

Abstract

This paper presents reliability and availability analyses of a mathematical model representing one robot and (n-1) standby safety units with a perfect switch to replace a failed safety unit. Robot and safety unit failure rates are assumed constant and the failed system repair times are assumed arbitrarily distributed. General expressions for state probabilities, system availabilities, reliability, and mean time to failure were obtained by using Markov and supplementary variable methods. General expressions for the robot - safety system steady state availability were developed for exponential, gamma, Weibull, Rayleigh, and lognormal failed system repair time distributions. Some plots of these expressions for special case models are shown. These plots shows that the robot-safety system availability, state probabilities decrease with time and robot-safety system steady state availability increases with the increasing values of the safety unit repair rate.

Introduction

Nowadays, the application of robots covers almost all aspects of our daily life and the population of robots is increasing at an impressive rate [1-4]. According to the findings of the United Nations Economic Commission for Europe (UN/ECE) and the International Federation of Robotics (IFR), the installations of industrial robots alone will be around 121,000 by 2008 throughout the world [5]. However, the negative effects of robotization must be considered seriously, since many people are injured and even killed by robots over the years [6-8]. As an unreliable robot may cause accidents, the inclusion of safety units in robot systems has become an important issue.

Thus, this paper presents a mathematical to perform reliability and availability analyses of a robot-safety system having one robot and (n-1) standby safety units with a perfect switch to replace a failed safety unit. Some examples of safety units are light curtains, flashing lights, and intelligent systems [6]. The block diagram of this robot-safety system is shown in Fig. 1 and its corresponding state space diagram is presented in Fig. 2. The numerals and letter n in the boxes of Fig. 2 denote system states.

As soon as a safety unit fails, it is replaced with one of the standby units. The overall robot-safety system can fail, the following three ways:

- The robot-safety system fails with an incident.
- The robot-safety system fails safely.
- The robot-safety system fails due to the malfunction of the robot.

The partly or fully failed robot-safety system is repaired.
The following assumptions are associated with this model:

- The robot-safety system is composed of one robot, n identical safety units (only one operates and the rest remain on standby) and a perfect switch.
- Robot, switch and one safety unit start operating simultaneously at time $t = 0$.
- The perfect switch means it never fail or the failure rate is too small to be considered compared with failure rates of other parts such as robot and safety unit.
- The completely failed robot-safety system and its individually failed units (i.e. robot, and safety unit) can be repaired. Failure and repair rates of robot and safety units are constant.
- The failed robot-safety system repair rates can be constant or non-constant.
- All failures are statistically independent.
- A repaired safety unit, robot or the total robot-safety system is as good as new.

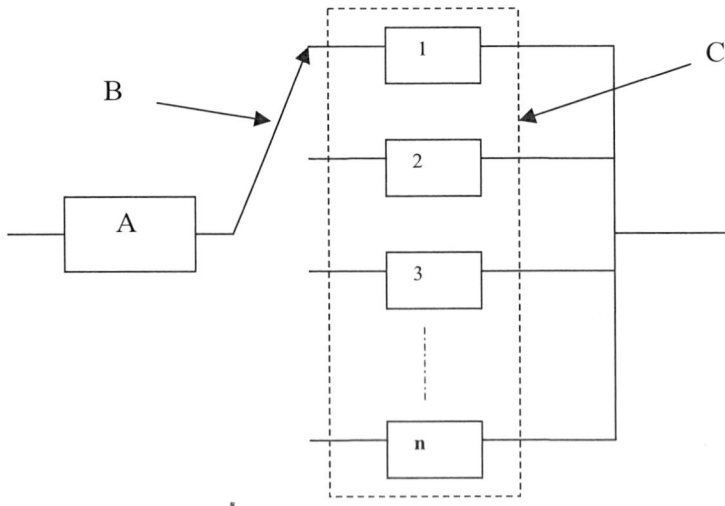

A : Robot
B : Switch for replacing a failed safety unit and it cannot fail.
C: n identical safety units (one operating and n-1 on standby)

Figure 1. The block diagram of the robot-safety system containing one robot and (n-1) standby safety units with a perfect switch.

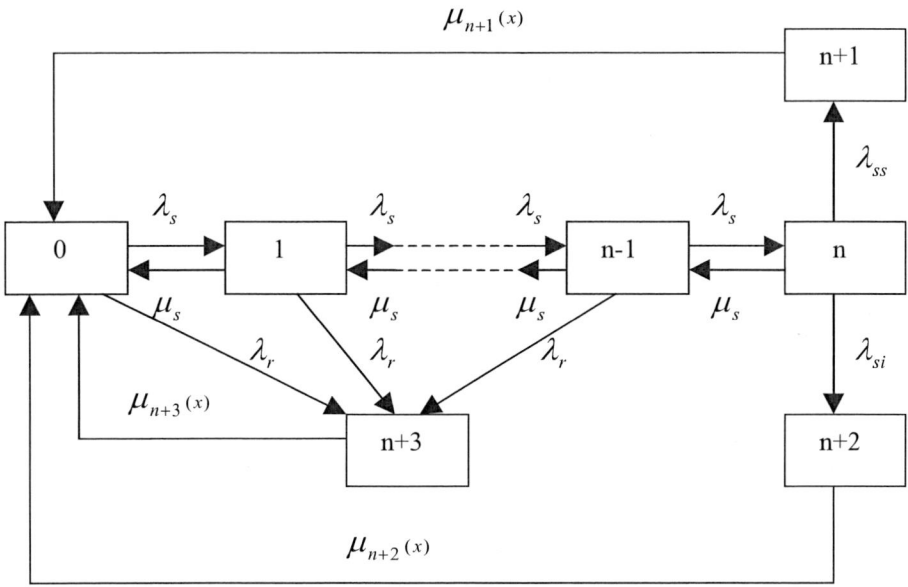

Figure 2. The state space diagram of the robot-safety system containing one robot and (n-1) standby safety units with a perfect switch.

Notation

The following symbols are associated with the model:

- i i^{th} state of the robot-safety system:
  - for $i = 0$, means the robot, the switch and one safety unit are working normally;
  - for $i = 1$, means the robot, the switch, one safety unit are working normally and one safety unit has failed;
  - for $i = k$, means the robot, the switch, one safety unit are working normally and k safety units have failed; (i.e., $k = 2, 3 \ldots \ldots n-1$);
  - for $i = n$, means the robot, the switch are working normally and all the safety units have failed.
- j j^{th} state of the robot–safety system:
  - for $j = n + 1$, the robot-safety system failed safely;
  - for $j = n + 2$, the robot-safety system failed with an incident;
  - for $j = n + 3$, the robot-safety system failed due to the malfunction of the robot;
- t Time.
- λ_s Constant failure rate of the safety unit.
- λ_r Constant failure rate of the robot.
- λ_{ss} Constant failure rate of the robot-safety system failing safely.
- λ_{si} Constant failure rate of the robot-safety system failing with an incident.

μ_s	Constant repair rate of the safety unit.
Δx:	Finite repair time interval.
$\mu_j(x)$	Time dependent repair rate when the failed robot-safety system is in State j and has an elapsed repair time of x; for j = n+1, n+2, n+3.
$P_j(x,t) \Delta x$	The probability that at time t, the failed robot-safety system is in state j and the elapsed repair time lies in the interval [x, x+Δx]; for j = n+1, n+2, n+3.
pdf	Probability density function.
$w_j(x)$	Pdf of repair time when the failed robot-safety system is in state j and has an elapsed time of x; for j = n+1, n+2, n+3.
$P_j(t)$	Probability that the robot-safety system is in state j at time t; for j = n+1, n+2, n+3.
$P_i(t)$	Probability that the robot-safety system is in state i at time t; for i = 0, 1,2…n.
P_i	Steady state probability that the robot-safety system is in state i; for i = 0, 1…n.
P_j	Steady state probability that robot-safety system is in state j; for j = n+1, n+2, n+3.
s	Laplace transform variable.
$P_i(s)$	Laplace transform of the probability that the robot-safety system is in state i; for i = 0,1,2…n.
$P_j(s)$	Laplace transform of the probability that the robot-safety system is in state j; for j = n+1, n+2, n+3.
AVrs(s)	Laplace transform of the robot-safety system availability with one normally working safety unit, the switch and the robot.
AVr(s)	Laplace transform of the robot-safety system availability with or without a normally working safety unit.
AVrs(t)	Robot-safety system time dependent availability with one normally working safety unit, the switch and the robot.
AVr(t)	Robot-safety system time dependent availability with or without one normally working safety units.
SSAVrs	Robot-safety system steady state availability with one normally working safety unit.
SSAVr	Robot-safety system steady state availability with or without one normally working safety unit.
Rrs(s)	Laplace transform of the robot-safety system reliability with one normally working safety unit.
Rr(s)	Laplace transform of the robot-safety system reliability with or without one normally working safety unit.
MTTFrs	Robot-safety system mean time to failure with one normally working safety unit.
MTTFr	Robot-safety system mean time to failure with or without one normally working safety unit.

Generalized Robot-Safety System Analysis

By using the supplementary method [9, 10], we write the following equations for the Figure 2 diagram:

$$\frac{dP_0(t)}{dt} + a_0 P_0(t) = \mu_s P_1(t) + \sum_{j=n+1}^{n+3} P_j(x,t)\mu_j(x)dx \tag{1}$$

$$\frac{dP_i(t)}{dt} + a_i P_i(t) = \mu_s P_{i+1}(t) + \lambda_s P_{i-1}(t) \tag{2}$$

(for $i = 1, 2, \ldots, n-1$)

$$\frac{dP_n(t)}{dt} + a_n P_n(t) = \lambda_s P_{n-1}(t) \tag{3}$$

$$\frac{\partial P_j(x,t)}{\partial t} + \frac{\partial P_j(x,t)}{\partial x} + \mu_j(x) P_j(x,t) = 0 \tag{4}$$

(for $j = n+1, n+2, n+3$)

where

$a_0 = \lambda_s + \lambda_r$

$a_i = \lambda_s + \lambda_r + \mu_s$ (for $i = 1, 2, \ldots, n-1$)

$a_n = \lambda_{ss} + \lambda_{si} + \mu_s$

The associated boundary conditions are as follows:

$$P_{n+1}(0,t) = \lambda_{ss} P_n(t) \tag{5}$$

$$P_{n+2}(0,t) = \lambda_{si} P_n(t) \tag{6}$$

$$P_{n+3}(0,t) = \lambda_r \sum_{i=0}^{n-1} P_i(t) \tag{7}$$

At time $t = 0$, $P_0(0) = 1$, and all other state probabilities are equal to zero.

Using Laplace transforms and solving Equations (1) – (7), we get:

$$(s + a_0) P_0(s) = 1 + \mu_1 P_1(s) + \sum_{j=n+1}^{n+3} \int_0^\infty P_j(x,s)\mu_j(x)dx \tag{8}$$

$$(s + a_i) P_i(s) = \mu_{i+1} P_{i+1}(s) + \lambda_i P_{i-1}(s) \quad \text{(for } i = 1, 2, \ldots, n-1\text{)} \tag{9}$$

$$(s + a_n) P_n(s) = \lambda_n P_{n-1}(s) \tag{10}$$

$$sP_j(x,s) + \frac{\partial P_j(x,s)}{\partial x} + \mu_j(x) P_j(x,s) = 0 \quad (\text{for } j = n+1, n+2, n+3) \tag{11}$$

The boundary conditions are as follows:

$$P_{n+1}(0, s) = \lambda_{ss} P_n(s) \tag{12}$$

$$P_{n+2}(0, s) = \lambda_{si} P_n(s) \tag{13}$$

$$P_{n+3}(0, s) = \lambda_r \sum_{i=0}^{n-1} P_i(s) \tag{14}$$

Solving Equation (11), we get:

$$P_j(s) = P_j(0,s) e^{-sx} \exp[-\int_0^x \mu_j(\delta) d\delta] \quad (\text{for } j = n+1, n+2, n+3) \tag{15}$$

Together with

$$P_j(s) = \int_0^\infty P_j(x,s) dx \quad (\text{for } j = n+1, n+2, n+3) \tag{16}$$

we obtain:

$$P_j(s) = P_j(0,s) \frac{1 - W_j(s)}{s} \quad (\text{for } j = n+1, n+2, n+3) \tag{17}$$

Where

$$\frac{1 - W_j(s)}{s} = P_j(0,s) \int_0^\infty e^{-sx} \exp[-\int_0^x \mu_j(\delta) d\delta] \tag{18}$$

$$(\text{for } j = n+1, n+2, n+3)$$

or

$$W_j(s) = \int_0^\infty e^{-sx} w_j(x) dx \quad (\text{for } j = n+1, n+2, n+3) \tag{19}$$

$$w_j(x) = \exp[-\int_0^x \mu_j(\delta) d\delta] \mu_j(x) \tag{20}$$

where $w_j(x)$ is the failed robot safety system repair time probability density function.

Using Equations (8) – (20), we get the following Laplace transforms of state probabilities:

$$P_0(s) = [s(1+\sum_{i=1}^{n} Y_i(s) + \sum_{j=n+1}^{n+3} a_j(s)\frac{1-W_j(s)}{s})]^{-1} = \frac{1}{G(s)} \qquad (21)$$

$$P_i(s) = Y_i(s)P_0(s) \qquad (22)$$
$$(\text{for } i = 1,2,\ldots\ldots n)$$

$$P_j(s) = a_j(s)\frac{1-W_j(s)}{s}P_0(s) \qquad (\text{for } j = n+1,n+2,n+3) \qquad (23)$$

where

$$Y_i(s) = \prod_{k=1}^{i} L_k(s) \quad (\text{for } i = 1,2,\ldots\ldots n)$$

$$a_{n+1}(s) = \lambda_{ss} Y_n(s)$$

$$a_{n+2}(s) = \lambda_{si} Y_n(s)$$

$$a_{n+3}(s) = \lambda_r [1+\sum_{i=1}^{n-1} Y_i(s)]$$

$$L_n(s) = \frac{\lambda_s}{s+a_n}$$

$$L_i(s) = \frac{\lambda_s}{(s+a_i)-\mu_s L_{i+1}(s)} \quad (\text{for } i = 1, 2\ldots\ldots n-1)$$

$$G(s) = s[1+\sum_{i=1}^{n} Y_i(s) + \sum_{j=n+1}^{n+3} a_j(s)\frac{1-W_j(s)}{s}]^{-1} \qquad (24)$$

$$W_j(s) = \int_0^\infty e^{-sx} w_j(x)dx \quad (\text{for } j = n+1,n+2,n+3) \qquad (25)$$

The Laplace transform of the robot-safety system availability with one normally working safety unit, the switch, and the robot is given by:

$$AV_{rs}(s) = \sum_{i=0}^{n-1} P_i(s) = \frac{1+\sum_{i=1}^{n-1} Y_i(s)}{G(s)} \qquad (26)$$

The Laplace transform of the robot-safety system availability with or without one normally working safety unit is given by:

$$AV_r(s) = \sum_{i=0}^{n} P_i(s) = \frac{1 + \sum_{i=1}^{n} Y_i(s)}{G(s)} \qquad (27)$$

Taking the inverse Laplace transforms of the above equations, we can get the time dependent state probabilities, $P_i(t)$ and $P_j(t)$, and robot-safety system availabilities, AVrs(t) and AVr(t).

Robot-Safety System Time Dependent Analysis for a Special Case

For three safety units (i.e. one working, others on standby), by substituting $n = 3$ into Equations (21) - (27), we obtain:

$$P_0(s) = \frac{1}{s[1 + \sum_{i=1}^{2} Y_i(s) + \sum_{j=3}^{5} a_j(s) \frac{1 - W_j(s)}{s}]} \qquad (28)$$

$$P_i(s) = Y_i(s) P_0(s) \qquad \text{(for } i = 1, 2, 3) \qquad (29)$$

$$P_j(s) = a_j(s) \frac{1 - W_j(s)}{s} P_0(s) \qquad \text{(for } j = 4, 5, 6) \qquad (30)$$

where

$$Y_i(s) = \prod_{k=1}^{i} L_k(s) \quad \text{(for } i = 1, 2, 3)$$

$$a_4(s) = \lambda_{ss} Y_n(s)$$

$$a_5(s) = \lambda_{si} Y_n(s)$$

$$a_6(s) = \lambda_r [1 + \sum_{i=1}^{2} Y_i(s)]$$

$$L_3(s) = \frac{\lambda_s}{s + a_3}$$

$$L_i(s) = \frac{\lambda_s}{(s + a_i) - \mu_s L_{i+1}(s)} \qquad \text{(for } i = 1, 2)$$

$$G(s) = s[1+\sum_{i=1}^{3}Y_i(s)+\sum_{j=4}^{6}a_j(s)\frac{1-W_j(s)}{s}] \qquad (31)$$

The Laplace transform of the robot-safety system availability with one normally working safety unit, the switch and the robot is given by:

$$AV_{rs}(s) = \sum_{i=0}^{2}P_i(s) = \frac{1+Y_1(s)}{G(s)} \qquad (32)$$

The Laplace transform of the robot-safety system availability with or without a normally working safety unit is given by:

$$AV_r(s) = \sum_{i=0}^{3}P_i(s) = \frac{1+\sum_{i=1}^{3}Y_i(s)}{G(s)} \qquad (33)$$

Taking the inverse Laplace transforms of the above equations, we can obtain the time dependent state probabilities, $P_i(t)$ and $P_j(t)$, and robot-safety system availabilities, AVrs(t) and AVr(t) respectively.

Thus, for the failed robot-safety system exponentially distributed repair times, the probability function is expressed by

$$w_j(x) = \mu_j e^{-\mu_j x} \qquad (\mu_j > 0, j = 4,5,6) \qquad (34)$$

where x is the repair time variable and μ_j is the constant repair rate of state j.

Substituting Equation (34) into Equation (19), we get

$$W_j(s) = \frac{\mu_j}{s+\mu_j} \qquad (\mu_j > 0, j = 4, 5, 6) \qquad (35)$$

By inserting Equation (35) into Equations (28) – (33), setting $\lambda_s = 0.001$, $\lambda_r = 0.0009$, $\lambda_{ss} = 0.0035$, $\lambda_{si} = 0.0015$, $\mu_s = 0.0002$, $\mu_4 = 0.0003$, $\mu_5 = 0.0001$, $\mu_6 = 0.00035$; and using Matlab computer program [11], the Figure 3 and Figure 4 were obtained. These plots show that state probabilities and system availabilities decrease and increase with varying time t.

$\lambda_s = 0.001$, $\lambda_r = 0.0009$, $\lambda_{ss} = 0.0035$, $\lambda_{si} = 0.0015$, $\mu_s = 0.0002$, $\mu_4 = 0.0003$, $\mu_5 = 0.0001$, $\mu_6 = 0.00035$

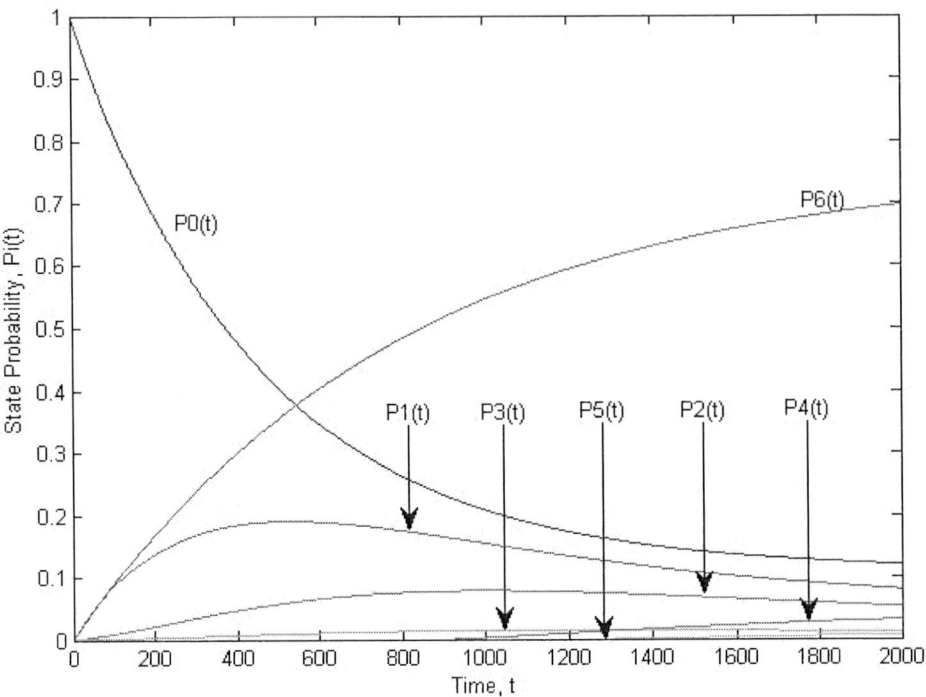

Figure 3. Time-dependent state probability plots for the special case robot-safety system with exponentially distributed failed system repair times.

Generalized Robot-Safety System Steady State Analysis

As time approaches infinity, all state probabilities reach the steady state. Thus, from Equations (1)-(7) we get:

$$a_0 P_0 = \mu_s P_1 + \sum_{j=n+1}^{n+3} \int_0^\infty \mu_j(x) P_j(x) dx \tag{36}$$

$$a_i P_i = \mu_s P_{i+1} + \lambda_s P_{i-1} \quad \text{(for } i = 1,2,3\ldots\ldots,n\text{-}1\text{)} \tag{37}$$

$\lambda_s = 0.001$, $\lambda_r = 0.0009$, $\lambda_{ss} = 0.0035$, $\lambda_{si} = 0.0015$, $\mu_s = 0.0002$, $\mu_4 = 0.0003$, $\mu_5 = 0.0001$, $\mu_6 = 0.00035$

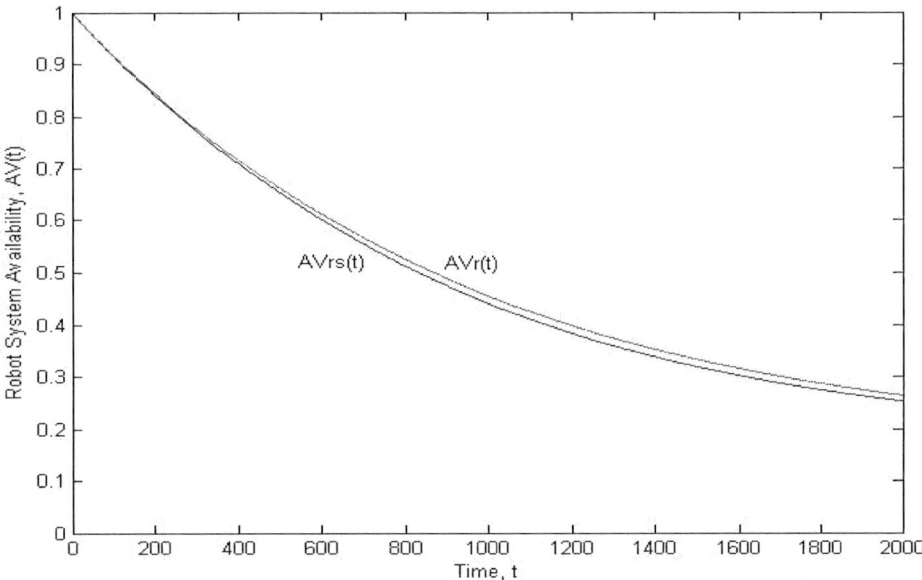

Figure 4. Time-dependent availability plots for the special case robot-safety system with exponentially distributed failed system repair times.

$$a_n P_n = \lambda_s P_{n-1} \tag{38}$$

$$\frac{dP_j(x)}{dx} + \mu_j(x) P_j(x) = 0 \qquad (\text{for } j = n+1, n+2, n+3) \tag{39}$$

The associated boundary conditions are as follows:

$$P_{n+1}(0) = \lambda_{ss} P_n \tag{40}$$

$$P_{n+2}(0) = \lambda_{si} P_n \tag{41}$$

$$P_{n+3}(0) = \lambda_r \sum_{i=0}^{n-1} P_i \tag{42}$$

By solving Equation (39), we get:

$$P_j(x) = P_j(0) \exp[-\int_0^x \mu_j(\delta) d\delta] \tag{43}$$

Together with:

$$P_j = \int_0^\infty P_j(x)dx \qquad \text{(for } j = 2n+2, 2n+3) \qquad (44)$$

We get:

$$P_j = P_j(0)E_j[x] \qquad \text{(for } j = 2n+2, 2n+3) \qquad (45)$$

where

$$E_j[x] = \int_0^\infty \exp[-\int_0^x \mu_j(\delta)d\delta]dx$$
$$= \int_0^\infty x w_j(x)dx \qquad \text{(for } j = 2n+2, 2n+3) \qquad (46)$$

where

$w_j(x)$ is the failed robot safety system repair time probability density function.

$E_j[x]$ is the mean time to robot safety system repair when the failed robot safety system is in state j and has an elapsed repair time x.

Solving Equations (36) – (46), together with

$$\sum_{i=0}^{n} P_i + \sum_{j=n+1}^{n+3} P_j = 1 \qquad (47)$$

We obtain:

$$P_0 = [1 + \sum_{i=1}^{n} Y_i + \sum_{j=n+1}^{n+3} a_j E_j[x]]^{-1} = \frac{1}{G} \qquad (48)$$

$$P_i = Y_i P_0 \qquad (49)$$
(for $i = 1, 2 \ldots \ldots n$)

$$P_j(s) = a_j E_j[x] P_0(s) \qquad \text{(for } j = n+1, n+2, n+3) \qquad (50)$$

where

$$Y_i = \lim_{s \to 0} Y_i(s) = \prod_{k=1}^{i} L_k \quad \text{(for } i = 1, 2 \ldots \ldots n)$$

$$a_{n+1} = \lambda_{ss} Y_n$$

$$a_{n+2} = \lambda_{si} Y_n$$

$$a_{n+3} = \lambda_r [1 + \sum_{i=0}^{n-1} Y_i]$$

$$L_n = \frac{\lambda_s}{a_n}$$

$$L_i = \frac{\lambda_s}{a_i - \mu_s L_{i+1}}$$

$$G = 1 + \sum_{i=1}^{n} Y_i + \sum_{j=n+1}^{n+3} a_j E_j[x] \tag{51}$$

$$E_j[x] = \int_0^\infty \exp[-\int_0^x \mu_j(\delta)d\delta]dx \tag{52}$$
$$= \int_0^\infty x w_j(x) dx \qquad (\text{for } j = n+1, n+2, n+3)$$

The generalized steady state availability of the robot safety system with one normally working safety unit, the switch and the robot is given by:

$$\text{SSAVrs} = \sum_{i=0}^{n-1} P_i = \frac{1 + \sum_{i=1}^{n-1} Y_i}{G} \tag{53}$$

Similarly, the generalized steady state availability of the robot safety system with or without a working safety unit is:

$$\text{SSAVr} = \sum_{i=0}^{n} P_i = \frac{1 + \sum_{i=1}^{n} Y_i}{G} \tag{54}$$

For different failed robot-safety system repair time distributions, we obtain expressions for G as follows:

) For the failed robot-safety system gamma distributed repair time x, the probability density function is expressed by

$$w_j(x) = \frac{\mu_j^\beta x^{\beta-1} e^{-\mu_j x}}{\Gamma(\beta)} \qquad (\beta > 0,\ j = n+1, n+2, n+3) \tag{55}$$

where x is the repair time variable, $\Gamma(\beta)$ is the gamma function, μ_j is the scale parameter, and β is the shape parameter.

Thus, the mean time to robot-safety system repair is given by

$$E_j(x) = \int_0^\infty x w_j(x) dx = \frac{\beta}{\mu_j} \qquad (\beta > 0,\ j = n+1, n+2, n+3) \tag{56}$$

Substituting Equation (56) into Equation (51), we get

$$G = 1 + \sum_{i=1}^{n} Y_i + \sum_{j=n+1}^{n+3} a_j \frac{\beta}{\mu_j} \tag{57}$$

) For the failed robot-safety system Weibull distributed repair time x, the probability density function is expressed by

$$w_j(x) = \mu_j \beta x^{\beta-1} e^{-\mu_j(x)^\beta} \qquad (\beta > 0,\ j = n+1, n+2, n+3) \tag{58}$$

where x is the repair time variable, μ_j is the scale parameter, and β is the shape parameter.

Thus, the mean time to robot-safety system repair is given by

$$E_j[x] = \int_0^\infty x W_j(x) dx = \left(\frac{1}{\mu_j}\right)^{1/\beta} \frac{1}{\beta} \Gamma\left(\frac{1}{\beta}\right) \qquad (\beta > 0, j = n+1, n+2, n+3) \tag{59}$$

Substituting (59) into Equation (51), we obtain

$$G = 1 + \sum_{i=1}^{n} Y_i + \sum_{j=n+1}^{n+3} a_j \left(\frac{1}{\mu_j}\right)^{1/\beta} \frac{1}{\beta} \Gamma\left(\frac{1}{\beta}\right) \tag{60}$$

) For the failed robot-safety system Rayleigh distributed repair time x, the probability density function is expressed by

$$w_j(x) = \mu_j x e^{-\mu_j x^2/2} \qquad (\mu_j > 0,\ j = n+1, n+2, n+3) \tag{61}$$

where x is the repair time variable, μ_j is the scale parameter.

Thus, the mean time to robot-safety system repair is given by

$$E_j(x) = \int_0^\infty x W_j(x) dx = \sqrt{\frac{\pi}{2\mu_j}} \qquad (\mu_j > 0, j = n+1, n+2, n+3) \tag{62}$$

Substituting Equation (62) into Equation (51), we get

$$G = 1 + \sum_{i=1}^{n} Y_i + \sum_{j=n+1}^{n+3} a_j \sqrt{\frac{\pi}{2\mu_j}} \tag{63}$$

) For the failed robot system Lognormal distributed repair time x, the probability density function is expressed by

$$w_j(x) = \frac{1}{\sqrt{2\pi} x \sigma_{y_j}} e^{[-\frac{(lnx - \mu_{y_j})^2}{2\sigma_{y_j}^2}]} \quad (\text{for } j = n+1, n+2, n+3) \quad (64)$$

where x is the repair time variable, lnx is the natural logarithm of x with a mean μ and variance σ^2. The conditions on parameters are:

$$\sigma_{y_j} = \ln\sqrt{1 + (\frac{\sigma_{x_j}}{\mu_{x_j}})^2} \quad (65)$$

and

$$\mu_{y_j} = \ln\sqrt{\frac{\mu_{x_j}^4}{\mu_{x_j}^2 + \sigma_{x_j}^2}} \quad (66)$$

Thus, the mean time to robot-safety system repair is given by

$$E_j(x) = e^{(\mu_{y_j} + \frac{\sigma_{y_j}^2}{2})} \quad (\text{for } j = n+1, n+2, n+3) \quad (67)$$

Substituting Equation (67) into Equation (51), we get

$$G = 1 + \sum_{i=1}^{n} Y_i + \sum_{j=n+1}^{n+3} a_j e^{(\mu_{y_j} + \frac{\sigma_{y_j}^2}{2})} \quad (68)$$

) For the failed robot system exponentially distributed repair time x, the probability density function is expressed by

$$w_j(x) = \mu_j e^{-\mu_j x} \quad (\mu_j > 0, j = n+1, n+2, n+3) \quad (69)$$

where x is the repair time variable and μ_j is the constant repair rate of state j.

Thus, the mean time to robot-safety system repair is given by

$$E_j(x) = \int_0^\infty x w_j(x) dx = \frac{1}{\mu_j} \quad (\beta > 0, j = n+1, n+2, n+3) \quad (70)$$

Substituting Equation (70) into Equation (51), we get

$$G = 1 + \sum_{i=1}^{n} Y_i + \sum_{j=n+1}^{n+3} a_j \frac{1}{\mu_j} \quad (71)$$

Robot-Safety System Steady State Analysis for a Special Case

For three safety units (i.e. one working, others on standby) by substituting n = 3 into Equations (48) - (54), we obtain:

$$P_0 = \frac{1}{1 + \sum_{i=1}^{3} Y_i(s) + \sum_{j=4}^{6} a_j E_j[x]} \tag{72}$$

$$P_i = Y_i P_0 \quad \text{(for } i = 1, 2, 3\text{)} \tag{73}$$

$$P_j(s) = a_j E_j[x] P_0 \quad \text{(for } j = 4, 5, 6\text{)} \tag{74}$$

where

$$Y_i = \prod_{k=1}^{i} L_k \quad \text{(for } i = 1, 2, 3\text{)}$$

$$a_4 = \lambda_{ss} Y_3$$

$$a_5 = \lambda_{si} Y_3$$

$$a_6 = \lambda_r [1 + \sum_{i=1}^{2} Y_i]$$

$$L_3 = \frac{\lambda_s}{a_3}$$

$$L_i = \frac{\lambda_s}{a_i - \mu_s L_{i+1}} \quad \text{(for } i = 1, 2\text{)}$$

$$G = 1 + \sum_{i=1}^{3} Y_i + \sum_{j=4}^{6} a_j E_j[x] \tag{75}$$

The steady state availability of the robot-safety system with one normally working safety unit, the switch, and the robot is given by:

$$\text{SSAV}_{rs} = \sum_{i=0}^{2} P_i = \frac{1 + \sum_{i=1}^{2} Y_i}{G} \tag{76}$$

The steady state availability of the robot-safety system with or without a normally working safety unit is given by:

$$\text{SSAV}_r = \sum_{i=0}^{3} P_i = \frac{1 + \sum_{i=1}^{3} Y_i}{G} \tag{77}$$

1) For exponentially distributed failed robot-safety system repair time, substituting Equation (71) into Equations (72) - (77), setting:

$\lambda_s = 0.001$, $\lambda_r = 0.0009$, $\lambda_{ss} = 0.0035$, $\lambda_{si} = 0.0015$, $\mu_4 = 0.0003$, $\mu_5 = 0.0001$, $\mu_6 = 0.00035$ and using Matlab computer program [11], the Figures 5 and 6 plots were obtained.

$\lambda_s = 0.001$, $\lambda_r = 0.0009$, $\lambda_{ss} = 0.0035$, $\lambda_{si} = 0.0015$, $\mu_4 = 0.0003$, $\mu_5 = 0.0001$, $\mu_6 = 0.00035$

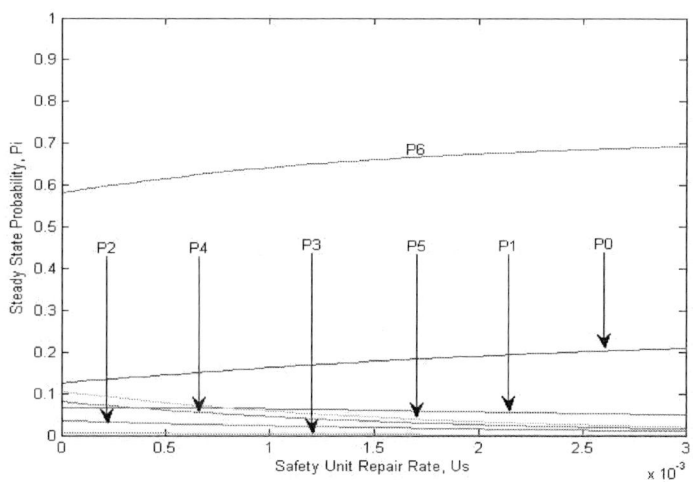

Figure 5. Robot-safety system steady state probability versus safety unit repair rate Us (means μ_s) plots for exponentially distributed failed system repair times.

$\lambda_s = 0.001$, $\lambda_r = 0.0009$, $\lambda_{ss} = 0.0035$, $\lambda_{si} = 0.0015$, $\mu_4 = 0.0003$, $\mu_5 = 0.0001$, $\mu_6 = 0.00035$

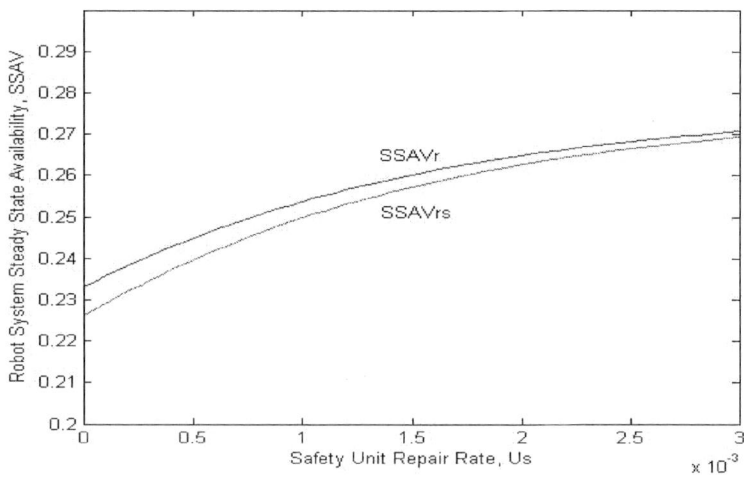

Figure 6. Robot-safety system steady state availability versus safety unit repair rate Us (means μ_s) plots for exponentially distributed failed system repair times.

2) For Rayleigh distributed failed robot-safety system repair time, substituting Equation (63) into Equations (72) - (77), setting:

$\lambda_s = 0.001$, $\lambda_r = 0.0009$, $\lambda_{ss} = 0.0035$, $\lambda_{si} = 0.0015$, $\mu_4 = 0.0000003$, $\mu_5 = 0.0000001$,

$\mu_6 = 0.00000035$ and using Matlab computer program [11], the Figure 7 was obtained.

$\lambda_s = 0.001$, $\lambda_r = 0.0009$, $\lambda_{ss} = 0.0035$, $\lambda_{si} = 0.0015$, $\mu_4 = 0.0000003$,

$\mu_5 = 0.0000001$, $\mu_6 = 0.00000035$

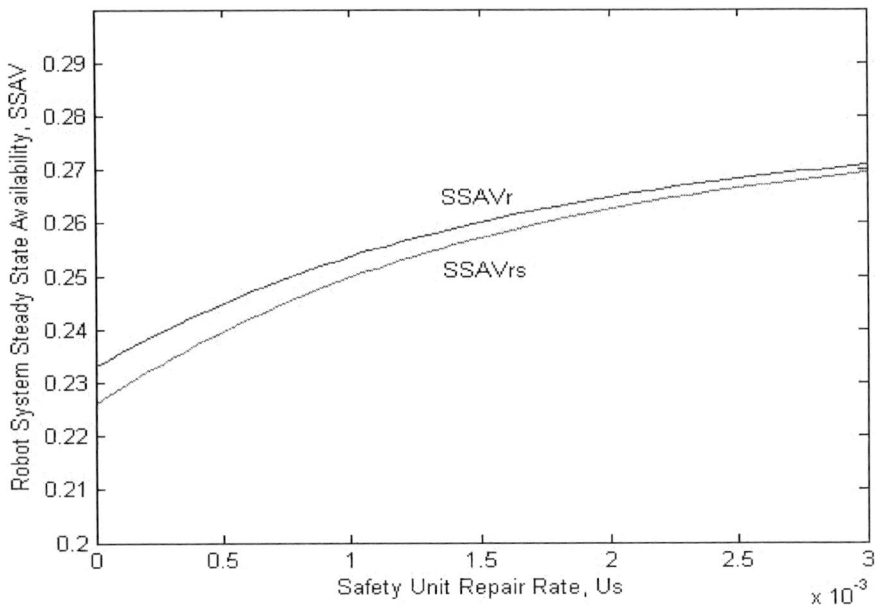

Figure 7. Robot-safety system steady state availability versus safety unit repair rate Us (means μ_s) plots for Rayleigh distributed failed system repair times.

3) For lognormally distributed failed robot-safety system repair time, substituting Equation (68) into Equations (72) - (77), setting:

$\lambda_s = 0.001$, $\lambda_r = 0.0009$, $\lambda_{ss} = 0.0035$, $\lambda_{si} = 0.0015$, $\mu_4 = 0.0009$, $\mu_5 = 0.0006$,

$\mu_6 = 0.0007$, $\sigma = 0.5$ and using Matlab computer program [11], the Figure 8 plots were obtained.

$\lambda_s = 0.001$, $\lambda_r = 0.0009$, $\lambda_{ss} = 0.0035$, $\lambda_{si} = 0.0015$, $\mu_4 = 0.0009$,

$\mu_5 = 0.0006$, $\mu_6 = 0.0007$, $\sigma = 0.5$

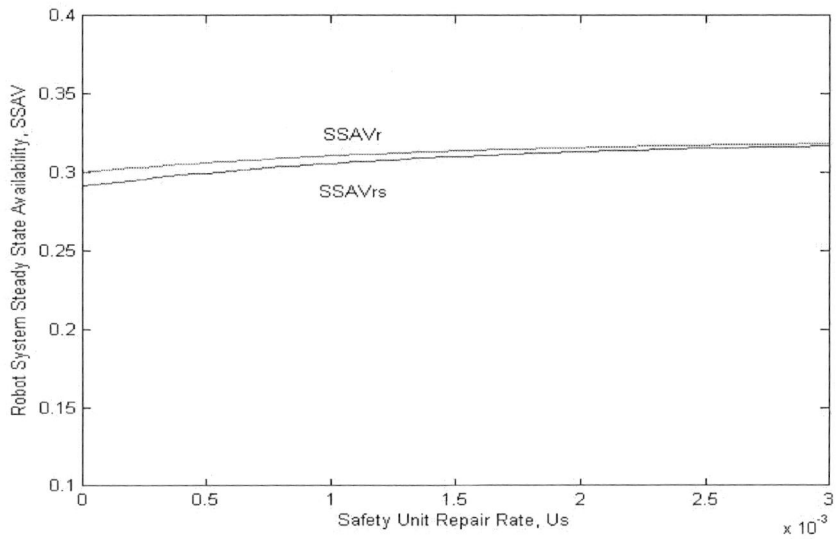

Figure 8. Robot- safety system steady state availability versus safety unit repair rate Us (means μ_s) plots for lognormally distributed failed system repair times.

Robot - Safety System Reliability and Mean Time to Failure Analysis

Setting $\mu_j = 0$, (for $j = n+1, n+2, n+3$), in Figure 2 and using the Markov method [12], we obtain the following set of differential equations:

$$\frac{dP_0(t)}{dt} + a_0 P_0(t) = \mu_s P_1(t) \tag{78}$$

$$\frac{dP_i(t)}{dt} + a_i P_i(t) = \mu_s P_{i+1}(t) + \lambda_s P_{i-1}(t) \tag{79}$$

(for $i = 1,2,\ldots,n-1$)

$$\frac{dP_n(t)}{dt} + a_n P_n(t) = \lambda_s P_{n-1}(t) \tag{80}$$

where

$a_0 = \lambda_s + \lambda_r$

$a_i = \lambda_s + \lambda_r + \mu_s$ (for $i = 1,2,\ldots,n-1$)

$a_n = \lambda_{ss} + \lambda_{si} + \mu_s$

$$\frac{dP_{n+1}(t)}{dt} = \lambda_{ss} P_n(t) \tag{81}$$

$$\frac{dP_{n+2}(t)}{dt} = \lambda_{si} P_n(t) \tag{82}$$

$$\frac{dP_{n+3}(t)}{dt} = \lambda_r \sum_{i=0}^{n-1} P_i(t) \tag{83}$$

At time t = 0, $P_0(0) = 1$, and other initial condition probabilities are equal to zero. Using Laplace transforms and solving Equations (78) – (83), we get:

$$sP_0(s) + a_0 P_0(s) = \mu_s P_1(s) + 1 \tag{84}$$

$$sP_i(s) + a_i P_i(s) = \mu_s P_{i+1}(s) + \lambda_s P_{i-1}(s) \qquad (\text{for } i = 1,2,\ldots\ldots n-1) \tag{85}$$

$$(s + a_n) P_n(s) = \lambda_n P_{n-1}(s) \tag{86}$$

$$sP_{n+1}(s) = \lambda_{ss} P_n(s) \tag{87}$$

$$sP_{n+2}(s) = \lambda_{si} P_n(s) \tag{88}$$

$$sP_{n+3}(s) = \lambda_r \sum_{i=0}^{n-1} P_i(s) \tag{89}$$

By solving Equations (84) - (89), together with

$$\sum_{i=0}^{n} P_i(s) + \sum_{j=n+1}^{n+3} P_j(s) = \frac{1}{s}, \tag{90}$$

we get the following general Laplace transform Equations of state probabilities:

$$P_0(s) = [s(1+\sum_{i=1}^{n} Y_i(s) + \sum_{j=n+1}^{n+3} \frac{a_j(s)}{s}]^{-1} = \frac{1}{G(s)} \tag{91}$$

$$P_i(s) = Y_i(s) P_0(s) \qquad (\text{for } i = 1,2,\ldots\ldots n) \tag{92}$$

$$P_j(s) = \frac{a_j(s)}{s} P_0(s) \qquad (\text{for } j = n+1, n+2, n+3) \tag{93}$$

where

$$G(s) = s[1+ \sum_{i=1}^{n} Y_i(s) + \sum_{j=n+1}^{n+3} \frac{a_j(s)}{s}] \qquad (94)$$

The Laplace transform of the robot-safety system reliability with one normally working safety unit, the switch and the robot is given by:

$$R_{rs}(s) = \sum_{i=0}^{n-1} P_i(s) = \frac{1+ \sum_{i=1}^{n-1} Y_i(s)}{G(s)} \qquad (95)$$

Similarly, the Laplace transform of the robot safety system reliability with or without a working safety unit is given by:

$$R(s) = \sum_{i=0}^{n} P_i(s) = \frac{1+ \sum_{i=1}^{n} Y_i(s)}{G(s)} \qquad (96)$$

Using Equation (95), the robot-safety system mean time to failure with one normally working safety unit, the switch and the robot is given by [12]:

$$MTTF_{rs} = \lim_{s \to 0} R_{rs}(s) = \frac{1+ \sum_{i=1}^{n-1} Y_i}{\sum_{j=n+1}^{n+3} a_j} \qquad (97)$$

Similarly, using Equation (96), the robot safety system mean time to failure with or without a working safety unit is given by [12]:

$$MTTF_r = \lim_{s \to 0} R_r(s) = \frac{1+ \sum_{i=1}^{n} Y_i}{\sum_{j=n+1}^{n+3} a_j} \qquad (98)$$

Robot-Safety System Mean Time to Failure Analysis for a Special Case

Substituting n = 3 into Equations (97) and (98), we get:

$$\text{MTTF}_{rs} = \lim_{s \to 0} R_{rs}(s) = \frac{1 + \sum_{i=1}^{2} Y_i}{\sum_{j=4}^{6} a_j} \qquad (99)$$

and

$$\text{MTTF}_{r} = \lim_{s \to 0} R_{r}(s) = \frac{1 + \sum_{i=1}^{3} Y_i}{\sum_{j=4}^{6} a_j} \qquad (100)$$

For $\lambda_r = 0.0009$, $\lambda_{ss} = 0.0035$, $\lambda_{si} = 0.0015$, $\lambda_s = 0.001$; and using Equations (99), (100), and Matlab computer program, the Figure 9 plots were obtained [11]:

$\lambda_r = 0.0009$, $\lambda_{ss} = 0.0035$, $\lambda_{si} = 0.0015$, $\lambda_s = 0.001$

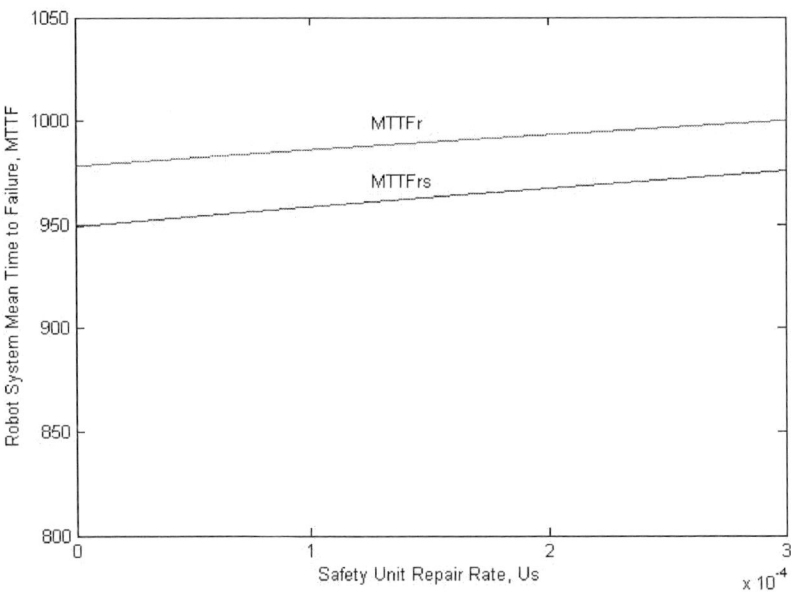

Figure 9. The robot-safety system mean time to failure plots for the increasing values of the safety unit repair rate Us (means μ_s).

Conclusion

This chapter has presented stochastic analysis of a repairable standby robot-safety system with a perfect switch to replace a failed safety unit. This study clearly demonstrates that the system availability increases significantly with the increase in safety units and safety unit repair rate.

References

[1] Dhillon, B.S., Fashandi, A.R.M., Safety and Reliability Assessment Techniques in Robotics, *Robotica*, Vol.15, pp. 701-708, 1997.

[2] Dhillon, B.S., Fashandi, A.R.M., Liu, K.L., Robot Systems Reliability and safety: A Review, *Journal of Quality in Maintenance Engineering*, Vol. 8, No.3, pp.170-212, 2002.

[3] Ruldall, B.H., *Automation and Robotics Worldwide: Reports and Survey*, Robotica, Vol.14, pp.243-251, 1996.

[4] Zeldman, M.I., *What Every Engineer Should Know About Robots*, Marcel Dekker, New York, 1984.

[5] United Nations Economic Commission for Europe (UN/ECE), *World Robotics 2005 – Statistics, Market Analysis, Forecasts, Case Studies and Profitability of Robot Investment*, United Nations, New York, 2005.

[6] Dhillon, B.S., *Robot Reliability and Safety*, Springer, NewYork, 1991.

[7] Nicolaisen, P., *Safety Problems Related to Robots*, Robotics, Vol.3, pp.205– 211, 1987.

[8] Nagamachi, M., *Ten Fatal Accidents Due Robots in Japan, in Ergonomics of Hybrid Automated Systems*, edited by H.R. Korwooski, M.R., Parsaei, M.R., Elsevier, Amsterdam, pp.391-396, 1998.

[9] Gaver, D.P., Time to failure and availability of paralleled system with repair, *IEEE Trans. Reliab.* Vol.12, pp.30-38, 1963.

[10] Grag, R.C., Dependability of a complex system having two types of components, *IEEE Trans Reliab.* Vol.12, pp.11-15, 1963.

[11] Hahn, Brian D., *Essential MATLAB for Scientists and Engineers*, Oxford: Butterworth - Heinemann, 2002.

[12] Dhillon, B.S., *Design Reliability: Fundamentals and Applications*, CRC, Boca Raton, Florida, 1999.

INDEX

A

A_β, 298, 299
abdomen, 232
access, 225, 246, 253, 254
Access Point, 85
accidents, 83, 325, 333
accuracy, 34, 50, 54, 101, 114, 149, 150, 194, 230, 272
activation, 258, 259, 260, 261
actuation, 80, 87, 95, 272
actuators, 9, 79, 80, 81, 94, 95, 96, 100, 101, 271, 313
adaptability, x, 309, 310
adaptation, viii, 10, 26, 29, 121, 290
adjustment, 82, 290, 311, 331
AFC, 20
age, ix, 155, 178, 180, 181, 182, 183, 184, 185, 253
agent, 130, 131, 138, 140, 150
aging society, 155
AIM, 188, 189
airship, vii, viii, 77, 78, 79, 80, 81, 82, 84, 85, 86, 87, 88, 90, 95, 96, 97, 98, 99, 100, 101, 102, 108, 109, 115, 116, 117, 118
algorithm, ix, 50, 60, 66, 83, 104, 107, 114, 117, 125, 131, 146, 172, 205, 219, 223, 225, 226, 227, 231, 239, 245, 256, 259, 261, 262, 271, 275, 296, 306
Alps, 118
alternative, 5, 10, 29, 50, 79, 141, 150, 204, 274
ambiguity, 44, 47, 53, 55, 109
AMD, 247
amortization, 251
Amsterdam, 73, 355
angular velocity, 88, 98, 276, 277, 296, 312, 313
application component, 253
Argentina, 220
argument, 5

artificial intelligence, 258
aspect ratio, 36, 46, 56
assessment, 115
assignment, 131
assumptions, 46, 214, 334
asymmetry, 188, 282
asymptotic, vii, viii, 9, 77, 81
asymptotically, vii, viii, 9, 12, 29, 77, 79, 81, 89, 92, 93, 94, 115, 296
atoms, 124
atrophy, 310
attention, 173, 178, 257, 268
Austria, 118, 119
authority, 95, 96
automata, ix, 193, 194, 195, 197, 198, 199, 200, 201, 202, 203, 206, 207, 208, 210, 219, 220, 221
automation, 4, 208, 251, 252, 262, 263, 264
automatization, 219
automobiles, 158, 159, 186
autonomous, 115, 116, 117, 118, 119, 151, 152, 153, 188, 189, 257, 263, 307, 308
autonomy, 130, 265, 266
availability, x, xi, 250, 333, 336, 339, 340, 341, 343, 345, 348, 349, 350, 351, 355
averaging, 107, 117, 230
avoidance, viii, ix, x, 121, 131, 134, 139, 140, 157, 158, 223, 224, 257, 258, 289, 290, 291, 294, 295, 296, 300, 305, 306, 307, 308

B

backlash, 272
bacteria, 4, 5
bandwidth, 269, 272
battery(ies), 82, 157, 158, 260
Bayesian, 248
behavior, viii, x, 79, 95, 97, 100, 101, 121, 122, 123, 124, 127, 128, 129, 130, 131, 133, 136, 137, 138,

139, 140, 146, 166, 200, 204, 205, 253, 263, 274, 289, 290, 307
behavior of children, 137
Belgium, 116, 118
bending, 313
benefits, 249
Bernstein- Bézier, x
binding, 310, 313
biodiversity, 78
biomedical, 248
birth, 4, 155
birth rate, 155
blocks, 136, 252, 262
body weight, 318
boundary conditions, 337, 338, 343
Brazil, 78, 118
breast, 227
bubbles, 225, 248
buffer, 165, 170, 175
building blocks, 136
bust, 320

C

C++, 83, 159, 251, 254, 256, 263
calibration, 33, 39, 44, 55, 101, 102, 103, 118, 119
Canada, 73, 333
candidates, 225
case study, ix, 29, 131, 193, 195, 208, 212, 219, 220
CDC, 286
cell, 35, 36, 112, 113, 208, 221, 312, 313, 317, 325, 327
changing environment, 139, 150
channels, 261
chemical, 145, 146
Chicago, 117
children, 136, 137, 138
chirality, 44
Cincinnati, 307
classes, 254, 256, 257, 263
classification, 33, 54
cleaning, 159
closed-loop, vii, 9, 11, 12, 13, 16, 18, 20, 25, 29, 79, 92, 94, 96
clustering, 80
Coast Guard, 257
coding, 253
cohesion, 126, 130, 131, 132, 133
collaboration, 173
collision-avoidance, x, 289, 291, 295, 296, 305, 306
collisions, 122, 125, 126, 127, 131, 132, 134, 272, 290
commercial, 83, 163, 166, 250

communication, 126, 145, 150, 157, 163, 256, 307
community, 4
compensation, 112, 275
competition, 258
complement, 101
complexity, 158, 195, 207, 219, 245, 253, 281
compliance, 9, 219, 255
components, x, 41, 51, 62, 63, 65, 66, 67, 69, 72, 81, 84, 94, 98, 101, 102, 104, 107, 109, 113, 124, 126, 145, 195, 196, 197, 202, 225, 253, 255, 267, 268, 270, 271, 275, 284, 294, 355
composite, 105
composition, 166
computation, 34, 47, 49, 54, 59, 60, 66, 68, 69, 70, 71, 72, 146, 239
computers, 255, 270, 283
computing, ix, 43, 47, 48, 49, 54, 55, 56, 61, 171, 190, 220, 224, 249, 255, 265
concentration, 146
concrete, 146
conditioning, 283
configuration, 21, 35, 45, 82, 95, 96, 123, 142, 143, 165, 250, 252, 254, 271, 272, 276, 277, 278, 279, 280, 281
Congress, 152, 188
connectivity, ix, 122, 125, 126, 127, 128, 129, 223, 226
consensus, 120
constraints, x, 40, 45, 46, 49, 80, 92, 98, 102, 130, 198, 199, 201, 218, 260, 274, 279, 289, 290, 294, 295, 296, 300, 304, 307, 308
construction, vii, 5, 206, 247, 254, 271, 285
consumption, 82, 322
continuity, 166, 167
control group, 263
convergence, 50, 306
conversion, 95, 258
convex, 12, 292
conviction, 320
Copenhagen, 73
CORBA, 83
correlation, 100, 109, 114
cosmic rays, 225, 248
costs, 250, 254
couples, viii, 121
coupling, 194, 197, 203, 208, 218
coverage, viii, 121, 132, 140, 257
covering, viii, 77, 79, 81, 115, 270
CPU, 82, 83, 85, 254, 255, 256, 260, 324
CRC, 355
credit, 131
critical state, 83
critical variables, 83

crystalline, 124
cues, 269
cybernetics, 116, 190, 307

D

damping, 88
danger, 176, 291, 322, 325
data gathering, 78
data set, 85, 109, 115, 227
data structure, 252
data transfer, 261, 262
database, 246
decisions, 212
decomposition, 47, 49, 54, 61, 94, 278
decoupling, x, 267, 271, 276, 285
defense, 257
definition, vii, 33, 47, 55, 67, 68, 92, 93, 94, 140, 200, 205, 206, 207, 211, 218, 225, 258
deformation, 51
degenerate, 47, 278
degradation, 78, 281
degrees of freedom, viii, 101, 268, 270, 278
delivery, 4
demand, 92, 93, 94, 100
Denmark, 73
denoising, 247
density, 88, 112, 133, 135, 139, 336, 338, 344, 345, 346, 347
Department of Defense, 115
derivatives, 87, 90
designers, 10
desire, 127
detection, 156, 258
determinism, 194, 202
deterministic, 194, 205, 251
developmental psychology, 136
deviation, 111, 113, 162, 180, 182, 227, 228, 234, 235, 236, 237, 238, 290, 302
DEVS, ix, 193, 194, 195, 196, 197, 198, 199, 201, 202, 203, 204, 205, 206, 207, 208, 209, 211, 212, 213, 215, 216, 217, 218, 219, 221
diamond(s), 109, 147, 148, 260
differential equations, 351
diffusion, 249
direct measure, 101
discrete event systems, 258
discrimination, 165, 170
disorder, 309
displacement, 113
distribution, 113, 122, 126, 181, 182, 184, 185, 225, 226, 231, 239, 240, 241, 244, 246, 251, 259
distribution function, 225

division of labor, 139
draft, 115
duration, vii, 77, 78, 200
dynamic control, 308
dynamic systems, 220

E

ears, 163, 283
earth, viii, 77, 87
eating, 158
ecology, 258
economics, 5
education, 330
ego, viii, 78
eigenvalue, 49, 50, 60, 63, 65, 69, 71, 72
eigenvector, 58, 59, 60, 63, 65, 69, 71, 72
elderly, 4, 180
electric current, 327
EMG, 330
employment, 4
energy, 82, 123, 140, 141, 142, 144, 313, 322
energy consumption, 82
entropy, ix, 223, 225, 226, 227, 229, 230, 232, 241, 245, 246, 247, 248
entropy maximization (EntMax), ix, 223, 225, 226, 227, 229, 230, 245, 246
environment, viii, x, 4, 5, 9, 80, 97, 102, 113, 115, 116, 121, 125, 127, 131, 132, 133, 135, 136, 137, 140, 146, 208, 220, 250, 253, 254, 257, 269, 270, 275, 278, 283, 290, 291, 309
environmental change, 131
environmental conditions, 131
epipolar geometry, vii, 33, 39, 51, 55
epipolar line, vii, 33, 41, 55
epipoles, vii, 33, 40, 41, 55
equality, 48
equilibrium, 12, 16, 18, 225
equipment, viii, 85, 155, 158, 160, 185, 309, 310
estimating, 102, 224
ethanol, 146
Euler, 80, 87, 97, 247, 276, 315
Euler equations, 247, 315
Euro, 73, 74
Europe, 258, 333, 355
European, 119, 190, 286
evidence, 127, 201
evolution, 125, 200, 257, 264
exaggeration, 3
exclusion, 258
excretion, 309
execution, 78, 170, 186, 200, 204, 205, 206, 251, 258, 259, 260, 261, 262, 284

exoskeleton, 286
exploitation, 136, 139, 257
extrinsic, vii, 33, 37, 55
eyes, 156, 166, 236, 241, 310

F

face recognition, 188
facial asymmetry, 188
facial expression, 156, 157
failure, x, 4, 78, 145, 309, 333, 334, 335, 336, 353, 354, 355
family, ix, 249, 250, 257, 258, 269
fatigue, 269
fear, 320, 327
feedback, 9, 10, 11, 25, 29, 80, 94, 130, 257, 269, 272, 283, 290, 296, 297, 300, 307, 308
feelings, 157
feet, 320
females, 181, 182, 185
filters, 231, 232, 233
fishing, 257
fitness, 126, 131, 132
flat-panel, ix, 155, 163, 166, 179, 180, 181, 182, 185
flexibility, ix, 10, 13, 79, 128, 195, 223, 250, 255, 292, 310
flexible joint, vii, 9, 10, 20, 24, 26, 29
flight, viii, 3, 77, 79, 80, 81, 82, 83, 86, 95, 96, 99, 101, 102, 109, 113, 115, 116, 117, 175
fluid, 118, 128, 129, 130, 156
focusing, 133, 156, 157, 181, 183, 194, 250, 253
forecasting, 186
France, 119, 266
freedom, viii, 44, 46, 59, 78, 101, 159, 185, 226, 232, 246, 268, 270, 278, 284
friction, 123, 124, 125, 269, 274, 275
funding, 4
fusion, ix, 118, 223, 226, 227, 247, 265
fuzzy logic, 10, 29
fuzzy sets, 11, 25

G

gases, viii, 121, 124
Gaussian, 126, 225, 226, 227, 228, 231, 232, 233, 234, 235, 236, 237, 238, 239, 241
gender, ix, 155, 178, 180, 181, 183, 184, 185
generalization, 123, 124, 225, 248
generation, 195, 219, 246, 247, 250, 251, 263
genetics, 139, 150
geology, 80
Germany, ix, 74, 78, 267, 286, 308
gestures, 34, 156, 157
Gibbs, 248
goals, 132, 133, 136, 140
GPS, 82, 83, 85, 101, 102, 108, 109, 110, 113, 115, 140, 149, 260, 261, 262
graduate students, 145
graph, 118, 128, 133, 258
grass, 146
gravitational constant, 125
gravitational force, 123, 150
gravity, 87, 101, 102, 103, 316, 317, 320, 322, 327
Greece, 73
grid generation, 247
grids, 113, 120
grouping, 119
groups, 137, 138, 180, 182, 185
growth, 249
guidance, vii, 77, 78, 80, 81, 85, 115, 258, 259, 260, 262

H

handicapped people, 158
hands, 282
haptic, ix, x, 267, 268, 269, 270, 271, 272, 274, 275, 276, 278, 279, 280, 281, 282, 283, 284, 285, 286, 287
harm, 3, 157
harmony, 3
head, 188, 227, 236, 237, 239, 246, 270, 283
height, 80, 102, 104, 107, 108, 109, 110, 111, 112, 113, 114, 115, 146, 160, 169, 173, 209, 272, 281, 318, 320, 322, 324
heterogeneity, 131
heuristic, 139
hexagonal lattice, 123
hip joint, 310, 313, 314
histogram, 262
homogeneity, 131
Honda, 3
hostile environment, 5
humane, 4
hybrid, 220
hydrology, 258

I

id, 127
identification, 29, 188
identity, 45, 52, 53, 87
illumination, 82
image analysis, 120

imagery, 80, 83, 110, 113, 115
images, vii, viii, ix, 33, 34, 39, 40, 41, 42, 43, 44, 45, 46, 47, 48, 49, 50, 52, 55, 66, 78, 80, 81, 82, 83, 85, 102, 103, 104, 105, 106, 107, 108, 109, 113, 115, 118, 223, 226, 227, 233, 239, 241, 242, 243, 244, 245, 246, 247, 248
imaging, vii, 33, 34, 38, 50, 51, 55, 105, 113, 157, 224
implementation, 86, 96, 106, 107, 115, 131, 138, 141, 144, 147, 194, 195, 201, 202, 203, 207, 208, 219, 221, 245, 250, 255, 259, 266, 275, 291, 300, 303, 306
incidence, 44, 99, 101
inclusion, 198, 201, 202, 221, 333
indeterminacy, 38, 46, 48, 55, 59
indexing, 33, 269
indication, 160, 166, 178, 182
indicators, 158
indices, 38
industry, 4, 250, 289
inequality, 10, 295
inertia, 21, 86, 88, 269, 275
infinite, 105, 193, 199, 200
Information System, 188
infrastructure, 250, 261, 309, 330
initial state, 199, 205, 206
injury(ies), 210, 310
insertion, 225
inspection, 78, 133
instability, 292
instruction, 164, 168, 170, 175, 176, 178, 186
integration, vii, 77, 80, 187, 224, 241, 245, 247, 248, 250, 258, 285
integrity, 253
intelligence, 139, 258
intelligent systems, 334
intensity, 96
intentions, 156, 157
interaction(s), ix, x, 122, 125, 126, 130, 131, 132, 133, 134, 156, 188, 189, 191, 223, 225, 226, 227, 232, 246, 253, 257, 267, 268, 270, 271, 274, 275, 276, 282, 284, 285, 287
interface, vii, ix, x, 83, 166, 191, 259, 267, 268, 269, 270, 271, 272, 274, 276, 281, 282, 283, 284, 285, 286, 287, 324
International Robot Exhibition, ix, 155, 156, 178
internet, vii, 266
interpretation, 36, 43, 47, 48, 54, 55, 181
interval, 36, 129, 146, 196, 197, 200, 204, 226, 229, 230, 231, 257, 298, 299, 336
intrinsic, vii, 33, 36, 39, 44, 46, 55, 60, 102, 107
intuition, 36
invariants, 199

inversion, 117
Iraq, 258
isolation, 83
Italy, 118, 249, 258, 263, 308
iteration, 50, 236

J

Jacobian, 278, 279
Japan, 3, 178, 186, 187, 188, 189, 190, 191, 286, 287, 308, 309, 324, 355
Jaynes, 247
joints, 9, 10, 29, 156, 271, 272, 273, 310, 313
joystick, 160, 164, 165, 175, 269
judgment, 61
Jung, 116, 118, 221

K

K^+, 22
Kalman filter, 109
kernel, 113, 250, 251, 252, 253, 254, 263, 264
kinematic model, 296
kinematics, viii, 77, 79, 80, 81, 87, 88, 271, 275, 276, 277, 278, 290, 291, 297
Korea, 9, 116

L

labeling, 33
labor, 139
LAN, 260, 324
land, vii, 41, 77, 78, 205
land use, 78
language, 156, 157, 198, 221
Laplace transforms, 337, 339, 340, 341, 352
laptop, 85, 255
laser, ix, 80, 155, 158, 163, 168, 169, 176, 180, 182, 183, 184, 185
laser pointer, ix, 155, 163, 168, 169, 176, 180, 182, 183, 184, 185
latency, 251, 252, 254
lateral motion, 80
lattices, 123
laws, 122, 123, 125, 126, 127, 128, 129, 131, 133, 150, 290
lead, 54, 94, 97, 130, 144, 146, 208, 241, 260, 274, 290, 310, 320
learning, 122, 131, 132, 134, 135, 138, 139, 150, 290
learning behavior, 139
lens, 141
levator, 100

lifetime, 197, 212, 216
lighter-than-air (LTA), vii, 77, 78
limitation, 82, 250, 277, 300
linear function, 21, 22
linear model, 297
links, 9
liquids, viii, 121, 124
literature, ix, 10, 79, 135, 139, 140, 193, 195, 203, 204, 224, 271, 284, 290
livestock, 78
localization, viii, ix, 80, 113, 121, 122, 140, 141, 142, 143, 149, 150, 151, 153, 223, 224
location, 46, 47, 59, 123, 125, 140, 141, 199, 224, 227, 228, 229, 230, 231, 232, 245, 314, 320, 321
locus, 175
lognormal, x, 333
London, 118, 222
long distance, 239
Lyapunov, vii, 9, 16, 18, 24, 79, 89, 90, 91, 92, 290
Lyapunov function, 16, 18, 89, 90, 91, 92

M

magnet, 210
magnetic sensor, viii, 77, 102
maintenance, 355
males, 181, 182, 184, 185
management, 256, 257, 258, 261, 262
mandible, 227
manifold(s), 9, 29, 235
manipulation, ix, x, 19, 142, 259, 267, 269, 270, 283
manufacturing, vii, 250, 251, 266
mapping, viii, 34, 38, 55, 78, 81, 101, 109, 113, 114, 115, 116, 118, 119, 153, 248, 276, 282
market, 250
Markov, x, 80, 118, 333, 351
Massachusetts, 153, 221
matrix, vii, 10, 12, 13, 16, 33, 36, 38, 39, 40, 41, 44, 45, 46, 47, 49, 50, 52, 53, 54, 55, 57, 58, 59, 60, 61, 62, 63, 64, 65, 66, 67, 68, 69, 70, 71, 72, 73, 80, 86, 87, 88, 89, 90, 92, 94, 102, 104, 106, 298, 299, 300
measurement, 12, 54, 96, 101, 102, 107, 109, 110, 111, 113, 114, 141, 227, 264, 319, 327
measures, 59, 85, 101, 105, 142, 226, 231, 246, 248, 278, 319
median, 59, 60, 248
membership, 11, 23, 25
memory, 163, 249, 253, 254, 259
mesh node, ix, 223, 226
metal oxide, 146
Mexico, 140, 252
Microsoft, 164

military, vii, 3, 78, 258
Ministry of Education, 330
Minnesota, 307
minors, 40
missions, 78, 115, 116, 257
MIT, 73, 152, 257, 258
MMP, 145
mobile phone, 4
mobile robots, viii, x, 3, 4, 5, 130, 155, 156, 185, 289, 290, 291, 294, 300, 301, 303, 304, 307, 308
mobility, 173, 208, 330
modeling, vii, 9, 10, 20, 24, 27, 29, 66, 79, 118, 193, 194, 197, 211, 219, 220, 221
models, xi, 12, 13, 27, 34, 51, 52, 53, 66, 79, 80, 101, 114, 124, 187, 194, 195, 196, 197, 198, 201, 202, 203, 204, 205, 206, 207, 218, 219, 220, 221, 224, 226, 227, 231, 247, 273, 333
modules, ix, x, 125, 145, 249, 253, 255, 258, 259, 260, 261, 269, 309, 310, 311, 314
molecular dynamics, 123
molecules, 124
momentum, 123
Moon, 221
mosaic, 80
motion, viii, x, 29, 34, 37, 38, 39, 43, 44, 50, 52, 53, 54, 55, 56, 61, 67, 78, 79, 80, 86, 95, 99, 101, 102, 117, 118, 182, 187, 188, 189, 259, 267, 268, 275, 276, 277, 281, 290, 308, 309, 310, 311, 314, 316, 317, 318, 319, 320, 321, 322, 323, 324, 325, 326, 327, 328, 329, 330
motion control, 29
motivation, 124
movement, viii, x, 81, 85, 101, 123, 155, 156, 157, 158, 159, 160, 162, 164, 165, 166, 167, 169, 170, 171, 172, 173, 174, 175, 176, 177, 178, 179, 180, 181, 182, 183, 184, 185, 209, 210, 284, 309, 310
MTS, 266
multimedia, 166
multiplication, 59, 109
multiplier, 38, 41, 229, 230
muscle atrophy, 310
mutation(s), 131, 133, 134, 136, 137, 138, 139
mutation rate, 136, 137, 138, 139

N

NASA, 3, 115
natural evolution, 125
natural laws, 122
navigation system, 79
Netherlands, 75
network, 130, 140, 143, 195, 203, 248, 251, 253, 254, 257, 290, 307

neural network(s), 29, 290
New Mexico, 140, 252
New Orleans, 30
New York, 73, 74, 220, 221, 355
Newton, 86, 315
Newtonian, 123, 124, 125, 127, 128, 129, 130, 140, 150
Newtonian physics, 140
next generation, 251, 263
nodes, ix, 203, 212, 223, 224, 225, 226, 235, 245, 254
noise, ix, 47, 59, 61, 80, 100, 130, 144, 223, 224, 225, 226, 227, 228, 231, 232, 233, 234, 235, 236, 237, 238, 239, 241, 245, 246, 248, 272, 274, 302, 304
nonholonomic robots, 289
non-linear, 247, 275, 283
nonlinear dynamics, viii, 77
nonlinear systems, 10, 20, 29, 30
nonverbal, 157

O

object recognition, 51, 54, 224, 247
observations, 187
off-the-shelf, ix, 249, 251
omni-directional, ix, 150, 155, 163, 165, 166, 179, 181, 182, 185
online learning, 134
opacity, 150
operating system, 249, 250, 251, 252, 253, 263, 273, 324
operator, x, 83, 168, 186, 257, 260, 267, 268, 269, 270, 273, 274, 275, 277, 278, 280, 281, 282, 283, 284, 285
optimization, x, 10, 12, 47, 106, 107, 125, 126, 224, 226, 228, 230, 232, 241, 245, 246, 247, 251, 267, 278, 281, 282, 285, 293, 294, 295, 300, 301, 303, 304
optimization method, 246, 300, 303
ORB, 118
organization, 122
orientation, viii, 45, 47, 51, 55, 59, 78, 80, 81, 85, 97, 101, 102, 103, 104, 111, 114, 115, 117, 162, 170, 172, 230, 232, 245, 269, 276, 280, 285, 292, 294, 302, 304
Ottawa, 333
outliers, 106, 113, 114
oxide, 146

P

pain, 156
parabolic, 141, 143, 150
Parallel distributed Compensation (PDC), vii, 9, 10, 11, 25
parameter, ix, 10, 12, 29, 36, 44, 46, 59, 60, 89, 102, 105, 106, 107, 123, 124, 125, 126, 223, 225, 226, 229, 230, 231, 232, 241, 259, 260, 261, 295, 345, 346
partition, 212
partnership, 78, 116
path planning, ix, 223, 224, 290, 291, 292, 300
pathways, 115
Pb, 292
PCM, 260
pedal, 269
penalty(ies), x, 126, 289, 291, 295, 296, 300, 301, 304, 305
pendulum, 29, 30, 310
perception, 3, 81, 115, 118, 134, 135, 136, 137, 138, 139, 269
performance, viii, ix, 4, 5, 10, 20, 77, 81, 98, 101, 115, 121, 122, 125, 126, 127, 129, 130, 131, 132, 135, 136, 137, 138, 146, 173, 221, 249, 250, 251, 253, 254, 264, 282, 287, 324
permit, 272
perturbation(s), 29, 50, 123, 131, 133
pheromones, 153
photocells, 35
photographs, 34
physicomimetics, viii, 121, 122
physics, viii, 121, 122, 123, 129, 130, 139, 140, 146, 150, 225, 251
pitch, 79, 97, 273
planning, ix, x, 78, 140, 223, 224, 246, 259, 289, 290, 291, 292, 300, 308
plants, 10
point of origin, 281
Poland, 117
pollution, 258
polygons, 176, 178
polymorphism, 3
polynomials, 96, 292
poor, 255
population, 125, 127, 131, 133, 139, 150, 181, 333
population size, 133
portability, 250, 255
ports, 197, 214, 260, 262
Portugal, 77, 117, 118, 119
posture, 306, 310, 322, 323
potential energy, 123, 140

power, ix, 78, 123, 125, 141, 142, 147, 150, 225, 249, 255, 260, 270, 274, 310, 311, 312, 313, 322
prediction, x, 78, 289, 290, 298
pressure, 95, 141
probability, 4, 112, 126, 225, 226, 336, 338, 341, 342, 344, 345, 346, 347, 349
probability density function, 338, 344
probability distribution, 225
problem solving, 50
production, 208, 221, 251
Production Cell, ix, 193, 195, 208
profit, 131
program, 163, 165, 210, 255, 266, 341, 349, 350, 354
programming, 253, 254
projector, ix, 155, 163, 173, 176, 180, 181, 182, 183, 184, 185
protocol(s), 144, 145
prototype, ix, 82, 84, 115, 155, 156, 163, 178, 185, 250
psychology, 136
Puerto Rico, 73, 74
pulse(s), 141, 142, 143, 144
P-value, 185

Q

quadrifocal tensor, vii, 33, 42, 55
quality control, 232, 246
quantization, 227
questionnaire(s), ix, 155, 156, 179, 181, 184, 185

R

race, 204
racism, 4
radar, 258
radiation, 168
radio, 141, 257
radius, 98, 125, 132, 134, 272, 279, 284, 295
range, ix, 79, 80, 95, 123, 130, 142, 143, 223, 224, 227, 232, 239, 241, 244, 247, 248, 256, 269, 270, 281, 312
range image registration, ix, 223
Rayleigh, x, 333, 346, 350
reading, 83, 261, 262
real numbers, 38, 199
real time, viii, 3, 121, 133, 139, 150, 168, 170, 171, 172, 179, 186, 219, 273, 289, 295, 307
realism, 224
reality, 48, 173, 269, 283
reasoning, 22, 197, 202, 308
recall, 86, 104, 105, 198

recognition, ix, 3, 33, 51, 54, 119, 160, 163, 187, 188, 191, 223, 224, 247
recombination, 131
reconstruction, vii, 33, 34, 43, 44, 45, 47, 48, 49, 52, 53, 54, 55, 56, 61, 66, 118, 223, 227, 239, 244, 245
recovery, 81, 101, 102, 113
reduction, 123, 163, 260
redundancy, 276
reference frame, 86
regulation, 27, 89
rehabilitation, 268, 310
Rehabilitation Center, 310
relationship(s), 4, 34, 38, 39, 42, 43, 44, 50, 52, 53, 61, 158, 258, 324
reliability, x, 250, 254, 333, 336, 353
repair, x, xi, 3, 122, 333, 334, 336, 338, 341, 342, 343, 344, 345, 346, 347, 348, 349, 350, 351, 354, 355
resolution, 82, 102, 227, 235, 236, 245, 252, 255, 263, 273, 275, 283
resources, 4, 201, 250, 253
response time, 27, 251
retina, 34
returns, 61, 214
revolutionary, 4
rewards, 131
RF, 141, 142, 143, 144, 145, 149, 150
rigidity, 127, 272
risk, 156, 176
rivers, 253
robotics, viii, ix, 3, 4, 5, 77, 123, 139, 150, 155, 173, 223, 249, 250, 251, 252, 257, 258, 263, 265, 266, 286, 290
robustness, 56, 98, 101, 117, 149, 221, 254, 306
rolling, 81, 95, 96, 308
Romeo ROV, 258
rotational inertia, 275
rotations, 66, 67, 95, 114, 128, 278
rovers, 3
ROVs, 250, 257, 263

S

safety, x, xi, 83, 126, 156, 157, 158, 186, 198, 199, 200, 203, 208, 210, 217, 218, 271, 289, 290, 291, 294, 296, 300, 301, 303, 304, 305, 309, 310, 320, 322, 323, 330, 333, 334, 335, 336, 338, 339, 340, 341, 342, 343, 344, 345, 346, 347, 348, 349, 350, 351, 353, 354, 355
sample, 120, 304
sampling, 227, 241, 257, 259, 262, 273, 298, 324
saturation, 79, 80, 92, 93, 94, 97, 100, 144, 290, 307

scalability, 250
scalar, 56, 58, 59, 79, 87, 89, 296
scaling, 142
scheduling, ix, 249, 252, 255, 256, 259, 262
schema, 307
science, vii, 10, 195, 265
scores, 182
search, 50, 54, 79, 140, 146, 207, 259, 261, 262
searching, 41, 96, 207, 290
security, 79
segmentation, 52, 80, 101, 118
self-organization, 122
self-repair, 122
self-worth, 131
semantic(s), 33, 193, 194, 197, 199, 200, 203, 204
semantic information, 33
semicircle, 281
sensing, viii, 78, 80, 101, 119, 121, 125, 134, 141, 158, 269
sensitivity, 144
sensors, vii, viii, 77, 80, 82, 85, 101, 115, 118, 119, 122, 144, 145, 150, 209, 210, 214, 216, 258, 261, 262, 282
separation, 79, 123, 125, 126
series, 197, 287
shape, 34, 44, 45, 53, 54, 55, 56, 67, 162, 169, 224, 227, 247, 259, 274, 292, 295, 345, 346
shaping, 81, 115
shares, 11, 25
sharing, 78, 131, 146, 156, 250, 256
shy, 48
sign(s), 55, 58, 61, 63, 65, 166, 175, 178, 181
signaling, 156, 159
signals, vii, 9, 12, 16, 18, 29, 79, 80, 85, 101, 143, 147, 164, 211, 254, 255, 268
similarity, 45, 113
simulation, vii, viii, x, 9, 20, 24, 25, 26, 29, 77, 81, 86, 97, 100, 101, 115, 118, 121, 123, 124, 125, 127, 130, 140, 156, 159, 160, 162, 164, 194, 197, 202, 203, 204, 205, 207, 217, 218, 219, 220, 222, 265, 289, 316, 318, 320, 322, 326, 327
Simultaneous Localization and Mapping (SLAM), 113, 115, 224, 248
skills, 156
slavery, 4
smoothing, ix, 113, 223, 224, 225, 226, 227, 229, 231, 232, 234, 235, 236, 237, 238, 239, 241, 242, 243, 244, 245, 246, 247, 248
smoothness, 224, 226, 245
society, 155
software, ix, 83, 131, 146, 156, 159, 202, 208, 210, 249, 250, 251, 253, 254, 259, 260, 261, 263
solar, 260

solar panels, 260
sounds, 83
space environment, 290, 291
Spain, 74, 117, 119, 152
species, 131
spectrum, vii, 77, 78
speculation, 191
speed, vii, viii, 4, 9, 77, 78, 81, 82, 85, 95, 115, 118, 141, 143, 146, 147, 155, 157, 158, 159, 160, 162, 163, 164, 165, 166, 168, 169, 170, 172, 173, 175, 176, 178, 179, 180, 181, 182, 183, 184, 185, 254, 273, 274, 277, 279, 284, 290, 312
spin, 247
spinal cord injury, 310
St. Louis, 29
stability, vii, viii, 9, 10, 11, 12, 16, 20, 29, 30, 77, 78, 81, 92, 253, 275, 307, 310, 324
stabilization, viii, 77, 79, 80, 81, 97, 98, 115, 117, 123, 290, 297, 307, 308
stages, 33, 34, 144, 194
standard deviation, 113, 147, 162, 180, 182, 227, 228, 234, 235, 236, 237, 238, 302
standard model, 34
standards, 250
stars, 113, 114
statistical mechanics, ix, 223, 225, 247
statistics, 248
stochastic, 101, 125, 335, 337, 339, 341, 343, 345, 347, 349, 351, 353, 354, 355
storage, 82, 83
strategies, x, 122, 267, 285
strength, 123, 124, 126, 150
stress, 4, 5, 253, 254, 309
stretching, 313
stroke, 95
structuring, 191
students, 145
success rate, 146
Sun, 120, 248
supervision, 4, 259
supply, 260, 311
surveillance, vii, viii, 77, 78, 81, 101, 115, 121
survivability, 134, 137, 138
survival, 133, 134, 135
survival rate, 133, 135
sustainability, 4
swarm(s), viii, 121, 122, 123, 126, 127, 128, 129, 130, 131, 132, 138, 139, 140, 150, 153
swarm intelligence, 139
switching, 80
Switzerland, 116, 248
symbols, 61, 199, 335
synchronization, 143, 203, 254, 255, 318

synthesis, 220
systems, vii, viii, ix, 9, 10, 12, 20, 29, 30, 34, 77, 79, 80, 81, 101, 115, 121, 122, 130, 131, 138, 139, 140, 143, 150, 157, 166, 186, 190, 193, 194, 195, 196, 197, 198, 200, 201, 202, 203, 204, 206, 207, 208, 220, 221, 223, 225, 249, 250, 251, 252, 253, 254, 258, 262, 263, 264, 265, 266, 269, 308, 333, 334

T

targets, 80, 81, 158
taxonomy, 256
TCC, 216, 217, 218
teaching, 4, 166
technology, viii, 3, 4, 5, 121, 122, 140, 146, 265
teens, 180
television, 4
temperature, 141, 157
temporal, ix, 193, 194, 202, 219
textbooks, 34, 36
theory, 4, 5, 79, 114, 194, 195, 198, 201, 207, 220, 225, 247, 324, 326
thermal properties, 272
thermodynamic(s), 225, 247
thinking, 5, 176
threat, 188
threshold(s), 144, 224, 252
time frame, 295, 298
timing, 144, 156, 159, 160, 162, 197, 250, 251, 252, 255, 256, 263, 325
Tokyo, 156, 178, 330
toxin, 146
Toyota, 158, 188
toys, 4
tracking, vii, viii, x, 9, 15, 16, 20, 27, 29, 77, 78, 79, 80, 81, 85, 90, 95, 97, 98, 101, 115, 117, 189, 269, 283, 289, 290, 296, 297, 298, 299, 300, 305, 306, 307, 308, 324, 327
trade-off, 122
traffic, 34, 164, 166, 176, 182
training, 125, 127, 135, 268, 284
trajectory, viii, x, 12, 16, 18, 20, 29, 52, 66, 78, 80, 101, 102, 103, 104, 108, 109, 110, 113, 114, 115, 146, 171, 172, 197, 200, 206, 274, 283, 284, 285, 289, 290, 296, 297, 298, 299, 300, 301, 304, 305, 306
transducer, 141, 142, 143, 150
transformation(s), 44, 45, 46, 53, 56, 59, 60, 87, 90, 104, 107, 143, 297
transformation matrix, 53, 90

transition(s), 5, 79, 95, 99, 162, 166, 167, 195, 196, 197, 198, 199, 200, 201, 202, 203, 204, 206, 207, 211, 212, 216, 218, 221, 327
translation, viii, 61, 67, 69, 71, 72, 78, 81, 102, 103, 104, 105, 106, 107, 108, 109, 111, 120, 159, 164, 168, 175, 176, 181, 208, 210
transmission, 78, 83, 145, 157
transmits, 85, 144
transport, 156, 214, 215
transportation, 158
trees, 146
trend, 5, 251
trial, 166
triangulation, 141
trifocal tensor, 33, 41, 55
turbulence, 98, 100, 246
turbulent flows, 225

U

UK, 118, 223, 263
ulcer, 4, 5
ultrasound, 150
uncertainty, 10, 12, 29, 113, 124, 226, 265
underwater robotics, 263, 265
underwater vehicles, 117, 257, 266
uniform, 241
United Nations (UN), 333, 355
universality, 198
Unmanned Aerial Vehicles, vii, 77, 78
updating, 16, 25
users, 250, 309, 330
UV, 61, 257

V

vacuum, 159
validation, 194, 195, 202, 203, 268
validity, 20, 29
values, xi, 48, 49, 52, 85, 92, 94, 99, 100, 123, 124, 126, 193, 195, 196, 199, 231, 232, 233, 236, 241, 259, 260, 298, 324, 333, 354
vapor, 150
variable(s), x, 93, 62, 79, 83, 90, 100, 113, 123, 124, 147, 193, 196, 198, 201, 205, 211, 212, 216, 258, 259, 260, 261, 292, 294, 295, 298, 301, 302, 333, 336, 341, 345, 346, 347
variance, 113, 181, 185
variation, 20, 162, 226, 228, 230, 246
vector, 12, 20, 25, 41, 48, 49, 50, 52, 59, 62, 63, 64, 65, 68, 87, 90, 92, 93, 96, 103, 104, 107, 109, 125, 248, 280, 292, 293, 294, 297, 298, 299

vehicles, vii, ix, 77, 78, 80, 113, 117, 122, 156, 158, 249, 250, 257, 258, 259, 262, 263, 265, 266, 290, 308
velocity, x, 79, 86, 87, 88, 90, 94, 95, 98, 99, 123, 125, 132, 139, 268, 273, 275, 276, 277, 278, 279, 280, 283, 289, 290, 292, 293, 294, 295, 296, 297, 300, 303, 304, 312, 313, 316, 324, 330
vertical take-off and landing, viii, 77, 81, 98
vessels, 258
vibration, 83
virtual reality (VR), 283
vision, vii, viii, 33, 34, 44, 48, 77, 78, 80, 81, 101, 102, 103, 115, 117, 119, 121, 140, 258
visual perception, 81, 115
voice, 157
voiding, 83

weapons, 265
web, 251, 254
websites, 246
Weibull, x, 333, 346
weight ratio, 78
welding, 5
wind, viii, 77, 80, 81, 82, 85, 86, 87, 88, 89, 90, 91, 94, 95, 96, 98, 99, 100, 101, 115, 118, 147
winter, 156
work environment, 4
workers, 5
writing, 253

X

X-axis, 175

W

Wales, 223
walking, 156, 283, 284
war, 4
water vapor, 150

Y

Y-axis, 175
yield, 4, 122, 127, 236